Keynes & Aidley's Nerve and Muscle

Fifth edition

This well-established and acclaimed textbook introducing the rapidly growing field of nerve and muscle function has been completely revised and updated. Written with undergraduate students in mind, it begins with the fundamental principles demonstrated by the pioneering electrophysiological experiments on cell excitability. This leads to more challenging material recounting recent discoveries from applying modern biochemical, genetic, physiological and biophysical, experimental and mathematical analysis. The resulting interdisciplinary approach conveys a unified contemporary understanding of nerve, and skeletal, cardiac and smooth muscle, function at the molecular, cellular and systems levels. Emphasis on important strategic experiments throughout clarifies the basis for our current scientific views, highlights the excitement and challenge of biomedical discovery, and suggests directions for future advance. These fundamental ideas are then translated into discussions of related disease conditions and their clinical management. Now including colour illustrations, it is an invaluable text for students of physiology, neuroscience, cell biology and biophysics.

Christopher L.-H. Huang is Professor of Cell Physiology at the University of Cambridge, UK. He made scientific contributions in excitation-contraction coupling, cell electrolyte homeostasis, migraine aura and cardiac arrhythmogenesis, whilst directing medical studies as Fellow of Murray Edwards College. He has been Editor of the *Journal of Physiology, Biological Reviews, Monographs of the Physiological Society* and *Europace*, and Director of Hutchison China Meditech and Hutchison Biofilm Medical Solutions. The first three editions of this book were authored by Professor R. D. Keynes (1919-2010), Professor of Physiology (1973-1987) and Fellow of Churchill College, Cambridge (1961-2010) and D. J. Aidley (1947-2000), Senior Lecturer and Fellow (1979-2000) in the School of Biological Sciences at the University of East Anglia, in the United Kingdom.

Keynes & Aidley's Nerve and Muscle

Fifth edition

Christopher L.-H. Huang
University of Cambridge

CAMBRIDGE
UNIVERSITY PRESS

CAMBRIDGE
UNIVERSITY PRESS

University Printing House, Cambridge CB2 8BS, United Kingdom

One Liberty Plaza, 20th Floor, New York, NY 10006, USA

477 Williamstown Road, Port Melbourne, VIC 3207, Australia

314–321, 3rd Floor, Plot 3, Splendor Forum, Jasola District Centre, New Delhi – 110025, India

79 Anson Road, #06–04/06, Singapore 079906

Cambridge University Press is part of the University of Cambridge.

It furthers the University's mission by disseminating knowledge in the pursuit of education, learning, and research at the highest international levels of excellence.

www.cambridge.org
Information on this title: www.cambridge.org/9781108495059
DOI: 10.1017/9781108860789

First published 1981
Second edition 1991
Third edition 2001
Fourth edition 2011
Fifth edition 2021

Printed in the United Kingdom by TJ Books Limited, Padstow Cornwall

A catalogue record for this publication is available from the British Library.

ISBN 978-1-108-49505-9 Hardback
ISBN 978-1-108-81687-8 Paperback

To friends and teachers: *In Memoriam Absentium, in Salutem Praesentium:*

Charles Michel and Morrin Acheson: The Queen's College, Oxford.

David Weatherall and John Ledingham: Nuffield Department of Medicine, University of Oxford.

Richard Adrian: Physiological Laboratory, University of Cambridge.

For at the first she will walk with him by crooked ways, and bring fear and dread upon him, and torment him with her discipline, until she may trust his soul, and try him by her laws. Then will she return the straight way unto him, and comfort him, and shew him her secrets.
Ecclesiasticus 4:17-18: King James Version (KJV)

Contents

Colour plates can be found between pages 140 and 141.

Preface

Initiation of movement, whether voluntary action by skeletal muscle or contraction of cardiac or smooth muscle, is the clearest observable physiological manifestation of animal life. It inevitably involves activation of contractile tissue initiated or modulated by altered activity in its nerve supply or signalling by its chemical modulators. An appreciation of structure and function in both nerve and muscle, and of the functional relationships between them, is fundamental to our physiological understanding. These processes, and their regulation and abnormalities, now also assume increasing applicability to the understanding and clinical management of disease processes.

This book introduces this important biological area in a form suitable for students taking university courses in physiology, cell biology or medicine. It gives a straightforward account of the fundamentals of this subject, whilst including some of the strategic classical and recent experimental evidence underpinning our current understanding.

Besides providing rewritten and reorganised chapters, this fifth edition covers major advances in this important and rapidly developing area of study. It includes contributions from recent molecular structural insights, opportunities arising from genetic manipulation, novel single-cell and multi-channel electrophysiological and optical recording techniques, and physical and mathematical analysis. It extends our appreciation of the implications of these molecular and cellular findings to the systems level. Many of these developments were prompted by their applicability to clinical medicine that itself has both inspired and become increasingly amenable to physiological analysis. They have led to major new insights from the resulting exciting and important discoveries concerning the molecules involved in electrical activity, activation of skeletal muscle and the function of cardiac and smooth muscle. This edition increases emphasis on new findings in excitation–contraction coupling, cardiac electrophysiology and arrhythmogenesis, and the cellular physiology of smooth muscle.

Nevertheless, in the spirit of previous editions, the earlier as well as introductory sections of the subsequent chapters in this revision first emphasise fundamental physiological principles prior to narrating more challenging recent material. In the course of this revision, I am particularly grateful to my current collaborators, Drs. Antony Jackson, Kamalan Jeevaratnam, Ming Lei, Hugh Matthews, James Fraser and Samantha Salvage, as well as undergraduate and postgraduate students in my college and laboratory, for stimulating pedagogical insights and continued scientific dialogue. I am also grateful for a visiting professorship generously awarded by the University of Surrey in the course of this revision.

Acknowledgements

The author is grateful for permission to reprint illustrative material, cited by the figure legends, conveyed through Copyright Clearance Center, Inc. or a Creative Commons Attribution 4.0 International License (CC–BY) in the course of preparing this new edition, to:

Prof. William F. Gilly, Hopkins Marine Station, CA, USA (Figure 5.4). American Physiological Society: *American Journal of Physiology* (Plate 19); *Physiological Reviews* (Plates 21 and 22, Figure 5.3). American Society for Biochemistry and Molecular Biology: *Journal of Biological Chemistry* (Plate 1); Company of Biologists Ltd.: *Journal of Cell Science* (Figure 14.7). Elsevier BV: *Advances in Surgery* (Figure 1.8); *Brain Research Reviews* (Figure 8.13); *Cardiovascular Research* (Plate 20); *Cell* (Plate 2); *Mechanisms of Ageing and Development* (Plate 18); *Neuroscience* (Figure 7.1); *Progress in Biophysics and Molecular Biology* (Plate 5). Federation of European Biochemical Societies (FEBS) Press: *FEBS Letters* (Plate 3). Frontiers Media SA: *Frontiers in Physiology* (Plates 11 and 16). John Wiley & Sons Inc: *Acta Physiologica* (Plates 20 and 24; Figure 14.6); *Biological Reviews of the Cambridge Philosophical Society* (Figure 8.13); *Clinical and Experimental Pharmacology and Physiology* (Plate 14); *Experimental Physiology* (Plate 24); *Journal of Anatomy* (Plate 4, Figure 10.7); *Journal of Cardiovascular Electrophysiology* (Figure 14.8); *Journal of Physiology* (Plate 5, 7, 10 and 17, and Figures 3.7, 3.13, 5.4, 7.3, 7.4, 7.7–7.9, 10.1–10.5, 10.10, 11.1, 11.3, 11.4, 11.12, 11.14, 12.16, 13.8, 14.2, 15.3, 15.5–15.7 and 15.9); *Protein Science* (Plates 8 and 9). Proceedings of the National Academy of Sciences USA (PNAS) (Plate 2). Rockefeller University Press: *Biophysical Journal* (Figure 10.6); *Journal of General Physiology* (Plate 7; Figures 10.12, 10.13). Society for Neuroscience: *Journal of Neuroscience* (Figure 7.1). Royal Society (Great Britain): *Open Biology* (Plate 18); *Bibliographical Memoirs of the Royal Society* (Figure 11.2). Springer-Nature: *Pflugers Archiv-European Journal of Physiology* (Plate 14, Figure 12.17); *Scientific Reports* (Figure 11.16).

Abbreviations used in the text

Notes:

(1) Units provided to give indication of unit dimensions are those frequently encountered in the physiological literature.
(2) Ion channel α-subunits, their currents and their encoding genes are summarised in Plates 6 and 13.

a	constant in the Hill equation
A	actin
αAR	α-adrenergic receptor
A-band	anisotropic band
AC	adenylyl cyclase
ACh	acetylcholine
AChR	acetylcholine receptor
A-curve	action potential restitution: dependence of APD on BCL
ADP	adenosine diphosphate
AF	atrial fibrillation
AgCl	silver chloride
α_h	Hodgkin–Huxley, h-variable, forward rate constant (/sec)
α_m	Hodgkin–Huxley, m-variable, forward rate constant (/sec)
α_n	Hodgkin–Huxley, n-variable, forward rate constant (/sec)
-AM	acetomethoxy (ester)
AMPA	α-amino-3-hydroxy-5-methyl-4-isoxazolepropionate
Ano1	gene encoding anoctamin-1 (Ca^{2+}-activated Cl^- channel)
2-APB	2-aminoethoxydiphenylborate (IP_3R blocker)
APD	action potential duration (ms)
APD_{90}	action potential duration to 90% full repolarisation (ms)
ARVC	arrhythmogenic right ventricular cardiomyopathy
ATP	adenosine triphosphate
ATP/ADP	ATP/ADP ratio
AV	atrioventricular
aVF	electrocardiogram recording between left leg and combined two remaining leads
aVL	electrocardiogram recording between left arm and combined two remaining leads
AVN	atrioventricular node

aVR	electrocardiogram recording between right arm and combined two remaining leads
Aα	peripheral nerve fibre subtype; conduction velocity ~100 m/s
Aβ	peripheral nerve fibre subtype; conduction velocity ~60 m/s
Aγ	peripheral nerve fibre subtype; conduction velocity ~40 m/s
b	constant in the Hill equation
B	peripheral nerve fibre subtype; conduction velocity ~10 m/s
Ba^{2+}	barium ion
BAPTA	1,2-bis(o-aminophenoxy)ethane-N,N,N',N'-tetraacetic acid tetrakis-acetoxymethyl ester
βAR	β-adrenergic receptor
BCL	basic cycle length (ms)
BDNF	brain-derived neurotrophic factor
β_h	Hodgkin–Huxley, *h*-variable, backward rate constant (/sec)
β_m	Hodgkin–Huxley, *m*-variable, backward rate constant (/sec)
β_n	Hodgkin–Huxley, *n*-variable, backward rate constant (/sec)
BK	large-conductance, Ca^{2+}-activated K^+ channel (maxiK)
BrS	Brugada syndrome
C	peripheral nerve fibre subtype; conduction velocity ~2 m/s
CACNA1C-G402S	L-type Ca^{2+} channel mutation involving the junction between DI/S6 and the I-II loop of Cav1.2
CACNA1C-G406R	L-type Ca2+ channel mutation involving the junction between DI/S6 and the I-II loop of Cav1.2
Cacna1g	gene encoding T-type $Ca_V3.2$ channel
Cacna1h	gene encoding T-type $Ca_V3.2$ channel
$[Ca^{2+}]_i$.	intracellular free Ca^{2+} concentration (mmol/L)
$[Ca^{2+}]_o$	extracellular Ca^{2+} concentration (mmol/L)
CaM	Ca^{2+}/calmodulin
CaMK	CaM-dependent kinase
CaMKII	calmodulin kinase II
cAMP	cyclic 3',5',-adenosine monophosphate
CASQ	calsequestrin
CASQ2	calsequestrin type 2, cardiac isoform
Cav1.1	L-type Ca^{2+} channel type 1, skeletal muscle isoform
Cav1.2	L-type Ca^{2+} channel type 2, cardiac muscle isoform
$Ca_V3.2$	T-type Ca^{2+} channel

$-CH_3$	methyl-
$CH_3-NH_3^+$	methylamine ion
Cl^-	chloride ion
$[Cl^-]_i$	intracellular Cl^- concentration (mMol/L)
$[Cl^-]_o$	extracellular Cl^- concentration (mMol/L)
ClC-1	Cl^- conducting channel
CLIC2	chloride intracellular channel protein 2
c_m	capacitance of unit length of fibre (μF/cm)
C_m	specific membrane capacitance of unit surface area (μF/cm^2)
CN	calcineurin
CN^-	cyanide ion
CPVT	catecholaminergic polymorphic ventricular tachycardia
8-CPT	8-(4-chlorophenylthio)adenosine-3',5'-cyclic monophosphate (Epac activator)
Cr	creatine
CrP	creatine phosphate
cryo-EM	cryo-electronmicroscope
Cs^+	caesium ion
CSD	cortical spreading depression
c_T	tubular membrane capacitance (μF/cm)
CT	crista terminalis
Cx	connexin
Cx40	connexin type 40
Cx43	connexin type 43
d	membrane thickness (nm)
D600	methoxyverapamil
DAD	delayed after-depolarisation
DAG	diacylglycerol
ΔAPD	transmural action potential duration gradient (ms)
ΔAPD$_{90}$	transmural repolarisation gradient in action potential duration to 90% recovery (ms)
DHPR	dihydropyridine receptor
DHPR1	dihyropyridine receptor type 1, skeletal muscle isoform, Cav1.1
DHPR2	dihyropyridine receptor type 2, cardiac muscle isoform, Cav1.2
DI	diastolic interval (ms)
di-4-ANEPPS	3-(4-{2-[6-(dibutylamino)naphthalen-2-yl]ethenyl}pyridinium-1-yl)propane-1-sulfonate
di-8-ANEPPS	3-[4-[(E)-2-[6-(dioctylamino)naphthalen-2-yl]ethenyl]pyridin-1-ium-1-yl]propane-1-sulfonate
DI$_{crit}$	critical DI where the restitution A-curve shows unity slope (ms)
DI	voltage-gated ion channel domain I
DII	voltage-gated ion channel domain II

DIII	voltage-gated ion channel domain III
DIV	voltage-gated ion channel domain IV
D-line	action potential restitution (consequence for DI of varying APD)
DNP	2,4-dinitrophenol
ΔV_c	cell volume change (μL)
dV/dt	action potential upstroke velocity (mV/ms)
$(dV/dt)_{max}$	maximum action potential upstroke velocity (mV/ms)
dV/dx	voltage drop with respect to distance along the intracellular space (mV/cm)
$\partial V/\partial x$	voltage gradient along length of fibre (mV/cm)
e	electron charge (C)
E	membrane potential (mV)
$\hat{E}$	transmembrane electric field (V/m)
ε	permittivity (F/m)
ε_r	dielectric constant
EAD	early after-depolarisation
ECG	electrocardiogram
EF-hand	Ca^{2+} binding protein motif
EGTA	ethylene glycol-bis(β-aminoethyl ether)-N,N,N', N'-tetraacetic acid
E_K	K^+ Nernst equilibrium potential (mV)
ELC	essential myosin light chain
E_m	membrane potential (mV)
EMF	electromotive force (mV)
E_{Na}	Na^+ Nernst equilibrium potential (mV)
Epac	exchange protein directly activated by cAMP
EPP	end-plate potential
EPSC	excitatory postsynaptic current
EPSP	excitatory postsynaptic potential
ERP	effective refractory period (ms)
F	Faraday constant (C/mol)
ζ	fraction of the total voltage drop V across the membrane
FDNB	1-fluoro-2,4-dinitrobenzene
FK506	tacrolimus; fujimycin
FKBP	FK506 binding protein
FKBP12	FK506 binding protein type 12
G-protein	guanosine triphosphate binding protein
$G_1(V)$	energy of channel resting state (J/mol)
$G_2(V)$	energy of channel activated state (J/mol)
$G_a(V)$	energy of transition state (J/mol)
GABA	gamma-amino butyric acid
g_{Ca}	membrane Ca^{2+} conductance (mS/cm^2)
GDP	guanosine diphosphate
g_f	membrane HCN channel conductance (mS/cm^2)
GFP	green fluorescent protein
GH	growth hormone

G_i	adenylyl cyclase inhibitory G-protein
G_{i2}	adenylyl cyclase inhibitory G-protein
$G_{i/o}$	adenylyl cyclase inhibitory G-protein
g_K	membrane K$^+$ conductance (mS/cm^2)
$\bar{g}_K$	maximum K$^+$ conductance (mS/cm^2)
g_{leak}	membrane leak conductance (mS/cm^2)
g_{Na}	membrane Na$^+$ conductance (mS/cm^2)
$\bar{g}_{Na}$	maximum Na$^+$ conductance (mS/cm^2)
G_q	phospholipase C activating G-protein
Group I	myelinated sensory fibre subtype, diameter 20 to 12 μm
Group II	myelinated sensory fibre subtype, diameter 12 to 4 μm
Group III	myelinated sensory fibre subtype, diameter <4 μm
G_s	adenylyl cyclase stimulatory G-protein
GTP	guanosine triphosphate
G_α	GTP binding subunit of trimeric G-protein
Gβγ	βγ component of trimeric G-protein
h	Hodgkin–Huxley, Na$^+$ conductance inactivation variable
H$^+$	hydrogen ion
HCN	hyperpolarisation-induced cyclic nucleotide-activated channel
HCS	hydrophobic constriction site, Na$^+$ channel
HEK293	Human embryonic kidney 293
HgCl	mercuric chloride
HMM	heavy meromyosin
5-HT	5-hydroxytryptamine
^{1}H-NMR	proton nuclear magnetic resonance
^{2}H-NMR	deuterium nuclear magnetic resonance
H-zone	Heller zone (German 'heller': brighter).
θ	action potential propagation velocity (m/s)
I	current
$I(V,t)$	charge movement (if normalised to background membrane capacitance: μA/μF)
I_0	current electrode; three-microelectrode voltage clamp (mA)
$I_0(t)$	current delivered to control voltage at V_1; three-microelectrode voltage clamp (mA)
i_a	axial current flow along the length, x, of a fibre (mA)
$i_a(t)$	axial intracellular current; three-microelectrode voltage clamp (mA)
I-band	isotropic band
I_{Ca}	Ca^{2+} current
I_{CaL}	voltage-gated L-type Ca^{2+} current carried by Cav1.1 or Cav1.2
I_{CaT}	voltage-gated T-type Ca^{2+} current carried by Cav3.1 or Cav3.2

I_{cat}	cationic current
ICC	interstitial cells of Cajal
ICC-CM	interstitial cells of Cajal within gastrointestinal circular muscle
ICC-DMP	interstitial cells of Cajal within deep muscular plexus between small intestinal inner and outer circular muscle sublayers
ICC-LM	interstitial cells of Cajal within gastrointestinal longitudinal muscle
ICC-MP or ICC MY	interstitial cells of Cajal between the circular and longitudinal muscle layers
ICC-SM	interstitial cells of Cajal between submucosa and circular muscle
ICC-SMP	submucosal interstitial cells of Cajal
ICC-SS	interstitial cells of Cajal within the subserosal connective tissue space
$I_{Cl(Ca)}$	Ca^{2+}-activated Cl^- current
I_{crac}	Ca^{2+}-release-activated (capacitative) current
I_{dr}	delayed rectifying K^+ current
IFMT	inactivation amino acid sequence, Na^+ channel
IGF-1	insulin-like growth factor 1
I_i	ionic current (mA/cm^2)
I_K	K^+ current
i_K	K^+ current per unit length of fibre (mA/cm)
$I_{K(ACh)}$	acetylcholine-gated K^+ current carried by Kir3.1
I_{K1}	inward ('anomalous') rectifying K^+ current carried by Kir2.1, Kir2.2 and Kir2.3
I_{K2p}	2-pore domain K^+ leak current carried by TWIK1 or TASK1
I_{KATP}	adenosine triphosphate (ATP)-sensitive K^+ current carried by Kir6.2
I_{KCa}	Ca^{2+} activated K^+ current carried by KCa1.1
I_{Kp}	two-pore domain K^+ channel mediated K^+ current carried by TWIK1 or TASK1
I_{Kr}	voltage-gated rapidly activating outward K^+ current carried by Kv11.1
I_{Ks}	voltage-gated slowly activating outward K^+ current carried by Kv7.1
I_{Kur}	voltage-gated ultra-rapidly activating atrial outward K^+ current carried by Kv1.5
I_m	membrane current (mA/cm^2)
i_m	transmembrane current per unit length of fibre (mA/cm^2)
$i_m(t)$	membrane current in the fibre segment extending a distance $3l/2$ from end; three-microelectrode voltage clamp (mA/cm); l, electrode distancing.
I_{maxK}	Ca^{2+}-activated K^+ current

I_{Na}	voltage-gated Na^+ current carried by Nav1.1, Nav1.4, Nav1.5
i_{Na}	Na^+ current per unit length of fibre (mA/cm)
I_{NaL}	late Na^+ current
I_{NCX}	Na^+–Ca^{2+} exchanger current
IP_3	inositol *tris*phosphate
IP_3R	inositol 1,4,5-*tris*phosphate receptor
IPSP	inhibitory postsynaptic potential
IQ domain	isoleucine-glutamine Ca^{2+}/calmodulin binding domain
I_{sac}	stretch-activated current
I_{ti}	transient inward current
I_{to}	early transient outward K^+ current
$I_{to,f}$	voltage-gated fast transient outward current carried by $K_v4.2$, Kv4.3
$I_{to,s}$	voltage-gated slow transient outward K^+ current carried by $K_v1,4$
IVC	inferior venacava
JNK	c-Jun N-terminal kinase
k	Boltzmann constant (J/K)
k	steepness factor, Boltzmann equation (mV)
K^+	potassium ion
$[K^+]_i$	intracellular K^+ concentration (mmol/L)
$[K^+]_o$	extracellular K^+ concentration (mmol/L)
K_{ATP}	ATP-sensitive K^+ channel
KChIP2	K^+ channel interacting protein 2
KCNH2	encoding gene for protein carrying I_{Kr}
KCNQ1	encoding gene for protein carrying I_{Ks}
K_{dr}	delayed rectifier K^+ channel
Kir	inwardly rectifying K^+ channel
Kir1.x, Kir4.x, Kir5.x, Kir7.x	inwardly rectifying K^+ channel variants
Kir2.1, Kir2.2, Kir2.3	K^+ channels carrying I_{K1}
Kir2.x	persistently active inwardly rectifying K^+ channel
Kir3.x	G-protein-receptor-coupled inwardly rectifying K^+ channel
Kir6.2	inward rectifying K^+ channel carrying I_{KATP}
Kir6.x	ATP- sensitive inwardly rectifying K^+ channel
KN-93	N-[2-[N-(4-chlorocinnamyl)-N-methylaminomethyl]phenyl]-N-(2-hydroxyethyl)-4-methoxybenzenesulfonamide (CaMK II inhibitor)
Kv4.2	K^+ channel, carrying $I_{to,f}$
Kv4.3	K^+ channel, carrying $I_{to,f}$
L	channel number density (channels/μm^2)
Λ	action potential wavelength (mm)
λ	dynamic space constant (mm)

$\lambda(\infty)$	steady-state space constant (mm)
Λ'	action potential wavelength (mm)
Λ_0'	action potential resting wavelength (mm)
$[(\text{lactate})^-]_i$	intracellular lactate ion concentration (mmol/L)
$[(\text{lactate})^-]_o$	extracellular lactate ion concentration (mmol/L)
Lead I	electrocardiogram recording between right and left arm
Lead II	electrocardiogram recording between right arm and left leg
Lead III	electrocardiogram recording between left arm and left leg
LMM	light meromyosin
LQTS	long QT syndrome
LQTS1	long QT syndrome, related to *KCNQ1*-mediated I_{Ks}
LQTS2	long QT syndrome, related to *KCNH2*-mediated I_{Kr}
LQTS3	long QT syndrome type 3 related to *SCN5A*-mediated I_{Na}
LQTS8	long QT syndrome type 8 related to *CACNA1c*-mediated I_{CaL} (Timothy syndrome)
LTD	long-term depression
LTP	long-term potentiation
LV	left ventricle
m	Hodgkin–Huxley, Na^+ conductance activation variable
M	myosin
m	quantal content (mean number of quanta released during an EPP)
M_2	muscarinic receptor type 2
M_3	muscarinic receptor type 3
MAP	monophasic action potential
maxiK	large-conductance, Ca^{2+}-activated K^+ channel (BK)
Mdg	muscular dysgenic
MEPP	miniature end-plate potential
$[Mg^{2+}]_o$	extracellular Mg^{2+} concentration (mmol/L)
mGluR	metabotropic glutamate receptor
Mkk4	mitogen-activated protein kinase kinase 4
Mkk4-acko	conditional *Mkk4* knockout
Mkk4-f/f	*Mkk4* control
MLC	myosin light chain
MLCK	myosin light-chain kinase
M-line	Mittelscheibe line (German: 'Mittel': middle; 'scheibe': disc)
Mn^{2+}	manganese ion
MRI	magnetic resonance imaging
mRNA	messenger ribonucleic acid

n	Hodgkin–Huxley, K^+ conductance activation variable
N	membrane density of a specified channel or transporter
N2B	cardiac-specific titin element
NA	noradrenaline
Na^+	sodium ion
$[Na^+]_i$	intracellular Na^+ concentration (mmol/L)
$[Na^+]_o$	extracellular Na^+ concentration (mmol/L)
NAD^+	oxidized nicotinamide adenine dinucleotide
NADH	reduced nicotinamide adenine dinucleotide
Nav1.1	α-subunit voltage-dependent Na^+ channel, neuronal isoform
$Na_v1.1$, $Na_v1.2$, $Na_v1.3$, $Na_v1.6$	central nervous system α-subunit voltage-dependent Na^+ channel isoforms
$Na_v1.4$	skeletal muscle α-subunit voltage-dependent Na^+ channel isoform
$Na_v1.5$	cardiac muscle α-subunit voltage-dependent Na^+ channel isoform
$Na_v1.7$, $Na_v1.8$, and $Na_v1.9$	peripheral nervous system α-subunit voltage-dependent Na^+ channel isoforms
NCX	Na^+–Ca^{2+} exchanger
NFAT	nuclear factor of activated T cells
$-NH_2$	amino-
NH_2–NH_3^+	hydrazine
NK_1	neurokinin receptor type 1
NK_3	neurokinin receptor type 3
NKCC1	Na^+-K^+-Cl^- cotransporter type 1
NMDA	N–methyl-D-aspartate
nNOS	neuronal nitric oxide synthase
NO	nitric oxide
-OH	hydroxyl-
OH–NH_3^+	hydroxylamine
P	force exerted by the muscle in the Hill equation (N)
P_0	isometric tension in the Hill equation (N)
P1	voltage-gated ion channel p-loop helix 1
P2	voltage-gated ion channel p-loop helix 2
P2XR	ionotropic P2X purinergic receptor
P2YR	metabotropic purinergic receptor
Pak1	p21 activated kinase-1
Pak1-/-	murine p21 activated kinase-1 knockout
Pak1-cko	murine p21 activated kinase-1 conditional knockout
P_{Cl}	membrane permeability to Cl^- (m/s)
PEVK	domains rich in proline (P), glutamate (E), valine (V), and lysine (K)
PGC-1α	peroxisome proliferator-activated receptor-γ coactivator-1α

PGC-1β	peroxisome proliferator-activated receptor-γ coactivator-1β
$Pgc\text{-}1\beta^{-/-}$	murine PGC-1β knockout
pH	negative logarithm to base 10 of [H$^+$]
Phase 0	cardiac action potential rapid depolarisation phase
Phase 1	cardiac action potential initial brief rapid repolarisation phase
Phase 2	cardiac action potential plateau phase
Phase 3	cardiac action potential terminal repolarisation phase
Phase 4	cardiac action potential electrical diastole
pH$_i$	logarithm to base 10 of intracellular [H$^+$]
PIP$_2$	phosphatidylinositol 4,5-*bis*phosphate
P_K	membrane permeability to K$^+$ (m/s)
P_k	probability of an EPP containing k quanta
pK_a	negative logarithm of dissociation constant
PKA	phosphokinase A
PKC	phosphokinase C
P_{Lac-}	membrane permeability to lactate ion (m/s)
P_{LacH}	membrane permeability to lactate acid (un-ionised) (m/s)
PLC	phospholipase C
PLN	phospholamban
PM	voltage-gated ion channel pore module
PMCA	sarcolemmal Ca^{2+}-ATPase
P_{Na}	membrane permeability to Na$^+$ (m/s)
PP1	protein phosphatase isoform 1
PP2A	protein phosphatase isoform 2
PPA1	protein phosphatase 1
[Pr^{z-}]	protein concentration (mmol/L)
P-wave	electrocardiogram recording, initial atrial depolarisation-related wave
<q>	microscopic charge movement (if normalised to background membrane capacitance: nC/μF)
$Q(V, t)$	charge movement (if normalised to background membrane capacitance: nC/μF)
$Q(V,\infty)$	steady state charge (if normalised to background membrane capacitance: nC/μF)
$Q_0(V)$	lipid bilayer charge (if normalised to background membrane capacitance: nC/μF)
Q_{max}	maximum charge (if normalised to background membrane capacitance: nC/μF)
qPCR	quantitative reverse transcriptase polymerase chain reaction
QRS-complex	major ventricular-related ECG deflection
R	universal gas constant (J/(K mol))
R_a	cytoplasmic resistivity (Ω cm)

r_a	intracellular resistance of unit fibre length (kΩ/cm)
r_a	intracellular resistance of unit length (kΩ.cm)
RA	right atrium
r_{ac}	tubular access resistance (kΩ cm)
RH237	(N-(4-sulfobutyl)-4-(6-(4-(dibutylamino)phenyl) hexatrienyl)pyridinium [styryl membrane voltage indicator]
Rhod-2	1-[2-amino-5-(3-dimethylamino-6-dimethyl-ammonio-9-xanthenyl)phenoxy-2-(2-amino-5-methylphenoxy)-ethane-N,N,N′,N′-tetraaacetic acid (BAPTA-derived Ca^{2+}-sensitive dye)
R_K	membrane resistance attributable to K^+ conductance (kΩ cm^2)
r_L	tubular luminal resistance (kΩ/cm)
RLC	regulatory myosin light chain
R_{leak}	membrane resistance attributable to leak conductance (kΩ cm^2)
r_m	resistance of unit length of fibre (kΩ cm)
R_m	specific membrane resistance (kΩ cm^2)
r_m	surface membrane resistance of unit length of fibre (kΩ cm)
R_{Na}	membrane resistance attributable to Na^+ conductance (kΩ cm^2)
RNA	ribonucleic acid
r_o	extracellular fluid resistance of unit length of nerve fibre (kΩ/cm)
ROCC	receptor-operated cation channel
ROS	reactive oxygen species
R_{patch}	resistance of membrane within the patch (MΩ); loose patch electrode recording
R_{pip}	loose patch clamp electrode resistance (MΩ); loose patch electrode recording
R_{seal}	seal resistance between loose patch electrode and external membrane surface (MΩ); loose patch electrode recording
r_T	tubular membrane resistance (kΩ/cm)
RV	right ventricle
RVOT	right ventricular outflow tract
RyR1	ryanodine receptor type 1, skeletal muscle isoform
RyR2	ryanodine receptor type 2, cardiac muscle isoform
RyR2-P2328S	RyR2 mutation exemplar
$RyR2^{S/+}$	murine heterozygotic RyR2-P2328S mutant
$RyR2^{S/S}$	murine homozygotic RyR2-P2328S mutant
RyR3	ryanodine receptor type 3, neuronal isoform
S1	myosin subfragment 2

S1	pacing stimulus
S1-S6	transmembrane segments in voltage-gated ion channel domains
S2	extrasystolic stimulus
S2	myosin subfragment 2
SAC	mechanosensitive, stretch-activated, channel
SAN	sino-atrial node
$sarcK_{ATP}$	sarcolemmal ATP-sensitive K^+ channel
SCF	stem cell factor
Scn5a+/−	heterozygotic Nav1.5 deficient genotype
Scn5a+/ΔKPQ	gain of function cardiac Na^+ channel genotype
Scn5a-1798insD	gain of function cardiac Na^+ channel genotype
SEM	standard error of the mean
SEP	atrial septum
SERCA	sarcoplasmic reticular Ca^{2+} ATPase
SERCA1	sarcoplasmic reticular Ca^{2+}-ATPase type 1, skeletal muscle isoform
SERCA2	sarcoplasmic reticular Ca^{2+}-ATPase type 2, cardiac muscle isoform
SF	ion channel selectivity filter
Slc12a2	gene encoding Na^+-K^+-Cl^- cotransporter type 1
SOCC	store-operated Ca^{2+} channel
SR	sarcoplasmic reticulum
Sr^{2+}	strontium ion
STD	spontaneous transient depolarization events/ unitary potentials
STIC	spontaneous transient inward current
SVC	superior vena cava
T	absolute temperature (K)
t	time (ms)
T−	transverse
τ	time constant (ms)
TdP	torsades de pointes (French: 'twisting of the peaks')
TGF-β_1	transforming growth factor β type 1
TM	tropomyosin
Tris	*tris*-hydroxyaminomethane
TRP	transient receptor potential channel
TRPC1	transient receptor potential channel protein type 1
TRPC3	transient receptor potential channel protein type 3
TRPC6	transient receptor potential channel protein type 6
TTCC	T-type Ca^{2+} channels
T-tubule	transverse tubule
TTX	tetrodotoxin
T-wave	electrocardiogram recording, ventricular-repolarisation-related wave

V	membrane voltage (mV)
V	velocity of shortening in the Hill equation (cm/s)
V	voltage (V)
$V(t)$	membrane voltage at time t (mV)
$V(x)$	membrane voltage at position x (mV)
$V(x, t)$	membrane voltage as a function of x and t (mV)
V^*	membrane voltage at which energies of G_1 and G_2 are equal (mV)
V1–V6	electrocardiogram recording: chest leads 1–6.
V_c	cell volume (μL)
VE	ventricular ectopic
VERP	ventricular ERP (ms)
VF	ventricular fibrillation
VIP	vasoactive intestinal polypeptide
VIP_1	vasoactive intestinal peptide type 1
V_{pip}	voltage at the back end of the pipette (mV); loose patch electrode recording
V_{rest}	external voltage of patched membrane (mV); loose patch electrode recording
VSM	voltage-sensing module, voltage-gated ion channel
VT	ventricular tachycardia
WT	wild type
x	position along length of a nerve (mm)
z	charge valency
Z-line	Zwischenscheibe line (German: 'Zwischen': spacer; 'scheibe': disk)
z_x	effective valency

Structural Organisation of the Nervous System

1.1 | Nervous Systems

An important characteristic of higher animals is their possession of a more or less elaborate system for rapid transfer of information through the body in the form of electrical signals or nervous impulses. At the bottom of the evolutionary scale, nervous systems of some primitive invertebrates consist simply of interconnected networks of undifferentiated nerve cells. The next step in complexity is a differentiation into *sensory* nerve cells responsible for gathering incoming information, and *motor* nerve cells that execute an appropriate response. The nerve cell bodies are grouped together to form *ganglia*. Different specialised receptor organs detect every kind of change in the external and internal environment. Likewise, different types of effector organ formed by muscles and glands receive and execute the outgoing instructions. In invertebrates, the ganglia that link the inputs and outputs remain to some extent anatomically separate. In vertebrates most of the nerve cell bodies are collected together in the *central nervous system*. The *peripheral nervous system* comprises *afferent* sensory nerves conveying information to the central nervous system, and *efferent* motor nerves conveying instructions from it. Within the central nervous system, the different pathways are connected up by large numbers of *interneurons* which have an integrative function. Fuller accounts of nervous system structure are summarised elsewhere (Brodal, 2016; Waxman, 2017).

Certain ganglia involved in internal homeostasis remain outside the central nervous system. Together with the preganglionic nerve trunks leading to them, and the postganglionic fibres arising from them, which innervate smooth muscle and gland cells in the animal's viscera and elsewhere, these constitute the *autonomic nervous system*. The preganglionic autonomic fibres leave the central nervous system in two distinct outflows. Those in the cranial and sacral nerves form the *parasympathetic* division of the autonomic system, while those coming from the thoracic and lumbar segments of the spinal cord form the *sympathetic* division.

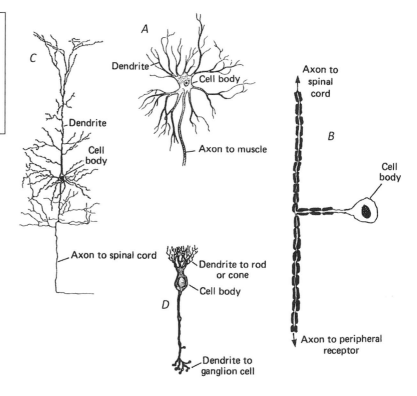

Figure 1.1 Schematic diagrams (not to scale) of the structure of: (A) a spinal motoneuron; (B) a spinal sensory neuron; (C) a pyramidal cell from the motor cortex of the brain; (D) a bipolar neuron in the vertebrate retina. (Adapted from Nicholls *et al.*, 2001.)

1.2 | The Anatomy of a Neuron

Each neuron has a cell body containing its nucleus and a number of processes or *dendrites* (Figure 1.1). One process, often much longer than the rest, is the *axon* or nerve fibre which carries outgoing impulses. The incoming signals from other neurons are passed on at junctional regions known as *synapses* scattered over the cell body and dendrites; discussion of their structure and of the special mechanisms involved in synaptic transmission is deferred to Chapters 7 and 8. The cell body is essential for long term maintenance of the axon (Section 1.5). However, it does not play an immediate role in axonal conduction of impulses. A nerve continues to function for a significant time after being severed from its cell body: electrophysiologists would have a harder time if this were not the case. We first concern ourselves with properties of peripheral nerves.

1.3 | Unmyelinated Nerve Fibres

Vertebrates have two main types of nerve fibre, larger fast-conducting *myelinated* axons, 1 to 25 μm in diameter, and small slowly conducting *unmyelinated* axons (under 1 µm). Most autonomic system fibres and peripheral sensory fibres subserving sensations like pain and temperature, where a rapid response is not required, are unmyelinated. Almost all invertebrates are equipped exclusively with unmyelinated fibres, but those showing particularly rapid conduction may have diameters as large as 500–1000 µm. Subsequent chapters will discuss the use of giant invertebrate axons in experiments clarifying the mechanism of conduction of the nervous impulse. The major fundamental advances made in electrophysiology have often depended heavily on technical possibilities opened up by opportunities offered by comparative zoology (Chapter 4).

All nerve fibres consist essentially of a long cylinder of cytoplasm, the *axoplasm*, surrounded by an electrically excitable *nerve membrane*. The electrical resistance of axoplasm is relatively low, by virtue of its appreciable concentrations of K^+ and other ions. That of the membrane is relatively high. The electrolyte-containing extracellular fluids outside the membrane are also good electrical conductors. Nerve fibre structure therefore parallels that of a shielded electric cable, with a central conducting core surrounded in turn by insulation and a further conducting layer. Many features of the behaviour of nerve fibres depend intimately on their *cable structure* (Section 6.1).

The layer analogous with the insulation of the cable does not, however, consist solely of the high-resistance nerve membrane, owing to the presence of *Schwann cells*. These are wrapped around the *axis cylinder* in a manner which varies with the different nerve fibre types. In the olfactory nerve (Figure 1.2), a single Schwann cell serves as a multi-channel supporting structure enveloping a short stretch of 30 or more tiny axons. Elsewhere, groups of individual axons may be associated with a single Schwann cell, with some deeply embedded within the Schwann cell, and others almost uncovered. In general, each Schwann cell supports a small group of up to half a dozen axons (Figure 1.3). In large invertebrate axons (Figure 1.4) the ratio is reversed, the whole surface of the axon being covered with a mosaic of many Schwann cells interdigitated with one another to form a layer several cells thick.

In all unmyelinated nerves, whether large or small, the axon membrane is separated from the Schwann cell membrane by a ~10 nm wide space sometimes termed the *mesaxon* by anatomists. This space freely communicates with the remaining tissue extracellular space. It provides a relatively uniform pathway for electric current flow during passage of an impulse. However, it can be tortuous and so ions entering it from the active nerve may temporarily accumulate within it, leading to events contributing to the *after-potential* (Section 6.5). Nevertheless, for the immediate purpose of describing nerve impulse propagation, unmyelinated fibres may be regarded as having a uniformly low external electrical resistance between different points on the outside of the membrane.

Figure 1.2 Electronmicrograph of a section through the olfactory nerve of a pike, showing a bundle of unmyelinated nerve fibres partially separated from other bundles by the basement membrane B. The mean diameter of the fibres is 0.2 μm, except where they are swollen by the presence of a mitochondrion (M). Magnification 54 800 ×. (Reproduced by courtesy of Prof. E. Weibel.)

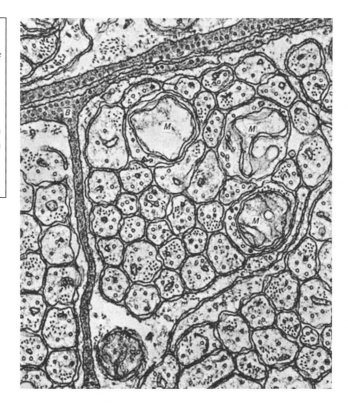

1.4 | Myelinated Nerve Fibres

In vertebrate myelinated nerve fibres, the axon is electrically insulated by the *myelin sheath* everywhere except at the *nodes of Ranvier* (Figures 1.5, 1.6, 1.7). In peripheral nerves, each stretch of myelin is laid down by a Schwann cell that repeatedly envelops the axis cylinder with many concentric layers of cell membrane (Figure 1.7). In the central nervous system, it is *oligodendroglial* cells that lay down the myelin. All cell membranes consist of a double layer of lipid molecules with which some proteins are associated (Section 3.1). This forms a structure that after appropriate staining appears under the electron-microscope as a pair of dark lines 2.5 nm across, separated by a 2.5 nm gap. In an adult myelinated fibre, the adjacent layers of Schwann cell membrane are partly fused together at their cytoplasmic surface, and the overall repeat distance of the double membrane, as determined by X-ray diffraction, is 17 nm. For a nerve fibre whose outside diameter is 10 μm, each stretch of myelin is about 1000 μm long and 1.3 μm thick, so that the myelin is built up of some 75 double layers of Schwann cell membrane. In larger fibres, the internodal distance, the thickness of the myelin and hence the number of layers, are all proportionately greater. Since myelin has a higher lipid content than cytoplasm, it also

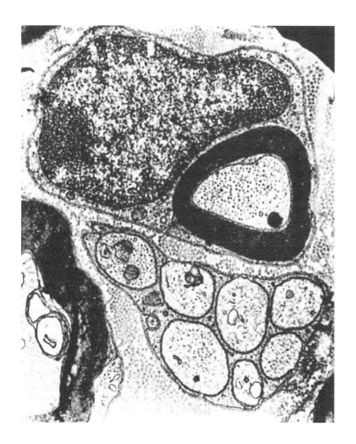

Figure 1.3 Electronmicrograph of a cross-section through a mammalian nerve showing unmyelinated fibres with their supporting Schwann cells and some small myelinated fibres. (Reproduced by courtesy of Professor J. D. Robertson.)

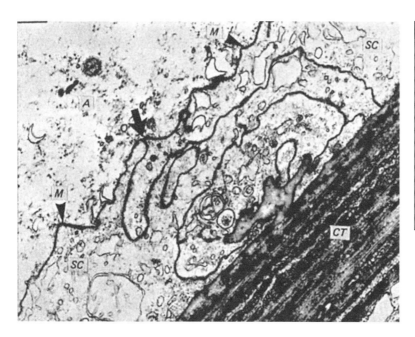

Figure 1.4 Electronmicrograph of the surface of a squid giant axon, showing the axoplasm (A), Schwann cell layer (SC) and connective tissue sheath (CT). Ions crossing the excitable membrane (M, arrowheads) must diffuse laterally to the junction between neighbouring Schwann cells marked with an arrow, and thence along the gap between the cells into the external medium. Magnification 22 600 ×. (Reproduced by courtesy of Dr F. B. P. Wooding.)

Figure 1.5 Electronmicrograph of a node of Ranvier in a single fibre dissected from a frog nerve. (Reproduced by courtesy of Professor R. Stämpfli.)

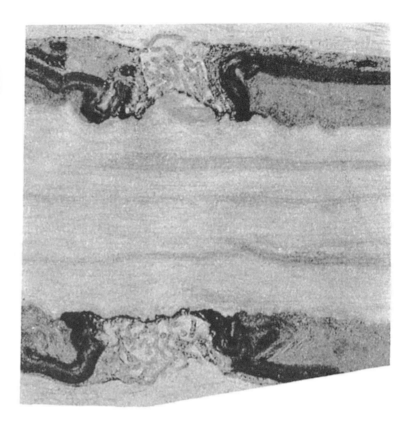

Figure 1.6 Schematic diagram of the structure of a vertebrate myelinated nerve fibre. The distance between neighbouring nodes is actually about 40 times greater relative to the fibre diameter than is shown here.

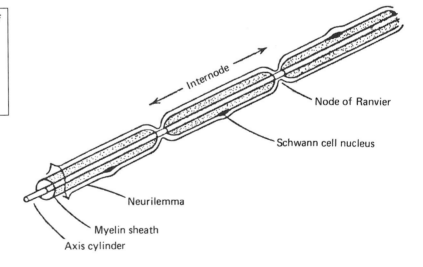

Internode

Node of Ranvier

Schwann cell nucleus

Neurilemma

Myelin sheath

Axis cylinder

has a greater refractive index. In unstained preparations it has a characteristic glistening white appearance. This accounts for the name given to the peripheral *white matter* of the spinal cord, consisting of columns of myelinated nerve fibres. In contrast, the central core of

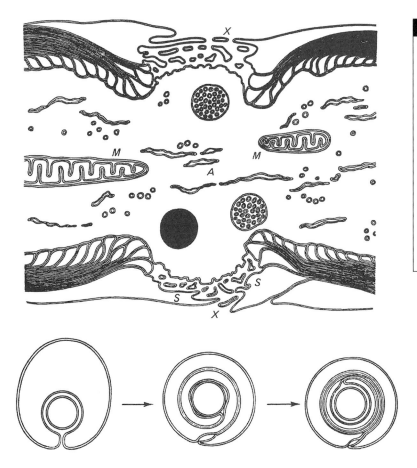

Figure 1.7 Drawing of a node of Ranvier made from an electronmicrograph. The axis cylinder (*A*) is continuous through the node; the axoplasm contains mitochondria (*M*) and other organelles. The myelin sheath, laid down as shown below by repeated envelopment of the axon by the Schwann cell on either side of the node, is discontinuous, leaving a narrow gap *X*, where the excitable membrane is accessible to the outside. Small tongues of Schwann cell cytoplasm (*S*) project into the gap, but do not close it entirely. (From Robertson, 1960.)

grey matter mainly comprises nerve cell bodies and supporting tissue. It also accounts for the difference between the white and grey rami of the autonomic system, containing respectively small myelinated nerve fibres and unmyelinated fibres.

At the node of Ranvier, the closely packed layers of Schwann cell terminate on either side as a series of small tongues of cytoplasm (Figure 1.7), leaving a gap ~1 μm in width where there is no obstacle between the axon membrane and the extracellular fluid. The external electrical resistance between neighbouring nodes of Ranvier is therefore relatively low, whereas the resistance between any two points on the internodal stretch of membrane is high because of the insulating effect of the myelin. The difference between the nodes and internodes in accessibility to the external medium is the basis for the *saltatory* mechanism of conduction in myelinated fibres (Section 6.3), which enables them to conduct impulses some 50 times faster than a non-myelinated fibre of the same overall diameter. Nerves may branch many times before terminating, and the branches always arise at nodes.

In peripheral myelinated nerves, the whole axon is covered by a thin, apparently structureless basement membrane, the *neurilemma*. The nuclei of the Schwann cells are found just beneath the neurilemma close to the midpoint of each internode. The fibrous connective tissue which separates individual fibres is known as the *endoneurium*. The fibres are bound together in bundles by the *perineurium*, and the several bundles which in turn form a whole nerve trunk are surrounded by the *epineurium*. The connective tissue sheaths in which the bundles of nerve fibres are wrapped also contain continuous sheets of cells which prevent extracellular ions in the spaces between the fibres from mixing freely with those outside the nerve trunk. The barrier to free diffusion offered by the sheath is probably responsible for some of the experimental discrepancies between the behaviour of fibres in an intact nerve and that of isolated single nerve fibres. The nerve fibres within the brain and spinal cord are packed together very closely, and are thought to lack a neurilemma. The individual fibres are difficult to tease apart, and the nodes of Ranvier are less easily demonstrated than in peripheral nerves by such histological techniques as staining with silver nitrate.

1.5 | Nerve Fibre Responses to Injury

Of major clinical importance is the limited capacity for nerve regeneration following injury leading to nerve transection. The cell bodies of the neurons involved undergo a chromatolysis reflecting dissolution of their contained *Nissl bodies* made up of rough endoplasmic reticulum. The nerve region distal to the injury shows a *Wallerian degeneration*. The myelinated sheaths retract from the nodes and break down into separate ellipsoidal structures (Figure 1.8). There is a loss of excitability in the distal nerve fibres within days after interruption. The distal axons are phagocytosed, leaving persistent tube-like Schwann cell sheaths. A process of regeneration commences from the nerve proximal to the lesion. There the fibre ends develop buds or sprouts that elongate into regenerating fibres. In the event that these reach and penetrate the sheaths left by the distal degenerating part of the nerve, further growth occurs along the sheaths at a rate of 3–4 mm/day. This can eventually lead to reinnervation of their effector structure. The regenerated axons recover their diameters and regain their myelination over the succeeding 4–12 months (Mulroy *et al.*, 1990).

It has been suggested that guidance of regenerating fibres into their distal stumps may involve chemotropic effects, but it is unlikely that there are long-range neurotrophic effects related to specific neuroeffector structures. Regenerating fibres appear to be directed in their growth by the degenerated sheaths, likely along their contained

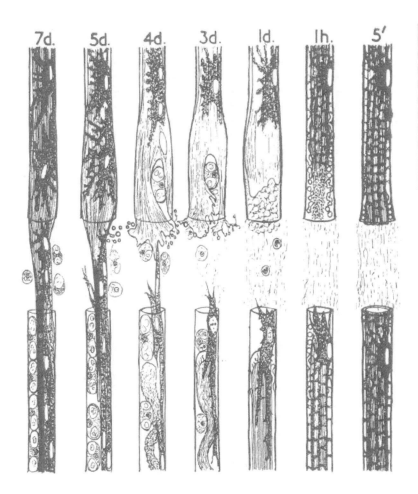

Figure 1.8 Wallerian degeneration and regeneration from 5 min to 7 days following nerve fibre transection. There is retraction of the central portion of the distal portion of the cut axon from 1 h, budding at the proximal portion from the 3rd and 4th day, and axon growth into the distal stump on the 5th and 7th days. (From Young, 1949.)

basement membrane material. The latter appears to support nerve growth even when derived from tissue other than nerve, such as basement membrane preparations from freeze-treated muscle tissue (Keynes *et al.*, 1984). In experimental surgical studies, use of appropriately oriented preparations promoted nerve regrowth and ultimate return of function (Glasby *et al.*, 1986a, 1986b). Such findings prompted explorations for future clinical applications in the surgical repair of nerve injury or transection (Gattuso *et al.*, 1988).

Resting and Action Potentials

2.1 | Electrophysiological Recording Methods

Electrical recording methods have classically provided much the most sensitive and convenient approach to studies of excitable activity, including the nervous impulse. This chapter provides a brief background to such measurements of both steady and rapidly changing electrical potentials. Later chapters will broaden the scope of these methods and additionally discuss radioactive tracer (e.g. Section 3.6), optical (Section 11.3), biochemical (Sections 5.1 and 5.2), genetic (Section 14.1) and mathematical modelling (Sections 3.7 and 4.7) techniques for studying nerve and muscle function. All these have proven important to our current understanding.

Recording potential differences between two points involves placing electrodes connected to a suitable amplifier and recording system at each of them. Investigations directed only at action potentials can employ fine platinum or tungsten wire electrodes. However, any bare metal surface has the disadvantage of becoming *polarised* by passage of electric current into or out of the solution which it contacts. Measurements of the absolute magnitude of a steady potential at the electrode tip require non-polarisable or reversible electrodes for which the unavoidable *contact potential* between metal and solution is both small and constant. The simplest type of reversible electrode is provided by electrolytically coating a silver wire with AgCl to produce an Ag/AgCl half-cell, but calomel (Hg/HgCl) half-cells are best employed for the most accurate measurements.

Recording the potential within a cell also requires the electrode to be well insulated, except where it directly contacts the intracellular fluid. It also needs to be fine enough to access the intracellular fluid with minimum cell damage to avoid electrical leaks across the membrane. The earliest intracellular recordings inserted a ~50 μm diameter glass capillary longitudinally down a 500 μm squid axon through a cannula tied into the cut end (Figure 2.1A). However, this method is not universally applicable to all excitable cells. The latter have often been studied using glass microelectrodes made from ~2 mm diameter

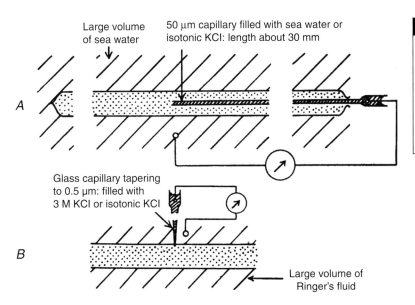

Large volume of sea water

50 μm capillary filled with sea water or isotonic KCl: length about 30 mm

A

Glass capillary tapering to 0.5 μm: filled with 3 M KCl or isotonic KCl

B

Large volume of Ringer's fluid

Figure 2.1 Methods for measuring absolute values of resting potential and action potential: (*A*) longitudinal insertion of 50 μm internal electrode into a squid giant axon; (*B*) transverse insertion of 0.5 μm internal electrode used for recording from muscle fibres and other cells. (From Hodgkin, 1951.)

hard glass tubing drawn out to produce a tapered micropipette of <0.5 μm tip diameter (Figure 2.1*B*). The microelectrode is filled with electrolyte solution, typically 3 M KCl, and an Ag/AgCl electrode inserted at the wide end for connection to an input amplifier. Variants of such microelectrodes have been used to measure membrane potentials both in single neurons and many other cell types (Sections 7.6, 11.4 and 14.9).

The potentials to be measured in electrophysiological experiments range from 150 mV down to a few μV. Faithful recording requires the recording system to have a flat frequency response from zero to ~50 kHz (1 Hz = 1 cycle/s). The amplifier requires a high input impedance and low electrical noise in the absence of input signal. These requirements are well fulfilled by currently available high-quality solid-state operational amplifiers, not available to the early investigators. The output is typically displayed on a monitor, classically a cathode-ray oscilloscope, ideally with storage capacity, and currently a computer screen. This permits detailed examination of the signals off-line. Permanent records were originally obtained by photographing the screen image, but are now typically digitally archived. Direct-writing recorders originally used to obtain continuous paper analogue records generally cannot follow the high frequencies required to faithfully reproduce individual action potentials. Current experimenters accomplish close examination of such signal timecourses by converting them into digital form, using an on-line computer for data storage and analysis (e.g. Figures 4.12 and 4.13).

Electrical recording studies typically accompany stimulation of electrophysiological activity. This can involve delivery of currents or voltages to the nerve exterior, or directly into the intracellular space.

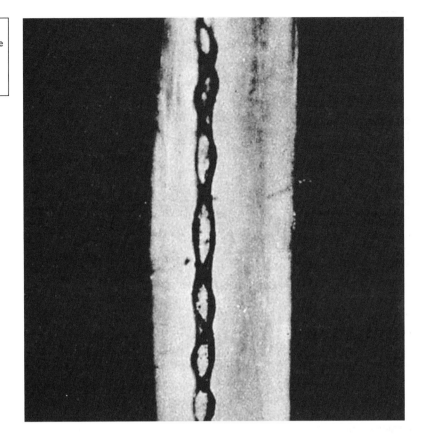

Figure 2.2 A squid giant axon into which a double spiral electrode has been inserted, photographed under a polarising microscope. Its diameter was 700 μm.

In the latter case, this requires introduction of two electrodes, for current delivery and voltage recording respectively, into the axon. Hodgkin and Huxley used an internal electrode system consisting of a double spiral of chloride-coated silver wire wound on a fine glass rod (Figure 2.2). Others used a glass microcapillary as a voltage electrode to which was glued an Ag/AgCl or platinised platinum wire to deliver current. The current electrode is connected to the output of an amplifier and pulse generator. This technique can be further extended by feeding the recorded voltage as a subtracting input to the current amplifier. The amplifier output can be controlled to produce just sufficient current to hold the potential at a desired command value, thereby achieving a *voltage clamp* of the membrane (Cole, 1941). This technique permitted measurements and analysis of membrane current in response to imposition of voltage steps to series of predetermined levels (Section 4.4; Hodgkin and Katz, 1949). Since its introduction, different variants of voltage clamping have played ever more important roles in investigating mechanisms of excitability in nerve (Section 6.5) and muscle (Sections 7.5, 7.6, 10.2, 11.4 and 13.6).

2.2 | Intracellular Recording of the Membrane Potential

Hodgkin and Huxley measured the absolute magnitude of the electrical potential within a living cell by introducing a 50 μm capillary electrode into a squid giant axon. When the tip of the electrode was far enough from the cut end it became up to 60 mV negative with respect to an electrode in the external solution. This *resting potential* across the membrane in the intact axon was about −60 mV, inside relative to outside. Stimulation of the axon by applying a shock at the far end elicited an *action potential* (Figure 2.3), or *spike*. Its amplitude was over 100 mV. At its peak the membrane potential was reversed by at least 40 mV. Typical values for isolated axons recorded with this type of electrode (Figure 2.4A) would be a resting potential of −60 mV and a spike of 110 mV. This would correspond to an internal potential of +50 mV at the peak of the spike. Records made with 0.5 μm diameter electrodes for undissected axons *in situ* in the squid's mantle gave slightly larger potentials, and did not show the underswing or *positive phase* at the tail of the spike (Figure 2.4B). At 20 °C the duration of the spike was about 0.5 ms; the records in Figure 2.4 were made at a lower temperature.

Every excitable tissue type, from mammalian motor nerve to muscle and electric organ, gave resting and action potentials of similar magnitudes (Figure 2.4C–H). The resting potential always fell between −60 and −95 mV and the potential at the peak of the spike between +20 and +50 mV. However, action potential shapes and durations varied. Their durations ranged from 0.5 ms in a mammalian myelinated fibre to 0.5 s in cardiac ventricular myocytes with their characteristically prolonged plateaus. Nevertheless, for a given

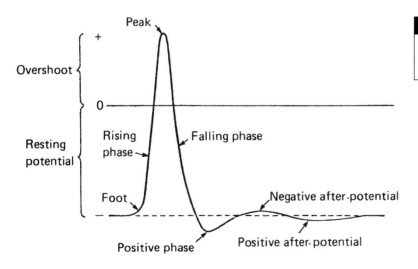

Figure 2.3 Nomenclature of the different parts of the action potential and the after-potentials that follow it.

Figure 2.4 Intracellular records of resting and action potentials. Horizontal lines (dashed in (A) and (B)) indicate zero potential; positive potential upwards. Marks on the voltage scales are 50 mV apart. The number against each time scale is its length (ms). In some cases the action potential is preceded by a stimulus artifact: (A) squid axon *in situ* at 8.5 °C, recorded with a 0.5 μm microelectrode; (B) squid axon isolated by dissection, at 12.5 °C, recorded with a 100 μm longitudinal microelectrode; (C) myelinated fibre from dorsal root of cat; (D) cell body of motor neuron in spinal cord of cat; (E) ventricular muscle fibre in frog's heart; (F) Purkinje fibre in sheep's heart; (G) electroplate in electric organ of *Electrophorus electricus*; (H) isolated fibre from frog sartorius muscle. ((A) and (B) recorded by A. L. Hodgkin and R. D. Keynes, from Hodgkin, 1958; (C) recorded by K. Krnjevic; (D) from Brock *et al.*, 1952; (E) recorded by B. F. Hoffman; (F) recorded by S. Weidmann, from Weidmann, 1956; (G) from Keynes and Martins-Ferreira, 1953; (H) from Hodgkin and Horowicz, 1957.)

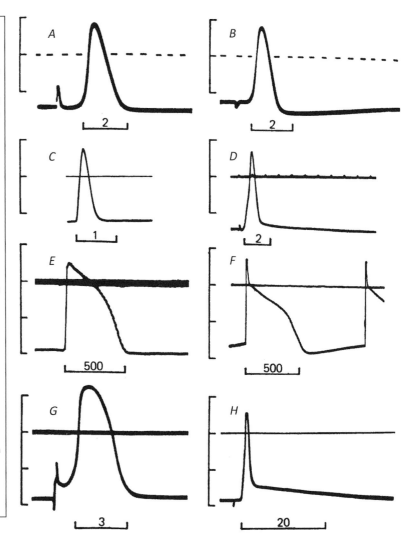

excitable cell the action potential remained exactly the same in both size and shape under constant external conditions such as temperature and bathing solution composition. This is an essential consequence of the *all-or-nothing* behaviour of the propagated impulse.

2.3 | Extracellular Recording of the Nervous Impulse

In many experimental situations it is impracticable to use intracellular electrodes. The passage of impulses can then only be studied using external electrodes. It is therefore important to relate findings obtained with such electrodes to potential changes at membrane level.

During the impulse the potential across the active membrane is reversed. This makes the membrane exterior negative with respect to

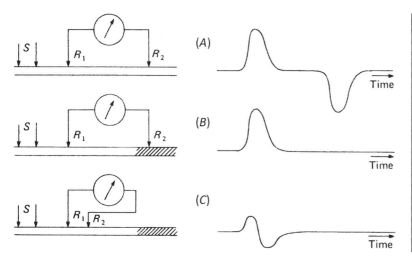

Figure 2.5 Electrical changes accompanying passage of a nerve impulse, as seen on an oscilloscope connected to external recording electrodes R_1 and R_2. S, stimulating electrodes. An upward deflection is obtained when R_1 is negative relative to R_2. (A) Diphasic recording seen when R_1 and R_2 are both on the intact portion of the nerve and are separated by an appreciable distance; (B) monophasic recording seen when the nerve is cut or crushed under R_2; (C) diphasic recording seen with R_2 moved back on to intact nerve, much closer to R_1.

the interior. Consequently, the active region of the nerve becomes electrically negative relative to the resting region. With two electrodes placed far apart on an intact nerve (Figure 2.5A), an impulse set up by stimulation at the left-hand end first reaches position R_1 and makes it temporarily negative. It next traverses the stretch between R_1 and R_2, and finally arrives under R_2, where it gives rise to a mirror-image deflection on the oscilloscope. This gives a *diphasic* record. If the nerve is cut or crushed under R_2, the impulse is extinguished when it reaches this point, and the record becomes *monophasic* (Figure 2.5B). However, it is sometimes difficult to obtain the classical diphasic action potential of Figure 2.5A because the electrodes cannot be separated by a great enough distance. In a frog nerve at room temperature, the duration of the action potential is of the order of 1.5 ms, and the conduction velocity is about 20 m/s. The active region therefore occupies 30 mm, and altogether some 50 mm of nerve must be dissected, requiring a rather large frog, to completely separate the upward and downward deflections. When the electrodes are closer together than the length of the active region, there is a partial overlap between the phases, and the diphasic recording has a reduced amplitude and no central flat portion (Figure 2.5C).

An entire nerve trunk contains a mixture of fibres with widely different diameters, spike durations and conduction velocities. Hence, even a monophasic spike recording may have a complicated appearance. When a frog sciatic nerve is stimulated strongly enough to excite all the fibres, an electrode placed near the point of stimulation records a monophasic action potential appearing as a single wave. However, a recording made at a greater distance reveals several waves because the spikes, conducted with different velocities, become dispersed with distance along the nerve. The three main groups of spikes are conventionally labelled A, B and C. Group A may be subdivided

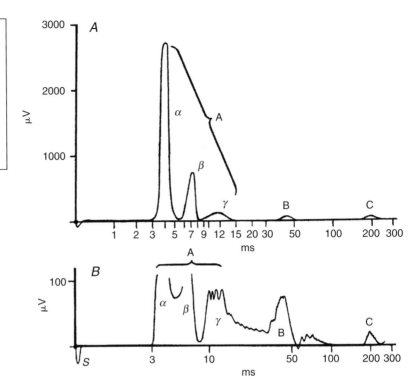

Figure 2.6 Monophasic recording of compound action potential from bullfrog peroneal nerve at a 13.1 cm conduction distance. Time in milliseconds on a logarithmic scale. Amplification for (B) is 10 times that for (A). S, stimulus artifact at zero time. (Redrawn after Erlanger and Gasser, 1937.)

into groups Aα, Aβ and Aγ. The experimental results shown in Figure 2.6 were obtained from a large American bullfrog studied at room temperature. The distance from stimulating to recording electrode was 131 mm. Reading the time for the foot of the wave to reach the recording electrode off the logarithmic scale (Figure 2.6A) gives conduction velocities of 41 mm/ms for Aα, 22 for Aβ, 14 for Aγ, 4 for B and 0.7 for C fibres. Mammalian nerve conduction velocities are somewhat greater (100 mm/ms for Aα, 60 for Aβ, 40 for Aγ, 10 for B and 2 for C fibres), partly because of their higher body temperatures and partly because their fibres are larger.

This wide conduction velocity distribution relates to correspondingly wide variations in fibre diameter. Large nerve fibres conduct impulses with higher velocities than small fibres, important features further discussed in Sections 6.2–6.5. Several other characteristics of nerve fibres depend markedly on their size. Smaller fibres need stronger shocks to excite them, so that the form of the volley recorded from a mixed nerve trunk is affected by stimulus strength. A weak shock, only elicits the Aα wave; a stronger shock elicits both Aα and Aβ waves, and so on. The amplitude of the voltage change picked up by an external recording electrode also varies with fibre diameter. It was suggested on theoretical grounds to vary with the square of diameter, but Gasser's reconstructions suggest that in practice the relationship is more nearly linear. In either case, this results in the observed electrical activity in a sensory nerve studied in situ being dominated by events in the largest as opposed to the small non-myelinated fibres.

While there is a wide range of fibre diameters in most nerve trunks, it is usually difficult to attribute particular functions to particular fibre sizes. Spinal cord sensory roots contain fibres giving A (that is Aα, Aβ and Aγ) and C waves. Motor roots show Aα, Aγ and B waves, the latter entering the white ramus. It is generally believed that B fibres occur only in preganglionic autonomic nerves, so that what is labelled B in Figure 2.6 might be better classified as Aδ. The grey rami contain sympathetic fibres and show mainly C waves. The fastest (Aα) fibres are either motor fibres activating voluntary muscles or afferent fibres conveying impulses from sensory receptors in these muscles. The γ motor fibres in mammals (Section 7.1) are connected to intrafusal muscle fibres in muscle spindles; in amphibia they innervate 'slow' as opposed to 'twitch' muscles (Section 12.2). At least some of the fibres of the unmyelinated C group convey pain impulses, but most belong to postganglionic autonomic nerves. The myelinated sensory fibres in peripheral nerves have also been classified according to diameter into group I (20 to 12 μm), group II (12 to 4 μm) and group III (less than 4 μm). Functionally, the group I fibres are found only in nerves from muscles. Subdivision IA is connected with annulo-spiral endings of muscle spindles. The more slowly conducting IB fibres carry impulses from Golgi tendon organs (Sections 8.1 and 8.2). The still slower group II and III fibres transmit other modes of sensation in both muscle and cutaneous nerves.

2.4 | Excitation

Before considering the ionic basis of the mechanism of conduction of the nervous impulse (Section 4.2), it is appropriate to outline some empirical features of excitation, particularly how impulses are set up in nerve and muscle fibres. This order of treatment reflects the historical sequence with which research on the subject developed. Progress towards fuller understanding of the fundamentals of the conduction mechanism was inevitably slow before introduction of intracellular recording techniques. Yet excitation could be investigated with comparatively simple methods, such as observing whether or not a muscle was induced to twitch.

A nerve can be *stimulated* by local application of a number of agents, including electric current, pressure, heat, solutions containing substances like KCl, or optical excitation of genetically introduced photosensitive molecules such as channelrhodopsin-2 (Ferenczi *et al.*, 2019). However, it is most easily and conveniently stimulated by applying electric shocks. The most effective electric current is one which flows outwards across the membrane and so *depolarises* it, that is to say reduces the size of the resting potential. The other agents

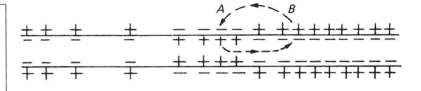

Figure 2.7 Diagrams illustrating the local circuit theory. The upper sketch represents an unmyelinated nerve fibre, the lower sketch a myelinated fibre. (From Hodgkin, 1958.)

listed above also cause depolarisation. Pressure and heat do so by damaging the membrane. A flow of current in the appropriate direction may be brought about either by applying a negative voltage pulse to a nearby electrode, making it *cathodal*, or through *local circuit* action when an impulse set up further along the fibre reaches the region of membrane under consideration.

It was suggested long ago that impulse *propagation* depends essentially on current flow in local circuits ahead of the active region. These depolarise the resting membrane thereby causing it to become active. This local circuit theory explains how current flowing from a given region *A* to region *B* in an unmyelinated fibre (Figure 2.7, top panel) results in movement of the active region towards the right. There are important differences to be discussed later (Sections 6.2–6.4) between current pathways in unmyelinated nerves or muscle fibres on the one hand, and myelinated fibres on the other (Figure 2.7, bottom panel), but the basic principle is similar. In studying effects of applied electric currents it is important to distinguish non-physiological physical events associated with and nevertheless important to the nervous impulse from physiological processes forming component events in normal excitation and propagation mechanisms.

The first key concept concerning *excitation* is that of a *threshold* stimulus. In a nerve–muscle preparation this is the smallest voltage which elicits a just-perceptible muscle twitch. It produces a voltage change just large enough to stimulate one of the nerve fibres, thereby causing contraction of the muscle fibres to which it is connected. If the nerve consisted only of a single fibre, further increases in the applied voltage would not make the twitch any stronger. This is because conduction is an *all-or-nothing* phenomenon: the stimulus either fails to set up an impulse if it is subthreshold or sets up a full-sized impulse if it is at or above threshold. No response of an intermediate size can be obtained by varying the stimulus strength, though of course occurrence of a response or otherwise

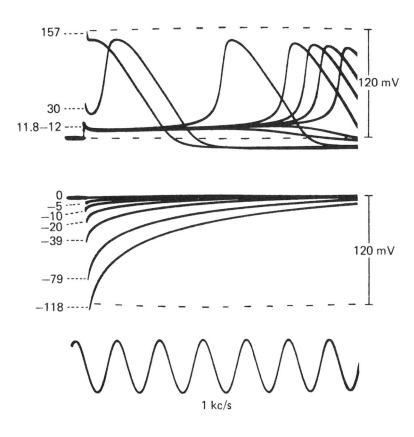

157 ----
30 -----
11.8–12 ----
120 mV

0
−5 ----
−10 ----
−20 ----
−39 ----
−79 -----
−118 -----
120 mV

1 kc/s

Figure 2.8 Threshold behaviour of membrane potential in squid giant axon at 6 °C. Shocks, whose strengths in nC/cm² membrane are shown against each trace, were applied to an internal wire electrode with a bare portion 15 mm long. The internal potential was recorded between a second wire 7 mm long opposite the centre of the stimulating wire and an electrode in the sea water outside. Depolarisation is shown upwards. The sine wave provides a 50 Hz time scale. (From Hodgkin et al., 1952.)

may be affected by changes in certain external conditions such as temperature or ionic environment.

An experiment exemplifying the threshold behaviour of a single nerve fibre stimulated an isolated squid giant axon over a 15 mm length by brief shocks applied between an axially inserted wire and an external electrode. A second wire with a bare portion opposite the central 7 mm of the axon recorded the internal membrane potential. The excitation threshold occurred with application of a 11.8–12 nC/(cm² membrane) depolarising shock to the stimulating wire. At this shock strength, the response arose after a latency of several milliseconds, during which the membrane was depolarised by about 10 mV and was in a metastable condition, sometimes giving a spike and sometimes simply reverting to its resting state (Figure 2.8, top panel). A larger shock reduced the latency but did not appreciably change spike size. Finally, reversing the direction of the shock gave an inward current which polarised the membrane beyond the resting level. The membrane potential simply decayed exponentially back to the resting value (Figure 2.8, bottom panel). The changes in the ionic permeability of the membrane that are responsible for this behaviour are explained in Chapter 3.

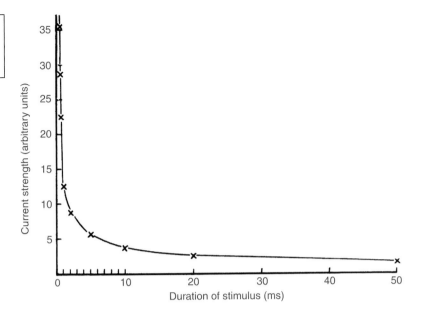

Figure 2.9 Strength–duration curve for direct stimulation of frog sartorius muscle. (From Rushton, 1933.)

A multi-fibre preparation like sciatic nerve contains hundreds of fibres with thresholds spread over a wide range of voltages. Hence increasing stimulus strength above that which just excites the lowest threshold fibre excites successively more higher-threshold fibres correspondingly increasing the size of the muscle twitch. The point at which the twitch ceases to increase any further reflects the point at which all the fibres in the nerve trunk are triggered. This corresponds to the *maximal* stimulus. A still larger, supra-maximal, shock does not produce a larger twitch.

A second key variable bearing on nerve excitability is the *duration* of the shock. Measurements of threshold demonstrate that for progressively longer shocks, the required applied current reaches an irreducible minimum known as the *rheobase*. In contrast, reducing shock duration results in a stronger shock becoming necessary to reach threshold. The resulting *strength–duration curve* demonstrates an inverse relation between shock strength and shock duration (Figure 2.9), with an essential requirement that eliciting the action potential requires the membrane to be depolarised to a critical level the existence of which is shown clearly in Figure 2.8. Reducing shock duration requires more current to flow outwards if the membrane potential is to attain this critical level before the end of the shock. It follows that for short shocks, a roughly constant total quantity of electricity has to be applied, and in Figure 2.8 the shock strength was therefore expressed in nC/cm^2 membrane.

For a short period after passage of an impulse, the stimulation threshold becomes raised. If a nerve is stimulated twice in quick succession, it may consequently fail to respond to the second stimulus. The *absolute refractory period* is the brief interval after a successful

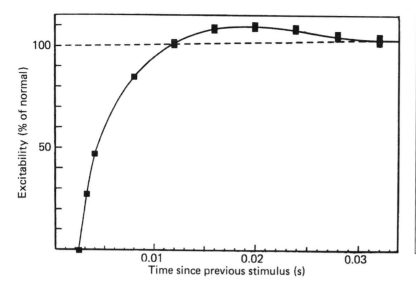

Figure 2.10 Timecourse of recovery of excitability (= 1/threshold) in frog sciatic nerve after passage of an impulse. Conditioning stimulus and test stimulus were applied at electrodes 15 mm apart, so that about 0.5 ms should be subtracted from each reading to obtain the course of recovery under the test electrode. The absolute refractory period lasted 2 ms, and the relative refractory period 10 ms; they were succeeded by a supernormal period lasting 20 ms. (From Adrian and Lucas, 1912.)

stimulus when no second shock, however large, can elicit another spike. Its duration roughly equals that of the spike, which in mammalian A fibres at body temperature is ~0.4 ms, or in frog nerve at 15 °C is ~2 ms. It is followed by the *relative refractory period*, during which a second response can be obtained if a strong enough shock is applied. This in turn is sometimes succeeded by a phase of supernormality when the excitability may be slightly greater than normal. Figure 2.10 illustrates the timecourse of the changes in excitability (= 1/threshold) in a frog sciatic nerve after the passage of an action potential.

The nerve refractoriness after conducting an impulse sets an upper limit to spike frequency. During the relative refractory period, both spike size and conduction velocity are subnormal, as well as excitability, so that two impulses traversing a long length of nerve must be separated by a minimum interval if the second one is to be full-sized. A mammalian group A fibre can conduct up to 1000 impulses/s, but the spikes would be small and would decline further during sustained stimulation. In group A fibres, recovery is complete after about 3 ms, giving a frequency limit for full-sized spikes of 300 impulses/s. Even this repetition rate is not often attained in the living animal, although certain sensory nerves may exceed it occasionally for short bursts of impulses.

Background Ionic Homeostasis of Excitable Cells

3.1 | Structure of the Cell Membrane

All living cells are surrounded by a plasma membrane comprising lipids and proteins. Its main function is to control passage of substances into and out of the cell. The ease with which a molecule crosses a cell membrane depends primarily on its charge and lipid solubility rather than its size. The lipid matrix is completely impermeable to all large water-soluble molecules, as well as small charged molecules and ions. It is permeable to water and small uncharged molecules like urea. The essential feature of membrane lipids that enables them to provide such an electrically insulating structure, acting as a barrier to free passage of ions, is their possession of hydrophilic (polar) head groups and hydrophobic (non-polar) tails. Such lipids spread on a water surface form a stable monolayer in which the polar ends contact the water and the non-polar hydrocarbon chains are oriented more or less at right angles to the plane of the surface. The cell membrane consists basically of two such lipid monolayers arranged back-to-back with the polar head groups facing outwards. The resulting sandwich interposes an uninterrupted hydrocarbon phase between the aqueous phases on either side with a thickness roughly twice the hydrocarbon chain length (Figure 3.1). Lipid *bilayers* of this type can readily be prepared artificially. Such 'black membranes' have been valuable as model systems in studies of properties of real cell membranes.

The chemical structure of the phospholipids of which cell membranes are mainly composed (Figure 3.2) comprises a glycerol backbone esterified to two fatty acids and phosphoric acid. This forms a phosphatidic acid with which alcohols like choline or ethanolamine combine through another ester linkage to give the neutral phospholipids lecithin and cephalin. Alternatively, an amino acid like serine is linked to give negatively charged phosphatidylserine. Another constituent of cell membranes is cholesterol, whose physical properties resemble those of a lipid through the –OH group attached to C-3. Spin-label and deuterium nuclear magnetic resonance (^{2}H-NMR) studies of

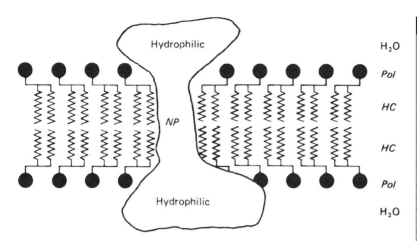

Figure 3.1 Schematic diagram of the structure of a cell membrane. Two layers of phospholipid molecules face one another with their fatty-acid chains forming a continuous hydrocarbon layer (*HC*) and their polar head groups (*Pol*) in the aqueous phase. The selective pathways for ion transport are provided by proteins extending across the membrane, which have a central hydrophobic section with non-polar side-chains (*NP*) and hydrophilic portions projecting on either side.

lipid bilayers show that the hydrocarbon chains are packed rather loosely so that the interior of the bilayer behaves like a liquid. With a chain length of 18 carbon atoms, the effective thickness of the hydrophobic region is about 3.0 nm. This is consistent with their observed electrical capacitance of 1 μF/cm^2 membrane and dielectric constant (ε_r) of 3.

The embedded proteins provide selective pathways for transport of ions and organic molecules both down and against prevailing electrochemical gradients. The particular nature of the transport pathways varies with the specific function of the cell under consideration. In the case of nerve and muscle, the pathways that are functionally important in connection with excitation and conduction are: (1) voltage-sensitive Na$^+$, K$^+$ and Ca^{2+} channels characteristic of electrically excitable membranes, (2) ligand-gated channels at synapses that transfer excitation onwards from the nerve terminal and (3) the ubiquitous Na$^+$, K$^+$-ATPase, responsible in all cell types for extruding Na$^+$ ions from the interior. cDNA sequencing studies (Section 5.1) have greatly advanced our understanding of the organisation of the membrane protein component.

The fluid mosaic model of membrane structure (Singer and Nicolson, 1972) describes a lipid bilayer containing integral membrane proteins traversing the entire membrane thickness and peripheral membrane proteins partially embedded within the membrane. High-resolution electronmicroscopy, permanganate- or osmic-acid-stained sections display the membrane in all cell types as two uniform lines separated by a space, the width of the whole structure being about 7.5 nm (Figure 3.3). This reflects uptake of the electron-dense stain by the polar groups of the phospholipids and their associated proteins. Additionally, electronmicroscope examination of freeze-fractured membranes demonstrates some of the integral membrane proteins traversing the bilayer to form specific ion-conducting or ion-pumping pathways as globular indentations or projections (Figure 3.4). Such

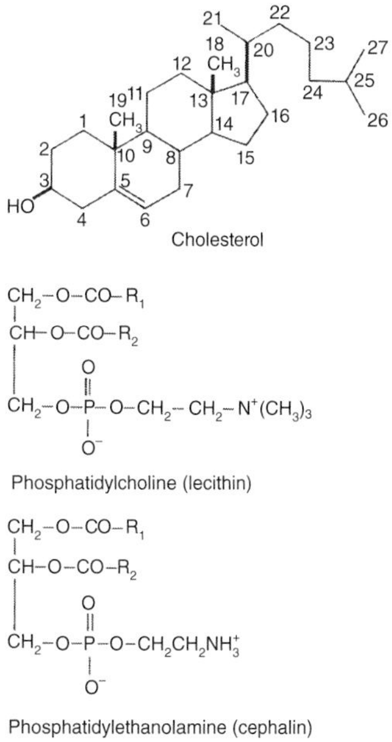

Cholesterol

Phosphatidylcholine (lecithin)

Phosphatidylethanolamine (cephalin)

Figure 3.2 The chemical structure of cholesterol and two neutral phospholipids.

Figure 3.3 Electronmicrograph at high magnification of the cell membrane stained with osmic acid. (Reproduced by courtesy of Professor J. D. Robertson.)

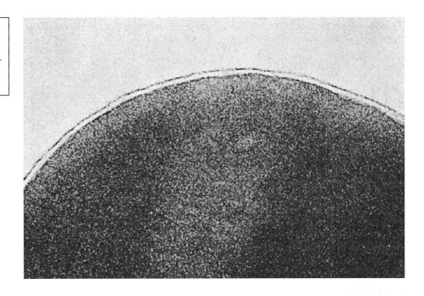

Figure 3.4 Electronmicrograph of a freeze-fracture preparation of a cell membrane. The proteins appear as globular indentations. (Reproduced by courtesy of Professor J. D. Robertson.)

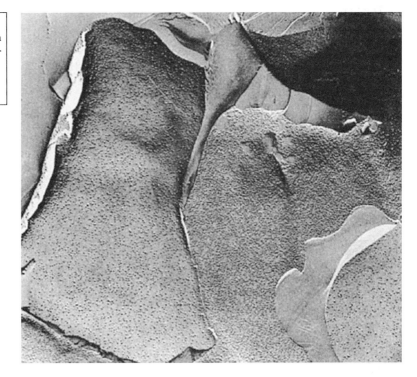

membrane proteins have a central non-polar section stabilised by the hydrophobic environment provided by the hydrocarbon chains of the lipids. They also show polar and often glycosidic portions extending into the aqueous medium both inside and outside. Whether they are held in a fixed position in the membrane by internal fibrils, or are free to rotate and move laterally, is not always clear, but it may well be that some freedom of movement is necessary for their normal functioning.

3.2 | Ion Distributions in Nerve and Muscle

Flame photometry and other microanalytical techniques have made it possible to determine quantities of ions present even in small tissue samples. To obtain the true intracellular concentrations, it is necessary to correct for the contents of the extracellular space. This may be done after measuring its size using a membrane-impermeant marker substance like inulin. Table 3.1 gives a simplified balance sheet of the ionic concentrations in frog muscle fibres and blood plasma determined in this way. In the case of the squid giant axon it is possible to extrude the axoplasm, just as toothpaste is squeezed from a tube, and so to obtain samples uncontaminated by extracellular ions. Table 3.2 shows the resulting ionic balance sheet.

The main features of the ion distributions common to all excitable tissues are that the intracellular [K^+], ([K^+]$_i$) is 20 to 50 times higher in cytoplasm than in blood, and for that of [Na^+] and [Cl^-] the situation is reversed. The total concentration of ions is, of course, about four times greater in a marine invertebrate like the squid, whose blood is isotonic with sea water, than it is in an amphibian like the frog, which lives in fresh water, but the concentration ratios are not very different. The principal anion in the external medium is Cl^-, but inside the cells its place is taken by a variety of non-penetrating organic anions (Section 3.4). The problem of achieving a balance between intracellular anions and cations is most severe in marine invertebrates, and is met by the presence either of large concentrations of aspartate and glutamate or, in squid, of isethionate.

Table 3.1	*Ionic concentrations in frog muscle fibres and plasma*	
	Concentration in fibre water (mM)	**Concentration in plasma water (mM)**
K^+	124	2.3
Na^+	3.6	108.8
Ca^{2+}	4.9	2.1
Mg^{2+}	14.0	1.3
Cl^-	1.5	77.9
HCO_3^-	12.4	26.6
Phosphocreatine	35.2	—
Organic anions	~45	~14

These figures are calculated from values given by Conway, 1957. At pH 7.0, phosphocreatine carries two negative charges; the remaining deficit in intracellular anions is made up by proteins.

Table 3.2	*Ionic concentrations in squid axoplasm and blood*	
	Concentration in axoplasm (mM)	**Concentration in blood (mM)**
K^+	400	20
Na^+	50	440
Ca^{2+}	0.4	10
Mg^{2+}	10	54
Cl^-	123	560
Arginine phosphate	5	–
Isethionate	250	–
Other organic anions	~110	~30

These values are taken from Hodgkin, 1958 and Keynes, 1963.

3.3 | The Genesis of Resting Potentials

When a membrane selectively permeable to a given ion separates two solutions containing different concentrations of that ion, an electrical potential difference is set up across it. In order to understand how this comes about, consider an intracellular compartment whose ionic concentrations are $[K^+]_i$ and $[Cl^-]_i$, and an extracellular compartment in which they are $[K^+]_o$ and $[Cl^-]_o$, bounded by a membrane that can discriminate perfectly between K^+ and Cl^- ions. This allows K^+ to pass freely but is totally impermeable to Cl^-. If $[K^+]_i$ is greater than $[K^+]_o$ there will initially be a net outward movement of K^+ down the concentration gradient, but each K^+ ion escaping from the compartment unaccompanied by a Cl^- ion will tend to make the outside of the membrane electrically positive. The direction of the electric field set up by this separation of charge will be such as to assist the entry of K^+ into the compartment and hinder their exit. A state of equilibrium will quickly be reached in which the opposed influences of concentration and electrical gradients on ionic movements will exactly balance one another. Although there will be a continuous flux of ions crossing the membrane in each direction, there will be no further net ion movement.

The argument may be placed on a quantitative basis by equating the chemical work involved in K^+ transfer from one concentration to the other with the electrical work involved in the transfer against the potential gradient. In order to move 1 gram-mole of the univalent K^+ from inside to outside, the chemical work that has to be done is $RT \log_e\{[K^+]_o/[K^+]_i\}$. The corresponding electrical work is $-E_m F$, where E_m is the membrane potential, inside relative to outside, R is the ideal gas constant and F is the charge carried by 1 gram-equivalent of ions or

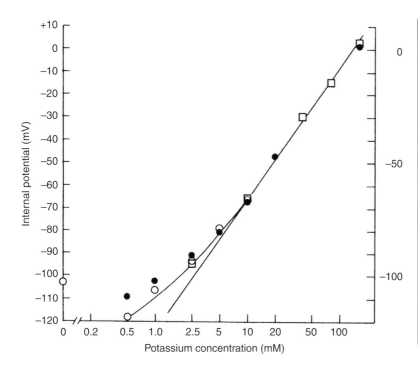

Figure 3.5 Variation in the resting potential of frog muscle fibres with the external K^+ concentration $[K^+]_o$. The measurements were made in a Cl^--free sulfate-Ringer solution containing 8 mM $CaSO_4$. Square symbols are potentials measured after equilibrating for 10 to 60 min; circles are potentials measured 20 to 60 s after a sudden change in concentration, filled symbols after increase in $[K^+]_o$, open symbols after decrease in $[K^+]_o$. For large values of $[K^+]_o$ the measured potentials agree well with the Nernst equation, $V = 58 \log_{10}\{[K^+]_o/[K^+]_i\}$, taking $[K^+]_i$ as 140 mM. The deviation at low $[K^+]_o$ can partly be explained by taking $P_{Na}/P_K = 0.01$, so that $V = 58 \log_{10}\{\{[K]_o - 0.01[Na]_o\}/140\}$. (From Hodgkin and Horowicz, 1959.)

Faraday constant. At equilibrium, no net work is done, and the sum of the two is zero, whence:

$$E_m = -\frac{RT}{F} \log_e \frac{[K^+]_o}{[K^+]_i} \qquad (3.1)$$

This relationship was first derived by the German physical chemist Nernst in the nineteenth century, and the equilibrium potential E_K for a membrane permeable exclusively to K^+ ions is known as the *Nernst potential* for K^+. The values of R and F are such that at room temperature the potential is given by:

$$E_K = -25 \log_e \frac{[K^+]_o}{[K^+]_i} = -58 \log_{10} \frac{[K^+]_o}{[K^+]_i} \qquad (3.2)$$

An e-fold change in concentration ratio corresponds to a 25 mV change in potential, or a 10-fold change to 58 mV.

Examination of the applicability of the Nernst relation to nerve and muscle demonstrates that it is well obeyed at high external K^+ concentrations, but for small values of $[K^+]_o$ the potential alters less steeply than predicted by Equation (3.2) (Figure 3.5). Thus, the membrane does not maintain a perfect selectivity for K^+ over the whole concentration range, and effects of other ions present must also be considered. The Goldman constant field equation provides a theoretical expression relating membrane potential to the permeabilities and concentrations of all the major ions in the system, whether positively or negatively charged. It involves assuming that the electric field is constant at all points across the membrane (Goldman, 1943):

$$E_m = -\frac{RT}{F} \log_e \frac{P_K[K^+]_o + P_{Na}[Na^+]_o + P_{Cl}[Cl^-]_i}{P_K[K^+]_i + P_{Na}[Na^+]_i + P_{Cl}[Cl^-]_o} \qquad (3.3)$$

The P terms are permeability coefficients for the various ions, and the suffixes o and i indicate external and internal concentrations respectively. Equation (3.3) fits rather well with experimental observation over a wide range of conditions, although it need not follow that the field is indeed truly constant. It is nevertheless empirically valuable for describing the behaviour of a membrane permeable to more than one species of ion. In the experiment of Figure 3.5, the deviation of the measured potential from a line with a slope of 58 mV is accounted for by taking P_{Na} to be 100 times smaller than P_K.

3.4 | The Donnan Equilibrium System in Muscle

We can therefore predict the resting potential across a membrane in an excitable cell of constant volume, and stable and defined intracellular and extracellular ionic concentrations. We can next explore how such background conditions arise in vivo. An initial contribution to the ionic concentration differences in Tables 3.1 and 3.2 might arise from the properties of the Donnan equilibrium. This was first studied experimentally in frog sartorius muscle cells (Boyle and Conway, 1941. A recent theoretical analysis (Fraser and Huang, 2007) considered a water-permeable membrane separating two, intracellular (i) and extracellular (o), compartments, each containing membrane-permeant K^+ and Cl^-. The intracellular compartment additionally contained membrane-impermeant, negatively charged, protein, Pr^{z-}, with valency, z, at concentration $[Pr^{z-}]$. Proteins contain amino acid residues with multiple titratable functional groups with which intracellular protons, H^+, associate or dissociate, with an overall logarithmic dissociation constant, pK_a, of typically ~6.2. The normal intracellular pH of ~7.2 would typically predict a net negative valency z of ~(−1.2).

An initial, hypothetical, situation balancing $[K^+]_o$ and $[K^+]_i$ and the concentration, $[Pr^{z-}]$, of intracellular impermeant Pr^{z-}, would predict an inequality of intracellular and extracellular ion concentrations, in which $[K^+]_i > [K^+]_o$, to ensure approximate intracellular and extracellular charge equality within each solution compartment (Figure 3.6A):

$$[K^+]_i = [K^+]_o + z[Pr^{z-}]$$

A more realistic situation in living cells would include not only the permeant K^+ and impermeant intracellular Pr^{z-}, but also permeant anions exemplified by Cl^-. Their equilibrium potential contributions to the resulting transmembrane potential can be computed from the

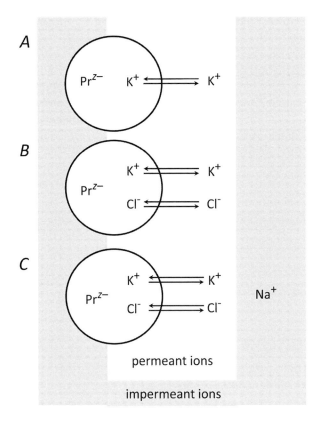

A

B

C

permeant ions

impermeant ions

Figure 3.6 Basis of the Donnan equilibrium. Simple cell systems considering intracellular and extracellular distributions of (A) hypothetical permeant K^+ and impermeant intracellular Pr^{z-} only, (B) permeant K^+ and Cl^-, and impermeant intracellular Pr^{z-} and (C) 'double' Donnan system containing permeant K^+ and Cl^- and intracellular impermeant Pr^{z-} and extracellular, relatively impermeant, Na^+.

respective Nernst potentials for K^+ and Cl^-, given by $E_K = (RT/F) \log_e\{[K^+]_o/[K^+]_i\}$ and $E_{Cl} = (RT/F) \log_e\{[Cl^-]_o/[Cl^-]_i\}$, respectively (Figure 3.6B). The equilibrium membrane potential is the voltage at which these contributions balance. There are then no net ion fluxes because the membrane potential equals the equilibrium potential for each diffusible ion. Since the membrane potential is common to both ions (Figure 3.6B):

$$E_m = -\frac{RT}{F} \log_e \frac{[K^+]_o}{[K^+]_i} = +\frac{RT}{F} \log_e \frac{[Cl^-]_o}{[Cl^-]_i}$$

The ratio for K^+ ions is the inverse of that for Cl^- ions because of their opposite charges:

$$\frac{[K^+]_o}{[K^+]_i} = \frac{[Cl^-]_i}{[Cl^-]_o} \tag{3.4}$$

The resulting Donnan condition for electrochemical equilibrium, Equation (3.4), can be written:

$$[K^+]_o[Cl^-]_o = [K^+]_i[Cl^-]_i, \tag{3.5}$$

Here E_m is driven towards the equilibrium potential of not only E_K, as discussed above in relationship to genesis of the resting potential, but also to E_{Cl}. The latter situation results in a further Cl^--mediated contribution to stabilisation of E_m in some excitable cells that show a significant Cl^- permeability (e.g. Section 10.1).

However, this situation results in an instability in cell volume, alterations in which would in turn further perturb the solute concentrations. Volume stability requires satisfaction of an additional condition of osmotic equilibrium in which intracellular and extracellular osmolarities are equal. In the presence of the negatively charged intracellular Pr^{z-}, this can be written:

$$[K^+]_o + [Cl^-]_o = [K^+]_i + [Cl^-]_i + [Pr^{z-}] \tag{3.6}$$

Since $[Pr^{z-}] > 0$, Equation (3.6) yields the inequality:

$$[K^+]_o + [Cl^-]_o \geq [K^+]_i + [Cl^-]_i \tag{3.7}$$

Finally, *electroneutrality* of the respective extracellular and intracellular compartments further requires that:

$$[K^+]_o - [Cl^-]_o = 0, \text{ or } [K^+]_o = [Cl^-]_o, \tag{3.8}$$

and

$$[K^+]_i - [Cl^-]_i + z[Pr^{z-}] = 0, \text{ or } [K^+]_i = [Cl^-]_i - z[Pr^{z-}] \tag{3.9}$$

Applying Equation (3.8), and substituting $[Cl^-]_o = [K^+]_o$ into, then squaring both sides of, Equation (3.7) gives:

$$4[K^+]_o^2 \geq [K^+]_i^2 + [Cl^-]_i^2 + 2[K^+]_i[Cl^-]_i \tag{3.10}$$

Since both $[K^+]_i$ and $[Cl^-]_i$ are positive, $([K^+]_i + [Cl^-]_i)^2 \geq 0$ and so:

$$[K^+]_i^2 + [Cl^-]_i^2 - 2[K^+]_i[Cl^-]_i \geq 0, \tag{3.11}$$

The right hand side of Equation (3.10), is:

$$[K^+]_i^2 + [Cl^-]_i^2 + 2[K^+]_i[Cl^-]_i \geq 4[K^+]_i[Cl^-]_i \tag{3.12}$$

From Equations (3.10) and (3.12),

$$4[K^+]_o^2 \geq 4[K^+]_i[Cl^-]_i \tag{3.13}$$

The osmotic stability condition therefore entails the inequality,

$$[K^+]_o[Cl^-]_o \geq [K^+]_i[Cl^-]_i \tag{3.14}$$

This is incompatible with, the *equality* requirement in Equation (3.5) representing the Donnan condition. The extent of the disparity between the conditions for the Donnan, electroneutrality and osmotic equilibria depends upon the magnitude of $z[Pr^{z-}]$ except where $[Cl^-]_o = [Cl^-]_i$. However, the latter corresponds to the situation in which $z[Pr^{z-}] = 0$, where Pr^{z-} is infinitely diluted by cell swelling resulting from water entry. The resulting inherent cell instability consequently is particularly marked with the large $z[Pr^{z-}]_i$ (~120–140 mM in amphibian skeletal muscle) known to exist in excitable cells (Usher-Smith et al., 2006a, 2009).

3.5 | Direct Tests of the Donnan Hypothesis

The osmotic activity of membrane-impermeant intracellular constituents $[Pr^{z-}]_i$ and their mean charge valency z emerge as centrally important in determining the background cell volume, V_c, and E_m. This was demonstrated in experiments investigating the effects of altering the effective valency z (Figure 3.7Aa) in amphibian muscle preparations known to show osmometric variations in V_c. Applied alterations in intracellular $[H^+]$ $([H^+]_i)$ were used to titrate the Pr^{z-} against H^+ within the physiological pH range, thereby neutralising its charge z. The experiments then assessed the effects of this cellular H^+ loading, on V_c and E_m. This tested the Donnan prediction

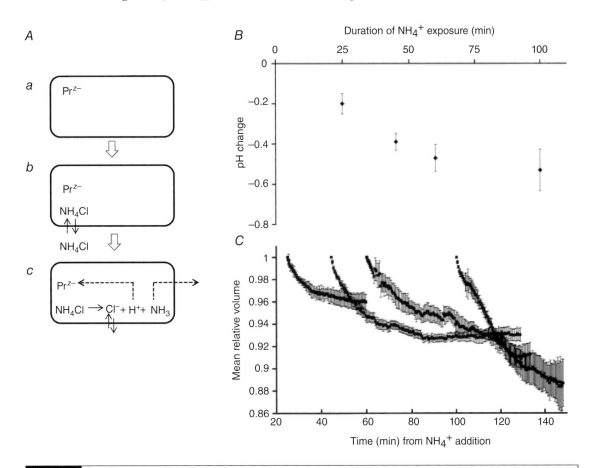

Figure 3.7 Effects of different durations of exposure to, followed by withdrawal of, extracellular 40 mM NH_4Cl on intracellular pH and cell volume V_c in amphibian skeletal muscle fibres. (A) Reactions in Pr^{z-}-containing cell (a) following NH_4Cl addition (b) and withdrawal (c). (B, C) Effects of NH_4Cl exposure duration on (B) mean changes in intracellular pH ($\pm$ SEM, $n \geq 6$ in each case) on return to normal Ringer's solution compared to intracellular pH preceding NH_4Cl withdrawal. (C) Relative cell volumes, V_c (means $\pm$ SEM, $n \geq 4$) plotted against time following NH_4Cl withdrawal following different exposure durations. Time zero is the point of NH_4Cl addition. Each plot begins at the point of NH_4Cl withdrawal, following 25, 45, 60 and 100 min exposures (left to right) respectively. (From Figure 3 of Fraser et al., 2005).

(Section 3.4) that intracellular acidification would decrease V_c, through reducing z.

The intracellular acidification was achieved by isosmotic additions of extracellular NH_4Cl followed by its withdrawal after graded exposure durations (Figure 3.7Ab). The added extracellular NH_4Cl would enter the intracellular space. Its subsequent withdrawal would increase the dissociation of the loaded intracellular NH_4Cl into acidic H^+, Cl^- and the basic and volatile NH_3 (Figure 3.7Ac). The membrane permeant and volatile NH_3 would then diffuse out of the intracellular space in preference to efflux of the charged, less membrane-permeant, NH_4^+. This would leave an increased $[H^+]_i$, providing protons available to titrate negative charges on the intracellular Pr^{z-}. This would reduce its valency, z, and permit assessment of its osmotic effects.

The extent of this intracellular acidification was varied by altering the duration of the preceding exposure to extracellular NH_4Cl. The latter would in turn alter the degree of intracellular NH_4Cl loading prior to its withdrawal from the extracellular space. pH electrode measurements demonstrated that NH_4Cl withdrawals following progressively lengthening these preceding exposures to extracellular NH_4Cl, produced progressively larger net intracellular acidifications of up to 0.53 ± 0.10 pH units (Figure 3.7B).

The corresponding cell volume (V_c) changes were continuously assessed using laser confocal microscope xz-plane scanning through these solution changes. These measured fibre cross-sectional areas with time in a whole muscle preparation of fixed length. In addition to the intracellular acidification indicated above, the NH_4Cl withdrawals were shown to induce *net* volume decreases. All the treated fibres reached final volumes that were significantly less than their initial pretreatment volumes. Furthermore, the observed net volume decrease became more marked with increases in the durations of the preceding cell loading exposures to NH_4Cl (Figure 3.7C; Fraser *et al.*, 2005).

The presence of intracellular impermeant charge therefore confers a tendency for the cell to swell and lyse. In vivo, the latter is balanced and cell stability regained, through the presence of charged, effectively membrane impermeant solute represented by Na^+ in the extracellular rather than the intracellular compartment. This would permit the existence of an osmotically stable *double* Donnan system (Figure 3.6C). This latter situation and the application of these physical principles to the ionic conditions occurring in the intact cell are next related to the operation of *active ion transport* processes described in the sections that follow.

3.6 | The Active Transport of Ions

In living cells, the intracellular impermeant Pr^{z-} is balanced by a relatively elevated concentration of the extracellular, relatively impermeant, Na^+. One might suggest that ion movements reflecting the consequently small Na^+ fluxes from extracellular to intracellular space are balanced by active processes driving Na^+ fluxes in the opposite direction. This would normally preserve or restore the relatively resting high $[Na^+]_o$ relative to $[Na^+]_i$. This scheme was first suggested by radioactive isotope ^{24}Na flux measurements. These demonstrated that about half of the intracellular Na^+ in frog sartorius muscle fibres was exchanged with extracellular Na^+ over one hour. Similarly, following dissection, squid and cuttlefish giant axons showed a steady Na^+ gain and K^+ loss. If not counteracted these would eventually have equalised their axoplasmic Na^+ and K^+ contents.

The resting cell membrane does have a finite Na^+ permeability. However, the inward Na^+ leakage is offset by operation of a *Na^+ pump*. This extrudes Na^+ at a rate which ensures that in the living animal $[Na^+]_i$ is roughly constant. As far as Na^+ and K^+ are concerned, this leads to a steady-state as opposed to an equilibrium situation. Since Na^+ expulsion from the cell takes place against both electrical and concentration gradients, it requires performance of electrochemical work, which necessitates an energy supply from cell metabolism. The process is termed *active transport*.

Cephalopod giant axons proved amenable to radioactive tracer studies on the mechanism of the Na^+ pump. Figure 3.8 shows results from measurements of Na^+ efflux from a ^{24}Na–loaded *Sepia* axon bathed in non-radioactive medium. These were made by counting samples of the bathing solution collected at 10 min intervals. The resting efflux, calculated in moles of Na^+ per unit area of membrane per unit time, was roughly constant at around 40 pmole/ (cm^2 s) at room temperature. The initial linear decline seen in the plots of Figures 3.8 and 3.9 reflects gradual dilution of the internal radioactivity by radioactively inactive Na^+ entering the axon as the experiment proceeds. However, adding the metabolic

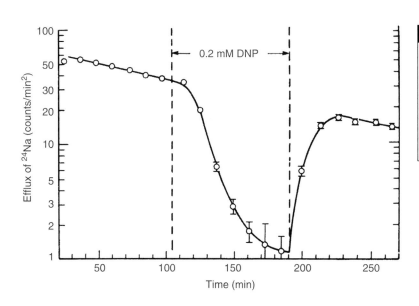

Figure 3.8 The effect on Na^+ efflux of blocking metabolism in a *Sepia* (cuttlefish) axon with dinitrophenol (DNP). At the beginning and end of the experiment the axon was in unpoisoned artificial sea water. Temperature 18 °C. (From Hodgkin and Keynes, 1955b.)

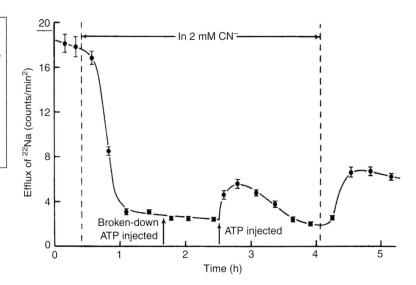

Figure 3.9 The rate of loss of radioactivity (per min) from a 780 μm squid axon initial loaded by micro-injection with 6700 counts/min of ^{22}Na, distributed over the 12 mm length of nerve. 32 nanomoles of ATP were injected over the same 12 mm length of nerve. Temperature 19 °C. (From Caldwell and Keynes, 1957.)

inhibitor 2,4-dinitrophenol (DNP) to the external medium caused a marked reduction of the counting rate to about one-thirtieth of its previous level. The effect was reversible: the efflux soon recovered on washing away the DNP. Axons treated with cyanide (CN$^-$) or azide yielded similar results. All these inhibitors are known to block production of the energy-rich compound adenosine triphosphate (ATP) by mitochondrial oxidative phosphorylation. The findings suggest that the Na$^+$ pump was driven by energy derived from hydrolysis of the terminal phosphate bond of ATP.

The role of ATP as immediate source of energy for Na$^+$ extrusion was further examined by testing its ability to restore Na$^+$ effluxes when injected into cyanide-poisoned axons. ATP injection brought about some recovery of the efflux (Figure 3.9). However, complete recovery required increasing the axoplasmic [ATP]/[ADP] ratio by injecting arginine phosphate, which serves as a high-energy phosphate reservoir in invertebrate tissues through the reaction (Figure 3.10):

ADP + arginine phosphate → ATP + arginine

A further important characteristic of the Na$^+$ pump activity is its dependence on the presence of extracellular K$^+$. When at the beginning of the experiment illustrated in Figure 3.10 [K$^+$]$_o$ had been reduced to zero, the unpoisoned Na$^+$ efflux fell to about quarter of its normal size. After the CN$^-$ had taken effect, a large amount of arginine phosphate was injected into the axon. This duly brought the efflux back to normal, but only during the first hour was it sensitive to removal of K$^+$. In a similar way, the efflux that reappeared on washing away the CN$^-$ only regained its K$^+$ sensitivity when sufficient time had been allowed for the [ATP]/[ADP] ratio to return to normal.

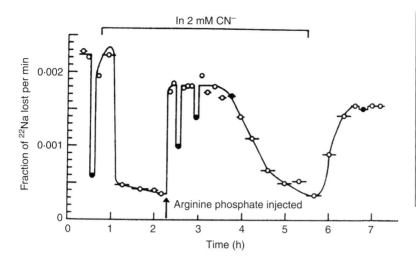

Figure 3.10 The effect on the efflux of labelled Na^+ from a squid giant axon of first blocking metabolism with cyanide and injecting a large quantity of arginine phosphate. Open circles show efflux with $[K^+]_o = 10$ mM; filled circles show efflux in a K^+-free solution. Immediately after the injection the mean internal concentration of arginine phosphate was 33 mM. Temperature 18 °C. (From Caldwell et al., 1960.)

The requirement of Na^+ pump activity for external K^+ suggested an obligatory coupling between Na^+ extrusion and K^+ uptake. Parallel measurements of ^{24}Na effluxes and ^{42}K influxes confirmed that this is the case. In a wide range of tissues the coupling ratio is normally 3:2. For every three Na^+ leaving the cell, two K^+ are taken up. Had the coupling ratio been exactly 1:1, the Na^+ pump would be electrically neutral in that its operation would bring about no net transfer of charge across the membrane. This coupling ratio greater than unity implies that the Na^+ pump is electrogenic. Its operation would cause a separation of charge tending to hyperpolarise the membrane. The after-potential that follows impulse firing in small unmyelinated nerve fibres may arise in this way from accelerated Na^+ pump function. The pumping of ions in many other situations has proved similarly to show some electrogenic properties.

The Na^+ pump occurs universally in cells of higher animals. It was identified with the Na^+, K^+-ATPase enzyme system first extracted from crab nerve by Skou at Aarhus University (Skou, 1998). Research on Na^+, K^+-ATPase chemistry depended heavily on exploitation of inhibitory actions of glycosides like ouabain and digoxin. In micromolar concentrations these block both the active Na^+ and K^+ fluxes in intact tissues, and ATPase activity by purified enzyme preparations. In addition, binding measurements using ^3H-labelled ouabain could estimate the number of Na^+ pumping sites in unit membrane area, assuming that each site binds one molecule of ouabain. The squid giant axon shows several thousand sites per square micrometre of membrane, while the smallest unmyelinated fibres show site densities about a tenth as great.

Although Na^+-pump-mediated Na^+ extrusion and K^+ uptake is quickly halted by ouabain or any metabolic inhibitor which deprives the pump of its ATP supply, neither treatment has any

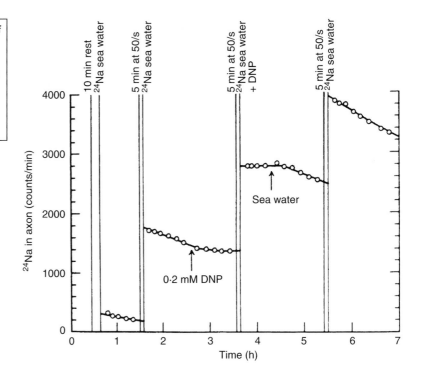

Figure 3.11 The lack of effect of dinitrophenol on the Na$^+$ entry during stimulation of a squid axon. The resting Na$^+$ influx for the first period of immersion in ^{24}Na sea water was 50 pmole/cm^2. Temperature 17 °C. (From Hodgkin and Keynes, 1955b.)

immediate effect on electrical excitability. Figure 3.11 shows the results of an experiment in which Na$^+$ influx into a squid giant axon was measured by soaking it for a few minutes in a solution containing ^{24}Na followed by mounting it above a Geiger counter in a stream of unlabelled artificial sea water. While Na$^+$ was being actively extruded from the axon, the counting rate fell steadily, but on the addition of DNP to the sea water bathing the axon, the counts remained constant. When the DNP was washed away, the Na$^+$ pump started up again. The rate of gain of radioactivity during the periods of exposure to ^{24}Na was increased by a factor of about 10 by stimulation at 50 shocks/s, but the extra entry of ^{24}Na was the same whether or not the Na$^+$ pump had been blocked. Washing out experiments showed that the accelerated outward movement of ^{24}Na during the impulse (Section 4.2) was affected equally little by DNP.

It follows from this evidence and further contrasts between electrical excitation and Na$^+$ pump properties and functions (Table 3.3) that the pathways for active and passive membrane ion transport likely function quite independently of one another (Figure 3.12). This was particularly clearly demonstrated in giant axons because of their large volume-to-surface ratio. A squid axon whose axoplasm had been extruded and replaced by a pure solution of K$_2$SO$_4$ was nevertheless capable of conducting over 400 000 impulses before becoming exhausted (Baker et al., 1962). In small unmyelinated nerve fibres

Table 3.3 | *Comparison of the properties of the Na^+ and K^+ channels with those of the Na^+ pump*

	Na^+ and K^+ channels	Na^+ pump
Direction of ion movements	Down the electrochemical gradient	Against the electrochemical gradient
Source of energy	Pre-existing electrochemical gradients	ATP
Voltage dependence	Regenerative link between potential and Na^+ conductance	Independent of membrane potential
Blocking agents	Tetrodotoxin blocks Na^+ channels at 10^{-8} M Tetramethylammonium blocks K^+ channels Ouabain has no effect	Tetrodotoxin has no effect Tetramethylammonium has no effect at 10^{-3} M Ouabain blocks at 10^{-7} M
External Ca^{2+}	Increase in $[Ca^{2+}]$ raises threshold for excitation; decrease in $[Ca^{2+}]$ lowers threshold	No effect
Selectivity	Li^+ is not distinguished from Na^+	Li^+ is pumped much more slowly than Na^+
Effect of temperature	Rate of opening and closing of channels has large temperature coefficient (Q_{10}), but maximum conductances have a small Q_{10}	Velocity of pumping has a large temperature coefficient
Density of distribution in the membrane	Squid axon has 290 TTX-binding sites per μm^2 Rabbit vagus has 100 TTX-binding sites per μm^2 of membrane	Squid axon has 4000 ouabain-binding sites per μm^2 Rabbit vagus has 750 ouabain-binding sites per μm^2
Maximum rate of movement of Na^+	100 000 pmol/(cm^{2}s) during rising phase of action potential	60 pmole/(cm^{2}s) of Na^+ at room temperature
Metabolic inhibitors	No effect; electrical activity is normal in axon perfused with pure salt solution	1 mM cyanide or 0.2 mM dinitrophenol block as soon as ATP is exhausted

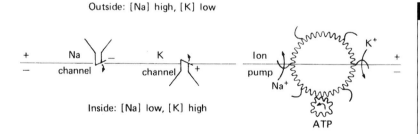

Outside: [Na] high, [K] low

Inside: [Na] low, [K] high

Figure 3.12 Two types of process mediating ion transport across the nerve membrane. The Na^+ pump responsible for transporting ions uphill and so creating the concentration gradients is shown as a bucket system driven by ATP. The Na^+ and K^+ channels involved in excitation are shown as funnel-shaped structures whose opening is controlled by the electric field across the membrane. In this diagram they are in the resting state with the charged gates held closed by the membrane potential. On depolarisation of the membrane the gates open and permit ions to flow downhill.

the downhill ionic movements during the nervous impulse are much larger in relation to the reservoir of ions built up by the Na^+ pump. Blockage of active transport does, after a relatively short time, affect the conduction mechanism indirectly by reducing the size of the ionic concentration gradients.

3.7 | Quantitative Reconstruction of Resting Cellular Ionic Homeostasis

It was possible to computationally assemble these distinct mechanisms underlying bioelectric potentials and their associated active and passive physiological processes to reconstruct the generation of the in vivo resting potential and associated parameters describing the quiescent cell. This employed experimentally established values of the individual ionic permeabilities, P_{Na}, P_K and P_{Cl}, the kinetic properties of Na^+, K^+-ATPase and their dependences on intracellular and extracellular $[Na^+]$ and $[K^+]$. The resulting E_m was derived from the total intracellular ionic charge that would be affected by their resulting transmembrane ionic fluxes, and the membrane capacitance, C_m, assuming extracellular neutrality in which the total extracellular concentrations of charged positive and negative ions were exactly equal:

$$E_m = \frac{F([Na^+]_i + [K^+]_i - [Cl^-]_i + z_x[Pr^{z-}]_i)}{C_m} \tag{3.15}$$

This reconstruction first recapitulated the existence of and values of a stable V_c and E_m. This involved both the presence of membrane impermeant intracellular ions, Pr^{z-}, and operation of a functional Na^+ pump. In the mathematical analysis, V_c and E_m converged to unique set points without requiring any explicitly V_c- or E_m -sensitive mechanisms. This involved a balance between two tendencies. On the one hand, the presence of Pr^{z-} prevents Cl^-, the major extracellular anion, from reaching chemical equilibrium. The resulting continued Cl^- influx in turn permits cation and therefore osmotic water influx. This drives cell swelling tending towards an infinite dilution of Pr^{z-}.

On the other hand, Na^+ pump activity tends to cause indefinite cell shrinkage in the absence of Pr^{z-}. Each Na^+ pump cycle tends to diminish $[Na^+]_i$ and increase $[K^+]_i$. The resulting outward K^+ gradient drives K^+ efflux causing cell hyperpolarisation. This also promotes Cl^-, accompanied by osmotic water, efflux, reducing V_c. However, the latter Cl^- efflux is limited by the consequently increased $[Pr^{z-}]_i$ relative to $[Cl^-]_i$. The resulting balance thereby sets V_c. The K^+ efflux drives membrane hyperpolarisation thereby setting the limiting E_m. The hyperpolarising E_m in turn opposes further K^+ efflux. These two factors balance to give unique and stable V_c and E_m.

Secondly, these mathematical computations also stably and consistently converged to unique values of $[Na^+]_i$, $[K^+]_i$ and $[Cl^-]_i$ as well as of E_m that were closely concordant with experimental findings obtained in the presence of normal $[Na^+]_o$, $[K^+]_o$ and $[Cl^-]_o$. This confirmed the validity of this charge difference approach (Fraser and Huang, 2004). The modelling could then proceed to clarify the contributions their component physiological processes made to E_m and V_c.

Thirdly, the modelling confirmed that maintenance of a stable V_c and E_m ultimately depended upon Na^+, K^+-ATPase mediated Na^+ pump activity. However, within defined limits, both V_c and E_m remained

A

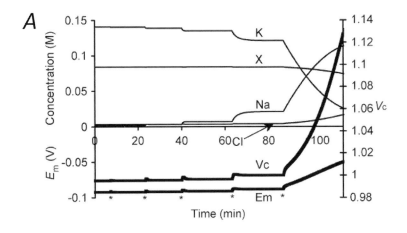

B

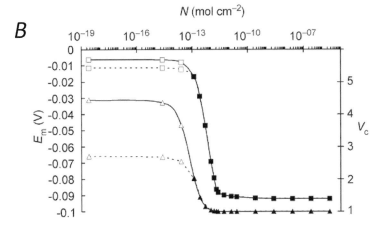

C

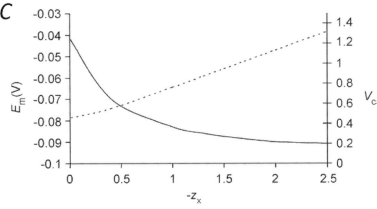

Figure 3.13 Systems reconstruction of the effects of Na^+ pump activity and of intracellular organic osmolyte and its valency upon intracellular ion homeostasis, cell volume stability and membrane potential.

(A, B) Na^+ pump activity underpins volume stability, but within limits, variations in its activity leave cell volume V_c and membrane potential E_m conserved. This is illustrated by: (A) Simulations of successive 10-fold reductions (each marked *) in Na^+ pump density (N) from an initial value of 5×10^{-8} mol cm^{-2}. Large changes in N through values sufficient to preserve stable cell volume exerted relatively little effect on V_c or E_m, Even a 10 000-fold reduction to $N = 5 \times 10^{-12}$ mol cm^{-2} produced only a $<1\%$ increase in V_c and $< 5\%$ depolarisation of E_m. However, a further 10-fold reduction in N destabilised cell volume.

(B) Overall relationship between N, V_c (triangles) and E_m (squares) demonstrates that for this particular set of modelled ion permeabilities, a pump density of $N \geq 2.5 \times 10^{-13}$ mol cm^{-2} was necessary to ensure stable V_c or E_m. For all values of N ($>3 \times 10^{-12}$ mol cm^{-2}) sufficient to support a value of E_m of at least −80 mV, variation in N has little influence on V_c or E_m. Filled symbols indicate final stable values. Open symbols denote values of E_m or V_c obtained either 50 min (dashed line) or 150 min (continuous line) after N was reduced from 5×10^{-12} mol cm^{-2} to the indicated value for N values insufficient to stabilise cell volume. (C) Effect of mean valency, z, of intracellular organic osmolyte on eventual stable values of both V_c and E_m. Cells modelled with initially identical intracellular organic osmolyte concentrations but different modelled cells studied with differing mean charge valency of osmolyte (z). Each cell reached stable but differing values of V_c, E_m, $[Na^+]_i$, $[K^+]_i$, $[Cl^-]_i$ or $[Pr^{z-}]_i$. Graph summarising influence of z upon E_m (continuous line) and V_c (dashed line). (From Figures 5 and 8C of Fraser and Huang, 2004.)

relatively conserved and stable despite wide variations in Na^+, K^+-ATPase membrane density (N) (Figure 13.13A). Successive stepwise 10-fold reductions in N from the reported normal values initially produced little alteration in either V_c or E_m, or of $[Na^+]_i$, $[K^+]_i$ or $[Pr^{z-}]$. However, this remained the case only until reductions in N to a critical value. Close to or below this, the parameters markedly altered with time, either becoming unstable or reaching altered steady state values (Figure 13.13A, B). Finally, such changes in N could not independently influence V_c and E_m but altered both these variables together. These findings suggest that once above a critical availability, N, variations in Na$^+$ pump density do not regulate or determine V_c or E_m in an excitable cell.

Fourthly, the analysis implicated the content of membrane-impermeant solute Pr^{z-} in determining V_c, but not E_m. For a given value of z, Pr^{z-} content linearly altered V_c as expected for the properties of an osmometer, but left E_m unchanged.

Fifthly, the charge z contrastingly influenced both V_c and E_m (Figure 13.13C). Weakly and strongly negative values of z would respectively permit low or high values of $[K^+]_i$, whilst still permitting approximate gross intracellular charge neutrality. For example, high $[K^+]_i$ would result in a more polarised E_m and imply a reduced K$^+$ efflux. Fig 13.13C summarises the resulting effects of z upon V_c (dashed line) and E_m (continuous line). For any given quantity of Pr^{z-}, V_c was minimal when z was zero, corresponding to the situation when Pr^{z-} was uncharged. V_c increased with increasing values of z, fulfilling expectations of its resulting in an increased effective intracellular osmotic pressure (Section 3.5). Conversely, for any given quantity of Pr^{z-}, increases in charge z shifted E_m to more negative potentials. Increasing the negative charge, z, on Pr^{z-} results in larger cells with larger, more negative, resting potentials, and correspondingly different $[Na^+]_i$, $[K^+]_i$, $[Cl^-]_i$ and $[Pr^{z-}]_i$.

Membrane Permeability Changes During Excitation

4.1 | Impedance Changes During the Spike

An important landmark in the development of theories about mechanisms of nerve excitation and conduction was the demonstration that passage of an impulse in the squid giant axon was accompanied by a substantial drop in its membrane electrical impedance (Cole and Curtis, 1939). An axon whose electrical properties are represented by resistance R_X and capacitance C_X was mounted in a trough between two plate electrodes connected in one arm of a Wheatstone bridge circuit for measurement of resistance and capacitance in parallel (Figure 4.1). The bridge output was displayed on a cathode-ray oscilloscope. The circuit elements R_v and C_v were adjusted to give a balance, and therefore zero output, with the resting axon. When the axon was stimulated at one end, the bridge went briefly out of balance (Figure 4.2A) with a timecourse similar to that of the action potential (Figure 4.2B). The change was attributable to a reduction in the membrane resistance from a ~1000 Ω cm^2 resting value to an active value around 25 Ω cm^2. There were no measurable changes in the membrane capacitance of ~1 μF/cm^2.

4.2 | The Sodium Hypothesis

Cole and Curtis's results were not wholly unexpected. It had previously been suggested by Bernstein that there was some kind of collapse in the membrane selectivity towards K$^+$ during the impulse. However, a year or two later both they, and Hodgkin and Huxley succeeded in recording internal potentials for the first time. This demonstrated that the membrane potential did not just fall towards zero at the peak of the spike, but instead reversed by quite a few millivolts (Figure 2.4). This unexpected overshoot could not be accounted for by reduced ionic selectivity of the nerve membrane, but required a radically different type of explanation.

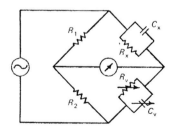

Figure 4.1 Wheatstone bridge circuit used for measurement of resistance, R_X, and capacitance, C_X, in parallel.

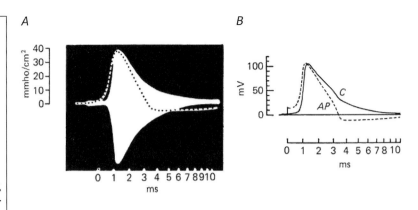

Figure 4.2 The timecourse of the impedance change during the conducted action potential in a squid giant axon (Cole and Curtis, 1939). (A) Double exposure of the unbalance of the impedance bridge and of the monophasic action potential at one of the impedance electrodes; the time marks at the bottom are 1 ms apart; (B) superimposed plots of the membrane conductance increase, C, and of the action potential, AP, after correction for amplifier response.

None was forthcoming until the suggestion of a *sodium hypothesis* of nervous conduction (Hodgkin *et al.*, 1949). This began from the observation that the external Na^+ concentration, $[Na^+]_o$, is greater than the internal concentration, $[Na^+]_i$, giving a Na^+ Nernst equilibrium potential (E_{Na}) reversed in polarity relative to E_K. This suggested a hypothesis that excitation involves a rapid and highly specific increase in membrane permeability to Na^+. This would shift the membrane potential from its resting level near E_K to a new value approaching E_{Na}. A first piece of evidence in support of this theory reported that nerves are indeed rendered inexcitable by Na^+-free extracellular solutions. Overton had showed previously for frog muscle that only Li^+ ions can fully replace Na^+, though several small organic cations like hydroxylamine subsequently proved to act as partial Na^+ substitutes (Section 5.6); and certain excitable tissues have Ca^{2+}-dependent spike mechanisms (Section 10.2). Replacement of part of the external Na^+ by glucose reduced both the rate of rise of the action potential and its height (Figure 4.3). This rate of rise, (dV/dt), was directly proportional to the extracellular Na^+ concentration $[Na^+]_o$. Furthermore, in accordance with Equation (3.2) applied to Na^+, and Equation (3.3), the slope of the line relating spike height to $\log_{10}[Na^+]_o$ was close to 58 mV until the point was reached where conduction failed. Subsequent experiments demonstrated a similar relationship when $[Na^+]_i$ was varied.

Altering the potential across a membrane of capacitance $\sim 1\ \mu F/cm^2$ from -60 mV at rest to $+50$ mV at the peak of the spike requires transfer of a total quantity of charge of $110\ nC/cm^2$. Such a charge would be carried by 1.1 picomoles of a monovalent ion crossing $1\ cm^2$ of membrane. A crucial test of the validity of the sodium hypothesis measured net Na^+ entry into and net K^+ loss from the fibre during passage of an impulse. Radioactivity analysis demonstrated that stimulated *Sepia* axons showed a net gain of 3.8 pmol Na/(cm^2 impulse) and a net loss of 3.6 pmol K/(cm^2 impulse). In squid axons the corresponding figures were 3.5 pmol Na and 3.0 pmol K (Keynes and Lewis, 1951). The measured ionic movements were thus more than large

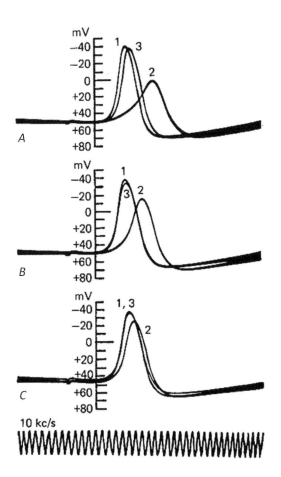

Figure 4.3 Effect of reducing external Na⁺ concentration on the action potential in a squid giant axon. In each set of records, record 1 shows the response with the axon in sea water, record 2 in the experimental solution and record 3 in sea water again. The solutions were prepared by mixing sea water and an isotonic dextrose solution, the proportions of sea water being: (A) 33%; (B) 50%; (C) 71%. (From Hodgkin and Katz, 1949.)

10 kc/s

enough to comply with the theory. This was not surprising in view of the likelihood of some exchange of K⁺ for Na⁺ over the top of the spike in addition to the net uptake of Na⁺ during its upstroke and the net loss of K⁺ during its falling phase. Such experiments with ^{24}Na (Figure 4.4) showed that there was an analogous exchange of labelled Na during the spike as well as a net entry. The extra inward movement of radioactive Na was estimated as 10 pmol/(cm^2 impulse), and the extra outward movement as about 6 pmol/(cm^2 impulse), the difference between the two figures being in good agreement with the analytical results.

4.3 | Predictions of the Sodium Hypothesis

The essential new property of the membrane envisaged by the sodium hypothesis was its possession of voltage-sensitive mechanisms providing appropriate control of its Na⁺ and K⁺ permeabilities. The sequence of events supposed to occur during the action potential may be

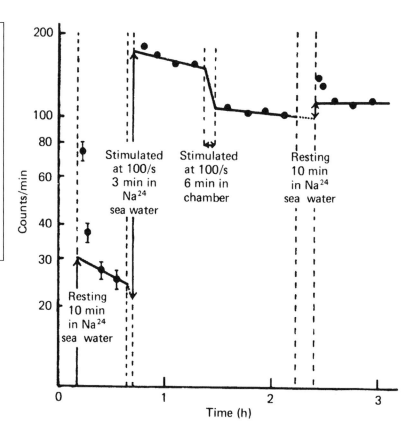

Figure 4.4 Movements of ^{24}Na in a stimulated *Sepia* axon of diameter 170 μm. The axon was alternately exposed to artificial sea water containing ^{24}Na, and mounted in a stream of inactive sea water above a Geiger counter to measure the amount of radioactivity taken up. The loss of counts during the first 10 min after exposure to ^{24}Na resulted from washing away extracellular Na$^+$, and was ignored. For the entry of ^{24}Na, I count/min was equivalent to 42.5×10^{-12} mol Na/(cm axon). Temperature 14 °C. (From Keynes, 1951.)

summarised as follows. Membrane depolarisation produced by outward current flow driven either by an applied cathode, or the proximity of an active region where the membrane potential is already reversed, rapidly increases its Na$^+$ permeability. There is a net inward movement of Na$^+$ ions, flowing down the Na$^+$ concentration gradient. If the initial depolarisation opens the Na$^+$ channels far enough, Na$^+$ enters faster than K$^+$ can leave. This causes the membrane potential to become further depolarised. The extra depolarisation increases Na$^+$ permeability even more, accelerating the change of membrane potential in a regenerative fashion. This linkage between Na$^+$ permeability and membrane potential forms a *positive feedback* mechanism (Figure 4.5).

The Na$^+$ entry does not continue indefinitely. First, it is halted partly because the membrane potential soon reaches a level close to E_{Na}, where the net inward driving force acting on Na$^+$ becomes zero. Secondly, the rise in Na$^+$ permeability decays inexorably with time from the moment when it is first triggered, a process termed *inactivation*. After the peak of the spike has been reached, the Na$^+$ channels begin to close, and the Na$^+$ permeability is soon completely inactivated. At the same time, the K$^+$ permeability of the membrane rises well above its resting value, and an outward movement of K$^+$

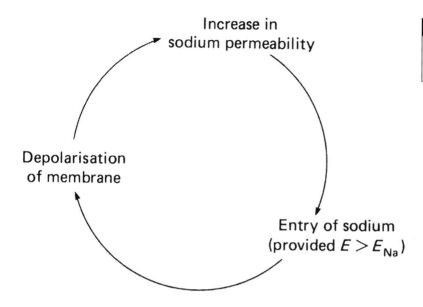

Increase in sodium permeability

Depolarisation of membrane

Entry of sodium (provided $E > E_{Na}$)

Figure 4.5 The regenerative linkage between membrane potential and Na^+ permeability. (From Hodgkin, 1951.)

takes place, eventually restoring the membrane potential to its original level.

At the end of the spike the membrane has returned to the normal resting potential, but its Na^+ permeability mechanism is still inactivated, resulting in refractoriness to further excitation. Lapse of further time allows the Na^+ permeability to recover from such inactivation, restoring it to the quiescent state in which it is still very low, as is characteristic of the resting membrane, but is now ready once more to increase regeneratively if the system is retriggered.

This scheme suggests two important features of the Na^+ channel. First, channel opening rapidly and steeply depends on membrane potential. A relatively small degree of depolarisation suffices to induce a large rise in Na^+ permeability (P_{Na}). Secondly, having opened quickly, they show a slower inactivation process. This occurs even when the potential has not returned to its resting level, and may still be reversed. In squid giant axon, K^+ channels are also strongly voltage dependent. However, their opening is delayed and they are not subsequently inactivated. The return of P_K to its resting value wholly depends on the membrane repolarisation taking place during the falling phase of the spike.

There is thus a separation in timing of the permeability changes. P_{Na} rises quickly and is cut off by inactivation. P_K rises with an appreciable lag. This minimises any energetically wasteful interchange of Na^+ and K^+ at the peak of the spike unaccompanied by useful alterations in membrane potential. It may also be noted that it is not essential to conduction that P_K should increase at all. In squid axon, the delayed rise of P_K nevertheless results in a more rapid return of the resting potential. It shortens the spike and subsequent refractory period. Nevertheless P_{Na} inactivation in conjunction with an unchanged K^+ efflux would also restore the resting potential, but would do so more slowly. Some nerve fibre types dispense with the rise of P_K.

4.4 | Voltage Clamp Experiments

The increase in membrane Na^+ permeability during the spike predicted by the sodium hypothesis can be measured with radioactive tracers (Figure 4.4). This approach has the advantage of specificity, providing unambiguous information about Na^+ movements. However, its time resolution is relatively poor. Its results refer only to cumulative effects of a large number of impulses. Detailed studies of membrane permeability changes through a single action potential resorted to measuring electric currents carried by the ions when they move across the membrane. This offered much greater sensitivity and time resolution. However, the amount learnt simply by recording the current flowing during the conducted action potential itself is limited, because the underlying permeability changes follow a fixed sequence determined by the nerve and not the experimenter.

To surmount the latter difficulty, Hodgkin and Huxley adopted an approach originally introduced by Cole and Marmont (Marmont, 1949) to measure ionic conductances in a nerve membrane whose potential was first 'clamped' at a chosen level, then subjected to a predetermined series of step changes. They thereby explored in detail the laws governing voltage-sensitive behaviour in Na^+ and K^+ channels. Such *voltage clamp* studies and their adaptations have provided much of our present-day knowledge of the permeability mechanisms underlying excitation and conduction not only in nerve, but also muscle (Section 10.2) and in synaptic transmission (Sections 7.5 and 7.6).

A typical experimental set-up for voltage clamping a squid giant axon involves introduction of two internal electrodes (Figure 4.6). The first monitors the potential at the centre of the stretch of axon to be clamped. The other passes current uniformly across the membrane over a somewhat greater length. In Hodgkin and Huxley's original apparatus, these electrodes were constructed by winding two spirals of AgCl-coated silver wire on a fine glass rod (Figure 2.2). Nowadays, the potential is recorded by a 50 μm micropipette filled with isotonic KCl, to which is glued a platinum wire 75 μm in diameter with a bare and platinised terminal portion that can pass current without undue polarisation (Figure 2.1). The external electrode consists of a platinum sheet in three sections: the current flowing to the central section is amplified and recorded, while the two outer sections ensure the uniformity of clamping over the fully controlled region.

After appropriate amplification, the internal potential is fed to a voltage comparator circuit, along with the square-wave signal to

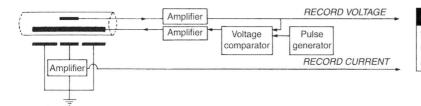

Figure 4.6 Schematic diagram of the arrangement for measuring membrane current in a voltage-clamped squid giant axon.

which the membrane potential is to be clamped. The output from the comparator is applied to the internal current wire. It is varied to increase or decrease the membrane current just enough to force the membrane potential to follow the square wave exactly. In electronic terms, this arrangement constitutes a negative feedback control system in which the potential across the membrane is determined by the externally generated command signal, and the resulting membrane current is measured. Various other electrode arrangements are required to voltage clamp smaller unmyelinated nerve fibres, single nerve cells, neuromuscular synapses (Section 7.5), muscle fibres (Section 10.2) or the isolated node of Ranvier in a myelinated nerve fibre. Nevertheless the basic principle of the circuit remains the same.

4.5 | Equivalent Electrical Circuit Description of the Nerve Membrane

The analysis of voltage clamp experimental results adopted an equivalent electrical circuit description of the nerve membrane (Figure 4.7). This consists of a capacitance C_m connected in parallel with three resistive ionic pathways each incorporating a resistance (R_K, R_{Na} and R_{leak}) in series with a battery. For a given ionic pathway, the driving forces acting on the ions are the membrane potential E_m and the concentration gradient for that species of ion. However, this concentration gradient may be equated to an electromotive force (EMF) calculated from the Nernst equilibrium potential (Section 3.3; Equation (3.2)). The appropriate values for the three battery potentials are E_K, E_{Na} and E_{leak} respectively. The net EMF acting on each ion is the difference between E_m and its Nernst potential. Ohm's law predicts that the ionic currents I_K, I_{Na} and I_{leak} are given by

$$I_K = \frac{(E_m - E_K)}{R_K} \tag{4.1}$$

and so on for I_{Na} and I_{leak}.

Although, in accordance with electrical convention, the ionic pathways are represented as resistances, it is often more convenient to think of them as reciprocal conductances, g_K, g_{Na} and g_{leak}. These represent the ease with which that particular ion passes across the

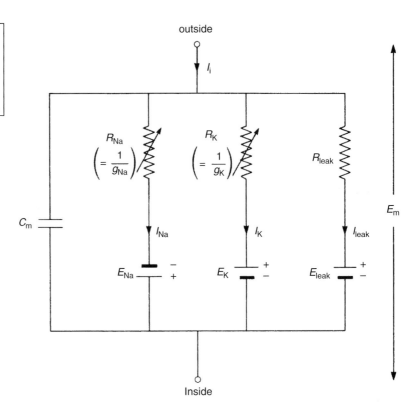

Figure 4.7 The equivalent electrical circuit of the nerve membrane according to Hodgkin and Huxley (1952). R_{Na} and R_K vary with membrane potential and time; the other components are constant.

membrane, and can be compared to the permeability coefficients in the constant field equation (Section 3.3), although measured in different units. In the equivalent circuit, R_K (= $1/g_K$) and R_{Na} (= $1/g_{Na}$) are indicated as being variable. The object of voltage clamp experiments is to investigate their dependence on membrane potential and time. R_{leak}, to which Cl⁻ is the main contributing ion, is constant.

In the absence of externally applied current, the electrical model predicts that the value of E_m is determined by the relative sizes of the ionic conductances. If g_K is, as in the resting condition, much larger than g_{Na}, E_m lies close to E_K. When Na⁺ channels are opened, g_{Na} rises, and E_m is driven towards E_{Na}. When the potential at which the membrane is clamped is suddenly altered by a step potential change, the current flowing across the membrane consists of the capacity current required to charge or discharge C_m plus the ionic current to be measured. The total current:

$$I = C_m \cdot \frac{dE_m}{dt} + I_i \tag{4.2}$$

The term I_i is the sum of the currents flowing through all three ionic pathways. With a well-designed voltage clamp system, (dE_m/dt) should have fallen to zero and the flow of capacity current should have ceased after no more than ~20 μs. All subsequent changes in the recorded

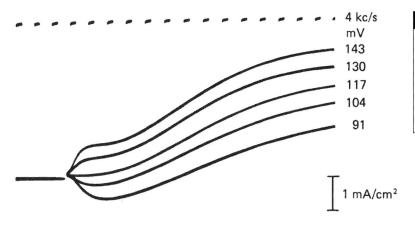

4 kc/s
mV
143
130
117
104
91

1 mA/cm²

Figure 4.8 Membrane currents for large depolarising voltage clamp pulses; outward current upwards. The figures on the right show the change in internal potential in mV. Temperature 3.5 °C. (From Hodgkin, 1958 after Hodgkin et al., 1952.)

current result from alterations in g_{Na} and g_K operative at the new membrane potential.

Figure 4.8 shows a typical family of superimposed current records for a squid giant axon subjected to increasingly larger depolarising voltage steps. The initial capacity transients were too fast to be photographed, so what is seen is purely the ionic current. These show an early phase flowing inwards for small depolarisations and outwards for large ones, and a late, consistently outward phase. This is consistent with the postulates of the sodium hypothesis. We next consider how the I_{Na} and I_K contributions were separated from one another.

4.6 | Separation of Ionic Current Components in Response to Voltage Change

Hodgkin and Huxley analysed their voltage clamp records following suppression of the inward Na⁺ current, I_{Na}, by substituting choline for Na⁺ in the external medium. The resulting records demonstrated that removal of external Na⁺ converted the initial hump of inward current into an outward one (Figure 4.9). However, it did not affect the late current, confirming that these inward and outward currents were carried by Na⁺ and K⁺ respectively. To eliminate I_{Na} completely, it was necessary to leave some Na⁺ in the external medium and to take E_m exactly to E_{Na}, where by definition $I_{Na} = 0$. In the experiment illustrated in Figure 4.10 [Na⁺]$_o$ was reduced to one-tenth its normal value and the depolarising step made to +56 mV. This yielded the isolated K⁺ current, I_K (Figure 4.10B). Subtraction of the resulting trace (B) from one recorded in normal sea water (Figure 4.10A), yielded the timecourse of I_{Na} (Figure 4.10C). The currents were finally converted into conductances by taking $g_K = I_K/(E_m - E_K)$ and $g_{Na} = I_{Na}/(E_m - E_{Na})$. Plots of g_K and g_{Na} against time show that,

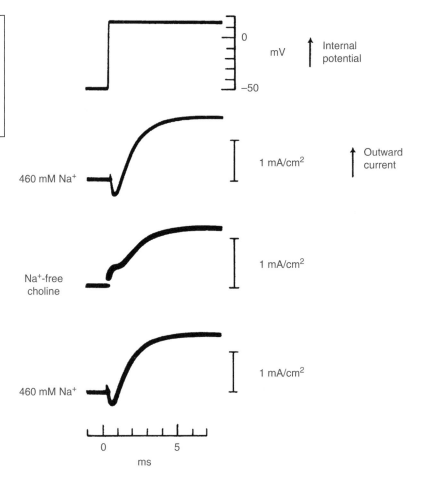

Figure 4.9 Membrane currents associated with depolarisation of 65 mV in the presence and absence of external Na$^+$. Outward current and internal potential shown upward. Temperature 11 °C. (From Hodgkin, 1958 after Hodgkin et al., 1952.)

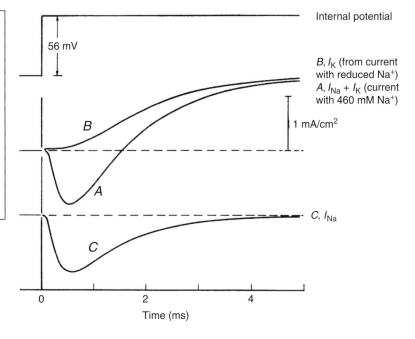

Figure 4.10 Analysis of the ionic current changes in a squid giant axon during a voltage clamp pulse that depolarised it by 56 mV. (A) (= I_{Na} + I_K) response with the axon in sea water containing 460 mM Na$^+$. (B) (= I_K) response with the axon in a solution made up of 10% sea water and 90% isotonic choline chloride solution. (C) (= I_{Na}) difference between traces (A) and (B). Temperature 8.5 °C. (From Hodgkin, 1958 after Hodgkin et al., 1952.)

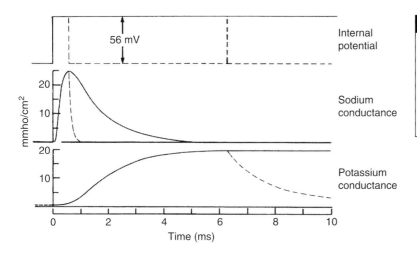

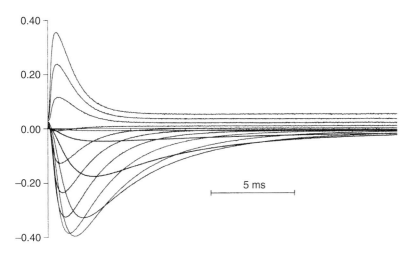

Figure 4.11 Timecourses of the ionic conductance changes during a voltage clamp pulse calculated from the current records shown in Figure 4.10. Dashed lines show the effect of repolarisation after 0.6 or 6.3 ms. (From Hodgkin, 1958 after Hodgkin *et al.*, 1952.)

Figure 4.12 Superimposed traces of Na$^+$ current in a voltage clamped squid axon whose K$^+$ channels were blocked by internal dialysis with 330 mM CsF + 20 mM NaF and which was bathed in a K$^+$-free artificial sea water containing 103 mM NaCl and 421 mM Tris buffer. The membrane was held at –70 mV and pulses were applied, taking the potential to levels varying between –40 and +80 mV in steps of 10 mV. Current scales mA/cm^2. For the smaller test pulses, the current flowed inward (downward), but above about +50 mV its direction reversed. For the largest pulses, inactivation was no longer complete, and the channels ended up in a non-inactivating open state. Temperature 5 °C. (Computer-acquired recording made by R. D. Keynes, N. G. Greeff, I. C. Forster and J. M. Bekkers.)

as explained above, g_{Na} rises quickly and is subsequently inactivated, while g_K rises with a definite lag and is not inactivated (Figure 4.11).

Separation of the two components of the ionic current is now achieved more easily by recording separated I_{Na} and I_K after complete pharmacological block of K$^+$ and Na$^+$ channels, respectively. A good method of abolishing I_K is by introducing Cs$^+$ into the axon by perfusion or dialysis. The Cs$^+$ enters the mouths of, thereby blocking, K$^+$ channels from the inside. Figure 4.12 shows a typical family of I_{Na} records for voltage clamp pulses of different sizes applied to a squid giant axon dialysed with CsF. For Na$^+$ channels, the Japanese puffer-fish poison tetrodotoxin (TTX) acts externally as an inhibitor with an affinity constant of ~3 nM. Figure 4.13 shows a family of I_K records for a squid axon dialysed with a KF solution and bathed in a Na$^+$-free solution containing 1 μM TTX. Such records give quantitative results confirming the findings and conclusions obtained by Hodgkin and Huxley. They provide convincing evidence for the assumption that Na$^+$ and K$^+$ channels are entirely separate entities related only through their strong dependence on a common potential gradient.

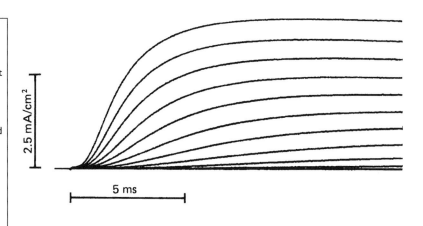

Figure 4.13 Superimposed traces of K$^+$ current in a voltage-clamped squid axon whose Na$^+$ channels were blocked by bathing it in artificial sea water containing 1 μM TTX, and which was dialysed internally with 350 mM KF. The membrane was held at −70 mV, and pulses were applied, taking the potential to levels varying between −60 and +40 mV steps. Outward current is upward. Temperature 4 °C. (Computer-acquired recording made by R. D. Keynes, J. E. Kimura and N. G. Greeff.)

4.7 | Mathematical Reconstruction of Ionic Current Properties

Hodgkin and Huxley next devised a set of mathematical equations empirically describing the behaviour of g_{Na} and g_K as functions of E_m and time. The Na$^+$ conductance obeyed the relationship:

$$g_{Na} = \bar{g}_{Na} m^3 h. \tag{4.3}$$

The constant term $\bar{g}_{Na}$ represents the peak conductance attainable, m is a dimensionless activation parameter varying between 0 and 1; h is a similar, inactivation, parameter varying between 1 and 0. The corresponding equation for the K$^+$ conductance was:

$$g_K = \bar{g}_K n^4 \tag{4.4}$$

The constant term $\bar{g}_K$ is the peak K$^+$ conductance; n is another dimensionless activation parameter. The quantities m, h and n described the variation of the conductances with E_m and time. They were determined by differential equations describing first-order, unimolecular transitions between inactive and active states:

$$\frac{dm}{dt} = \alpha_m(1 - m) - \beta_m m \tag{4.5}$$

$$\frac{dh}{dt} = \alpha_h(1 - h) - \beta_h h \tag{4.6}$$

And:

$$\frac{dn}{dt} = \alpha_n(1 - n) - \beta_n n \tag{4.7}$$

The α and β terms are voltage-dependent forward and backward rate constants whose dimensions are time^{-1}.

The precise details of the voltage-dependence of the six rate constants need not concern us further. Equations (4.3) to (4.7) are mainly cited to

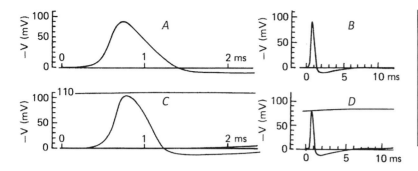

Figure 4.14 Comparison of computed (A, B) and experimentally recorded (C, D) action potentials propagated in a squid giant axon at 18.5 °C, plotted on fast and slow time scales. The calculated conduction velocity was 18.8 m/s; that actually observed was 21.2 m/s. (From Hodgkin and Huxley, 1952.)

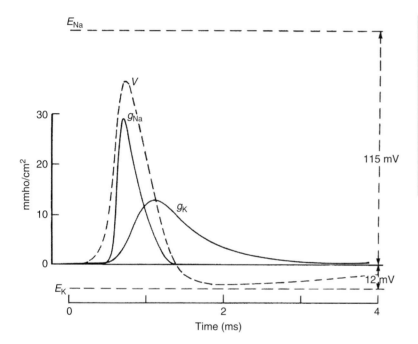

Figure 4.15 Timecourses of the propagated action potential and underlying ionic conductance changes computed by Hodgkin and Huxley from their voltage clamp data. The constants used were appropriate to a temperature of 18.5 °C. The calculated net entry of Na^+ was 4.33 pmol/cm^2, and the net exit of K^+ was 4.26 pmol/cm^2. Conduction velocity = 18.8 m/s. (From Hodgkin and Huxley, 1952.)

help the mathematically minded reader to follow the steps necessary for achievement of Hodgkin and Huxley's primary objective of testing their description of the permeability system by calculating from their equations the shape of the propagated action potential.

Finally, Hodgkin and Huxley used the mathematical formulation to compute the expected properties of the conducted action potential using parameters derived from the voltage clamp data. First, the predicted timecourse of the propagated action potential at 18.5 °C closely agreed with experimental observations at the same temperature (Figure 4.14). Secondly, the expected net exchange of Na^+ and K^+ could be calculated from the extents and degree of overlap of the changes in g_{Na} and g_K during the spike (Figures 4.15). The total entry of Na^+ and the exit of K^+ in a single impulse each worked out to be

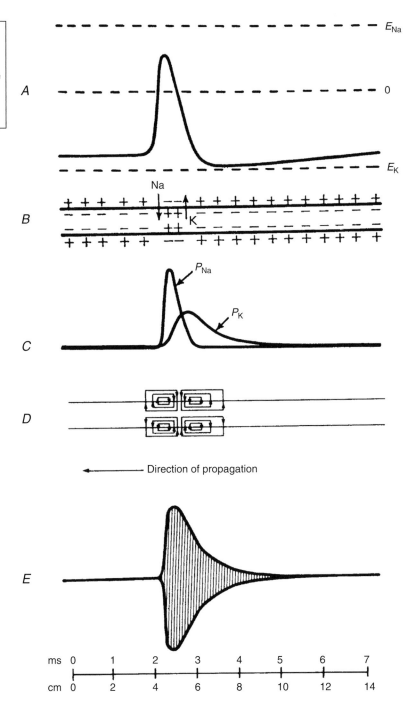

Figure 4.16 Time relations of the events during the conducted impulse: (A) membrane potential; (B) ionic movements; (C) membrane permeability; (D) local circuit current flow; (E) membrane impedance.

~4.3 pmol/cm^2 membrane, in agreement with the results of the analytical and tracer experiments discussed in Section 4.2. Finally, the computed action potential conduction velocity showed close agreement with observed values (Figure 4.16).

5

Voltage-Gated Ion Channels

Both voltage- and ligand-gated ion channels, as well as membrane transport proteins such as the Na^+, K^+-ATPase Na^+ pump are large intramembrane protein molecules. Following the classical findings described in previous chapters, their structure and function have been studied at the molecular level. This has included some of their primary and higher order structures. Biophysical studies have then contributed substantial new understanding of the mechanisms underlying their function.

5.1 | cDNA Sequencing Studies

A protein comprises one or more long chains built up of 20 different amino acids (Table 5.1), complexly folded on themselves. Its properties depend critically on this folding pattern. The latter in turn is determined by the exact sequence with which its constituent amino acids are assembled and interact. This is specified by the nucleotide base sequence of the corresponding DNA molecules in the cell genome. There are only four different bases. Each of the 20 known amino acids corresponds to a specific group of bases within a universally obeyed triplet code. The information embodied in the base sequence of a DNA molecule is transcribed onto an intermediary messenger RNA. This is translated during protein synthesis to yield the correct sequence of amino acids. Rapid nucleotide sequencing methods perfected by Sanger and his colleagues and modern recombinant DNA technology has made possible cloning of DNA so that the required quantity for the determination can be prepared from a single gene. Hence the amino-acid sequences of proteins are readily determined indirectly from the base sequences of the cDNA in which they are encoded.

5.2 | The Structure of Voltage-Gated Ion Channels

The substantial voltages generated by the electric organ of the electric eel (*Electrophorus*) depend on additive discharge of a large number of cells derived embryonically from muscle (Figure 2.4G). Their electrical

Table 5.1 | *The amino acids found in proteins*

Type	Amino acid	Side-chain	Abbreviations		Hydropathy index
Non-polar	Isoleucine	$-CH(CH_3)CH_2.CH_3$	Ile	I	4.5
	Valine	$-CH(CH_3)_2$	Val	V	4.2
	Leucine	$-CH_2.CH(CH_3)_2$	Leu	L	3.8
	Phenylalanine	$-CH_2.C_6H_5$	Phe	F	2.8
	Methionine	$-CH_2.CH_2.SCH_3$	Met	M	1.9
	Alanine	$-CH_3$	Ala	A	1.8
	Tryptophan	$-CH_2C(CHNH)C_6H_4$	Trp	W	−0.9
	Proline	$-CH_2.CH_2.CH_2-$	Pro	P	−1.6
Uncharged polar	Cysteine/cystine	$-CH_2SH$	Cys	C	2.5
	Glycine	$-H$	Gly	G	−0.4
	Threonine	$-CH(OH)CH_3$	Thr	T	−0.7
	Serine	$-CH_2OH$	Ser	S	−0.8
	Tyrosine	$-CH_2C_6H_4OH$	Tyr	Y	−1.3
	Histidine	$-CH_2C(NHCHNCH)$	His	H	−3.2
	Glutamine	$-CH_2.CH_2.CO.NH_2$	Gln	Q	−3.5
	Asparagine	$-CH_2.CO.NH_2$	Asn	N	−3.5
Acidic	Aspartic acid	$-CH_2.COO^-$	Asp	D	−3.5
	Glutamic acid	$-CH_2CH_2COO^-$	Glu	E	−3.5
Basic	Lysine	$-(CH_2)_4.NH_3^+$	Lys	K	−3.9
	Arginine	$-(CH_2)_3NH.C(NH_2)=NH_3^+$	Arg	R	−4.5

Amino acids have the general formula $R-CH(NH_2)COOH$, where R is the side-chain or residue. Proline is actually an imino acid, while cystine is made up of two cysteines linked by a disulfide bridge. The standard abbreviations are given in three- and one-letter codes. The hydropathy index is taken from (Kyte & Doolittle, 1982).

excitability involves an increase of Na^+ membrane permeability in common with the wide range of examples illustrated (Figure 2.4). This particular electric organ provided ideal material for both isolating, then purifying, Na^+ channel protein and determining its amino acid sequence. The initial biochemical work was greatly facilitated by the fact that the protein could be specifically labelled by the Japanese puffer fish poison tetrodotoxin (TTX). The Na^+ channel proved to be a single large peptide with a molecular mass of about 260 kDa, glycosylated at several points on incorporation into the membrane.

A team of scientists at Kyoto University (Noda *et al.*, 1984) successfully cloned and sequenced Na^+ channel cDNA from both *Electrophorus* electric organ and rat brain. This was quickly followed elsewhere by cloning of voltage-gated K^+ channels, most notably the *Shaker* gene of the fruit fly *Drosophila*, and voltage-gated Ca^{2+} channels such as the muscle dihydropyridine receptor (DHPR) (Section 11.6). It is now clear that there is a large family of K^+-, Na^+- or Ca^{2+}-selective membrane

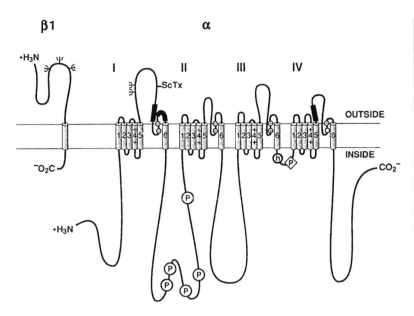

Figure 5.1 Transmembrane folding diagrams illustrating primary structures of α- and $β_1$-subunits of a Na^+ channel. The bold lines are polypeptide chains with the length of each segment roughly proportional to its true length in rat brain Na^+ channel. The cylinders represent probable transmembrane α-helices. Parts of the external links between transmembrane segments S5 and S6 are shown tucked back into the membrane to form the external pore. Sites are indicated of experimentally demonstrated glycosylation $ψ$, cAMP-dependent phosphorylation (P in a circle), protein kinase C phosphorylation (P in a diamond), amino acid residues required for TTX binding (ScTx), and of the inactivation particle (h in a circle). (From Catterall, 1992 with permission of the American Physiological Society.)

proteins with closely related primary structures, gated not only by voltage, but also in a variety of other ways. In *Electrophorus* Na^+ channels, the primary structure of the protein is a large monomer containing 1820 amino acid residues. The α-subunit consists of four linked homologous membrane-spanning domains labelled I, II, III and IV, with closely similar amino acid sequences (Figure 5.1).

Expression of the ion-conducting core α-subunits alone appears sufficient for channel voltage-dependent gating and ion permeation. The proteins whose primary structures were determined were expressed by injecting their corresponding messenger RNA (mRNA) into oocytes of the African clawed toad *Xenopus*. These large cells about to develop into mature eggs possess the normal translation machinery. Injection of mRNA induces synthesis of its encoded protein and its incorporation into the membrane. After synthesising the corresponding mRNA, most voltage-gated and ligand-gated channel proteins so far isolated have been successfully expressed in such oocytes. They demonstrated normal functional characteristics on either patch clamp or macroscopic single-cell current recordings (Section 7.6). This technique could be extended by altering the sequence of amino acid residues by site-directed mutagenesis. This made it possible to explore detailed effects of artificial modifications of protein structure.

cDNA sequencing studies provide accurate information about the primary structure of membrane proteins. These permit limited insights into their configuration within the membranes in which they are expressed. Of the 20 possible amino acids making up proteins (Table 5.1), eight types of residue are non-polar, seven are polar but uncharged, two are acidic and negatively charged, and three are basic and positively charged. The non-polar residues are hydrophobic, and tend to be located in the centre of the molecule within the lipid core of the membrane. The polar or charged residues are hydrophilic and are more likely to occur in the aqueous cytoplasmic environment or at the membrane outer surface. Studies of the hydropathy index of different stretches of the amino acid chain suggest that each homologous domain comprises six largely hydrophobic segments, S1–6. These form α-helices crossing the membrane from one side to the other. These segments form the transmembrane cylinders represented in Figure 5.1. They are connected through short or moderate-length extracellular and intracellular loops. Amino acid sequences of the S2, S3 and S4 segments in voltage-gated Na^+ and K^+ channels of some typical species are compared in Figure 5.2. Those of voltage-gated Ca^{2+} channels are similar.

The cDNA sequencing studies offer only limited insight into molecular folding details key to understanding conformational changes underlying and regulating channel opening and closing. However, cryo-electronmicroscopic (cryo-EM) methods have clarified such higher order structure for both electric eel and human skeletal muscle, Nav1.4, Na^+ channel variants (Yan *et al.*, 2017; Pan *et al.*, 2018). As indicated above, the single Nav1.4 polypeptide chain has four homologous domains (DI–IV). Each comprises six transmembrane segments, S1–S6. The domains are assembled around a central ion-selective pore (Plate 1*A*), the protein traversing the entire membrane thickness (Plate 1*B*). The pore module (PM) contains helices S5 and S6. Sequences between S5 and S6 form extracellular domains, the selectivity filter (SF) ensuring permeation only of Na^+ and the SF-supporting p-loop helices P1 and P2 (Plate 1*A*). The S4 helices from each domain, DI–DIV, form voltage-sensing modules (VSMs).

The domains together form a pseudotetrameric unit in which PMs from each domain line the central ion-conducting pore. The VSM in any given domain closely contacts the clockwise PM from the adjacent domain. Within each domain, the VSMs containing the S4 helices lie on the perimeter likely in close contact with the membrane lipid component. Their sequences are conserved through a wide variety of species (Figure 5.2). Each S4 helix carries between four and eight positively charged arginine or lysine residues, always separated by a pair of non-polar or in a few cases uncharged polar residues. This is consistent with a voltage-sensing function involving an ability to respond to alterations in electric field produced by changes in membrane potential.

Voltage-gated Ca^{2+} channels show a similar size and structure as Na^+ channels, as do voltage-gated K^+ channels, except that the latter are built up of four identical unconnected domains. Na^+ channels belong to a superfamily of 4×6 transmembrane, voltage-gated cation channels containing 24 transmembrane segments, variously Na^+, Ca^{2+} or non-selective. They occur widely through the animal kingdom, ranging from single-cell eukaryotes, through invertebrates, to vertebrates. There are likely evolutionary kinships amongst Na^+ (Nav2 and Nav1) and between Ca^{2+} channels (Cav1 and Cav2) within the inside

```
K+   S2     C S A F R I L L (E) M T F W I I C G T (E) I I      Loligo sKv1A
            C A L F R V T L (E) F T F W I I C L T (E) I L      Drosophila Shaker
            C A F F R V V L (E) F S F W I I C L T (E) V I      Rat rKv1.1
Na+  IS2    R A F L K I L A (E) C T Y I G T F I W (E) S V      Loligo sNa2
            K A I I K I V M (E) I S Y I A L F I Y (E) A (E)    Drosophila dNa1
            R A L I K I L S (E) F T Y I G T F T Y (E) V N      Rat rNaB2
     IIS2   L A L I K L F A (E) A A F V A T F V Y N G I        Loligo sNa2
            L A M L K V I C (E) F T F I S T F V K N G V        Drosophila dNa1
            M A I I K L F M (E) A T F I G T F V L N G V        Rat rNaB2
     IIIS2  F A F W K I F M (E) G I F I V T F C K (D) M Y      Loligo sNa2
            L A L W K L I M (E) V V F I L C F S F N I W        Drosophila dNa1
            Y A V W K L L M (E) L I F I Y T F V K (D) A Y      Rat rNaB2
     IVS2   L G M L K M V C (E) G T F G I F V M N I Y          Loligo sNa2
            L G V I K V I A (E) L G F V T T F F A N S V        Drosophila dNa1
            L S I L K L V C (E) G T F L V I F V L N I W        Rat rNaB2


K+   S3     S M N A I (D) V V S I M P Y F I T L G T V I        Loligo sKv1A
            V M N V I (D) I I A I I P Y F I T L A T V V        Drosophila Shaker
            I M N F I (D) I V A I I P Y F I T L G T (E) I      Rat rKv1.1
Na+  IS3    A W N W L (D) F V V I G L A Y L T (E) V V (D) L    Loligo sNa2
            P W N W L (D) F V V I T M R Y A T I G M (E) V      Drosophila dNa1
            P W N W L (D) F T V I T F A Y V T (E) F V N L      Rat rNaB2
     IIS3   P W N V F (D) S F I V F L S M L (E) L G L G G      Loligo sNa2
            G W N I F (D) L L I V T A S L L (D) I I F (E) L    Drosophila dNa1
            G W N I F (D) G F I V S L S L M (E) L G L A N      Rat rNaB2
     IIIS3  A W C W L (D) F L I V A V S I I M L A A (E) S      Loligo sNa2
            F W T I L (D) F I I V F V S V F S L L I (E) (E)    Drosophila dNa1
            A W C W L (D) F L I V D V S L V S L T A N A        Rat rNaB2
     IVS3   P W N I F (D) F V V V V L S I L G I A L S (D)      Loligo sNa2
            P W N S P (D) F L L V L A S I L G I L M (E) (D)    Drosophila dNa1
            G W N I F (D) F V V V I L S I V G M F L A (E)      Rat rNaB2


K+   S4     Q L G K S H R S L K F I R F V R V L R I V R L      Loligo sKv1A
            Q L G K S H R S L K F I R F V R V L R I V R L      Drosophila Shaker
            Q L G K S H R S L K F I R F V R V L R I V R L      Rat rKv1.1
Na+  IS4    K L G P I V A V T K L A R L V R F T R L A S L      Loligo sNa2
            K L G P M I S V T K L A R L V R F T R L G A L      Drosophila dNa1
            K L G P I V S I T K L A R L V R F T R L A S V      Rat rNaB2
     IIS4   N L T P W S K A L K F V R L L R F S R L V S L      Loligo sNa2
            K M T T W S K A L K L A R L L R L G R L V S L      Drosophila dNa1
            N L T P W S K A L K F V R L L R F S R L V S L      Rat rNaB2
     IIIS4  R M G (E) S R S V A R L P R L A R L T R M S R F    Loligo sNa2
            R M G Q W R S I A R L P R L A R L T R L S R L      Drosophila dNa1
            R M G (E) F R S L A R L P R L A R L T R L S K I    Rat rNaB2
     IVS4   R I G K A S K V L R L V R G V R F G V R V R L      Loligo sNa2
            R I G K A A K I L R L I R G I R F V R V V L        Drosophila dNa1
            R I G K A G K I L R L I R G I R A L R I V R F      Rat rNaB2
```

outside

Figure 5.2 Amino acid sequences of the charge-carrying S2, S3 and S4 transmembrane segments of all four domains of voltage-gated K^+ channels, and of individual domains of Na^+ channels, for the squid *Loligo opalescens*, the fly *Drosophila* and rat brain. Positively charged arginine residues (**R**) and lysine residues (**K**) are in bold type and the stretches over which they are separated by two non-polar residues are underlined. Negatively charged residues of aspartate (**D**) and glutamate (**E**) are ringed. (Data selected from Figures 1, 3 and 4 of Keynes and Elinder, 1999.)

superfamily of voltage-gated 4×6 transmembrane cation channels. Two rounds of domain duplications generated their common 24 transmembrane segment template, with asymmetrically arranged pores that allow for differing ion selectivity through a single DII or DIII lysine residue in the ion selectivity filter. The appearance of voltage-gated, Na^+-selective channels with rapid gating kinetics was likely a limiting factor in the evolution of nervous systems.

Na^+ and Ca^{2+} channels both show a number of similar regulatory or binding sites likely of physiological (Section 13.19) and clinical importance (Sections 14.14 and 14.17). Firstly, cytosolic $[Ca^{2+}]$ can modify channel function through direct Ca^{2+} binding at one or more Ca^{2+}-binding EF hand motifs. In addition, a isoleucine-glutamine (IQ) domain binds calmodulin (CaM) after its own EF-hand motifs have bound Ca^{2+}. Both these sites occur in the C-terminal domain. Na^+ channels also contain a III–IV loop Ca^{2+}–CaM binding site. Secondly, in common with a wide range of protein molecules, Na^+ and Ca^{2+} channels can undergo kinase-mediated phosphorylation at multiple sites that exert strategic regulatory effects. These include sites phosphorylatable by calmodulin kinase II (CaMKII) themselves regulated by Ca^{2+}–CaM (Takla *et al.*, 2020), and a phosphorylation site for PKC.

Thirdly, α-subunits of the different Na^+, Ca^{2+} and K^+ channel subtypes coassemble with auxiliary β-subunits. For example, accessory Na^+ channel β-subunits normally comprise an amino-terminal immunoglobulin connected by a short 'neck' to a single transmembrane domain. They may associate with α-subunits either non-covalently, as with $β_1$ and $β_3$, or through disulfide linkages, as with $β_2$ and $β_4$. They enhance Na^+ channel trafficking to the surface membrane, thereby modifying peak I_{Na}. They also modulate gating behaviour, particularly the voltages at which channel inactivation takes place, as well as the kinetics of such modulation (Namadurai *et al.*, 2015). This may influence the key conformational changes shown by Na^+ channels through the action potential cycle. They may also interact with, producing *trans* couplings between, β-subunits of adjacent cells (Salvage *et al.*, 2020a). Finally, ion channel α-subunits may go on to form oligomeric complexes within the plasma membrane; the organisation of those of Na^+ channels are further modified by the presence of β-subunits (Salvage *et al.*, 2020b). At even higher levels of organisation, ion channels may selectively occur with heterogeneous densities in different membrane regions (e.g. Section 10.2) and specialised structures in both nerve and muscle cells (e.g. Section 11.6).

Primary sequencing studies suggest at least nine human Nav channel subtypes. Nav1.1, Nav1.2, Nav1.3 and Nav1.6 may mainly occur in central nervous system. Nav1.4 and Nav1.5 occur in skeletal and cardiac muscle, respectively. Nav1.7, Nav1.8, and Nav1.9 mainly occur in peripheral nervous system. Mutations in voltage-gated channels are associated with a number of inherited clinical pathologies. Over 1000 point mutations have been identified in human Nav channels. Some of these are associated with neurological, cardiovascular, muscular and psychiatric disorders that include epilepsy and migraine (Section 8.9), muscle dystonias (Section 10.9), cardiac arrhythmias (Sections 14.6–14.9), pain syndromes and autism spectrum disorder.

5.3 | Biophysical Measurements of Channel Protein Conformational Changes

Ion channels thus form specialised protein structures embedded within the membrane bilayer matrix that include specialised ion permeation and charged, voltage-sensing, regions central to their function (Section 5.2). Insights from the classical studies on channel function (Sections 4.5 and 4.6) were limited to the effects of voltage on

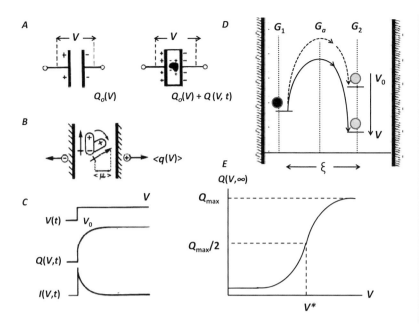

Figure 5.3 Charge movements reflecting conformational changes in intramembrane protein molecules. (A) Left: Capacitative charge $Q_0(V)$ of membrane of thickness d separating voltage V producing uniform transmembrane electric field $\hat{E}$. Right: Embedded polar molecules at density L within the lipid bilayer membrane producing increase in charge $Q(V, t)$. (B) Underlying charge movement along the direction of $\hat{E}$ displacing microscopic charge $<q(V)>$. (C) Voltage step $V(t)$ from reference voltage V_0 to V producing charge redistribution $Q(V, t)$, measurable as its derivative, the charge movement $I(V,t)$. (D) Energetics of charge transitions between resting state of energy $G_1(V)$ and active state of energy $G_2(V)$, via transition state of energy $G_a(V)$ with voltage change between V_0 and V. (E) The resulting steady-state dependence of charge movement $Q(V, \infty)$ upon membrane potential V. (From Figure 1 of Huang, 1988.)

eventual channel opening. Subsequent biophysical studies analysed the full underlying molecular mechanisms starting from the initial voltage-sensing processes in a wide range of ion channels.

A membrane with voltage V acting across its lipid bilayer thickness d (Figure 5.3A, Left panel) subjects its membrane protein components to an electric field $\hat{E} = -V/d$ (Figure 5.3A, Right panel), in turn alterable by imposing an activating voltage step $V(t)$ from V_0 to V. The latter would displace a charge $Q_0(V)$, reflecting the capacitative properties of the membrane lipid component. However, the same altered $\hat{E}$ also acts upon charged component groups in proteins embedded within the membrane. It moves or reorients such charge with consequent molecular conformational changes in the channel molecules from resting to activated states (Figure 5.3B). Their extent and the proportion of intramembrane molecules affected will vary with the change in $\hat{E}$ and therefore the voltage, V (Huang, 1988, 1993). The resulting transition in turn electrostatically displaces charge, $<q(V,t)>$, from the membrane surface (Figure 5.3B). An ensemble of such molecules occurring with density L within the membrane will produce an overall charge displacement $Q(V, t)$, producing a charge movement $I(V,t) = d[Q(V,t)]/dt$. This ultimately displaces a steady-state ($t = \infty$) charge given by $Q(V,\infty) = L<q(V,\infty)>$, measurable if this is separated from the contribution $Q_0(V)$ from the lipid bilayer component (Figure 5.3C).

The steady-state dependence $Q(V,\infty)$ throws light on the energetics of this underlying molecular transition. The simplest, activated and resting, two-state model, considers a charge of valence, z, traversing fraction $\xi \approx 1$ of the total transmembrane voltage drop. The energy of the activated, $G_2(V)$, relative to the resting state $G_1(V)$, is the charge-voltage product $(ze) \times [\xi(V - V^*)]$, where V^* is the voltage at which $G_1(V^*) = G_2(V^*)$ and e is the electron charge (Figure 5.3D). $Q(V,\infty)$ reflects the proportion of charge existing in the activated state. It can be quantified by a Boltzmann relationship between the maximum charge Q_{max}, V^* and steepness factor $k = ze/kT$:

$$Q(V,\infty) = \frac{Q_{max}}{\left\{1 + \exp\dfrac{-(V - V^*)ze}{kT}\right\}} \quad (5.1)$$

where k is the Boltzmann constant and T the absolute temperature. Successively larger voltage steps give correspondingly increasing values of $Q(V,\infty)$ up to the saturating maximum, Q_{max}, that reflects the total number of molecules available to undergo this transition, for which z defines the amount of charge involved, and V^* defines the voltage about which the transition takes place (Figure 5.3E).

The detailed timecourse of the charge movement $I(V,t)$ throws light on the kinetics of the molecular transition. The simplest two-state case predicts forward and backward unimolecular, first-order transitions across a single energy barrier $G_a(V)$ separating $G_1(V)$ and $G_2(V)$. The resulting voltage-dependent activation energies $[G_a(V) - G_1(V)]$ and $[G_a(V) - G_2(V)]$ determine the forward, $\alpha(V)$, and backward, $\beta(V)$, rate constants describing the net charge movement produced by the underlying functional groups of the protein undergoing the transition:

$$\frac{dQ(V,t)}{dt} = \alpha(V)[Q(V,\infty) - Q(V,t)] - \beta(V)Q(V,t) \quad (5.2)$$

This gives:

$$Q(V,t) = Q(V,\infty)\left\{1 - \exp\left(\frac{-t}{\alpha(V) + \beta(V)}\right)\right\} \quad (5.3)$$

The derivative of $Q(V,t)$ is the corresponding charge movement $I(V,t) = d[Q(V,t)]/dt$ and is the exponential decay (Figure 5.3C):

$$I(V,t) = Q(V,\infty)\left(\frac{1}{\alpha(V) + \beta(V)}\right)\left\{\exp\left(\frac{-t}{\alpha(V) + \beta(V)}\right)\right\} \quad (5.4)$$

5.4 | Charge Movements in Excitable Cell Membranes

Such charge movements, also termed *asymmetry* or *gating currents*, reflect transitions involving charged functional groups in single molecules within the cell membrane. These had escaped detection in the original ion channel studies (Hodgkin and Huxley, 1952). This reflects their expectedly considerably smaller amplitudes compared to those of the corresponding ion currents. They were subsequently demonstrated under experimental conditions abolishing these larger ion currents by pharmacological block, or intracellular and/or extracellular substitutions of their charge carriers with impermeant ions. Their contributions to the resulting electrophysiological records required separation of accompanying linear capacitative currents arising from the embedding membrane lipid bilayer matrix. This involved subtracting from the total capacitative currents produced by the test voltage clamp steps, from resting V_0 to test potentials V, control records from steps of identical or scaleable size traversing more negative voltages where the charge movements would be absent.

The first experimental demonstration of charge movements were made in amphibian fast twitch skeletal muscle (Figure 5.4A; Schneider and Chandler, 1973; Adrian and Almers, 1976a, 1976b; Chandler *et al.*, 1976) and cephalopod giant axons (Armstrong and Bezanilla, 1973, 1974, 1977; Keynes and Rojas, 1973, 1974; Figure 5.4B). The resulting records, illustrated here first for fast twitch amphibian muscle (Figure 5.4A), conformed to expectations for charge transitions *within* the membrane as opposed to ion currents flowing *across* the membrane. The currents attained peak amplitudes shortly following imposition of the test voltage steps (Figure 5.4Aa). They then declined to steady values reflecting residual membrane leak current which could be subtracted. Their amplitudes and timecourses both varied steeply with test potential (Figure 5.4Ab). The steady state charge $Q(V, \infty)$ derived from the area beneath the transient part of the current, increased with depolarisation, crossed an inflexion at half-maximal voltage V^*, and saturated at a finite maximum value Q_{max} (Section 5.3). These characteristics suggested a finite quantity of available charge moving within the membrane, as opposed to ionic currents crossing the membrane (Section 5.3). Furthermore, the 'on' charge displaced in one direction was always restored by the 'off' ending of the pulse (Figure 5.4Ac). Subsequent studies implicated charge movements in muscle, at least in part, to voltage-sensitive processes triggering contraction (Section 11.7).

Squid giant axon gating currents were similarly revealed after minimising I_{Na} and I_{K}. Na^+ permeation through open Na^+ channels was blocked by bathing the axon in Na^+-free TTX-containing solutions. K^+ channels were blocked by perfusion or internal dialysis with a Cs^+- or tetramethylammonium fluoride containing solution. Alternatively currents were compared before (Figure 5.4Ba) and following TTX-mediated I_{Na} block (Figure 5.4Bb). The gating current in response to suprathreshold depolarising steps (Figure 5.4Bb) was outward, rising to a peak within $\sim$100 µs. It was $>$50-fold smaller in amplitude than the corresponding I_{Na} observed under normal $[\mathrm{Na}^+]_{\mathrm{o}}$. Examination at faster sweep speeds demonstrated that it decayed to baseline well before the peak of the corresponding I_{Na} (Figure 5.4Bb'). Restoring the resting potential produced an exponential 'off' current decay in the opposite, inward, direction resembling the timecourse of the corresponding I_{Na} decay, restoring the previous outward movement of charge.

Subsequent explorations demonstrated charge movements in all excitable cell types through all species explored (Huang, 1993). They occurred in slow-twitch amphibian muscle, mammalian skeletal fast- (Figure 5.4C) and slow-twitch, and cardiac ventricular muscle

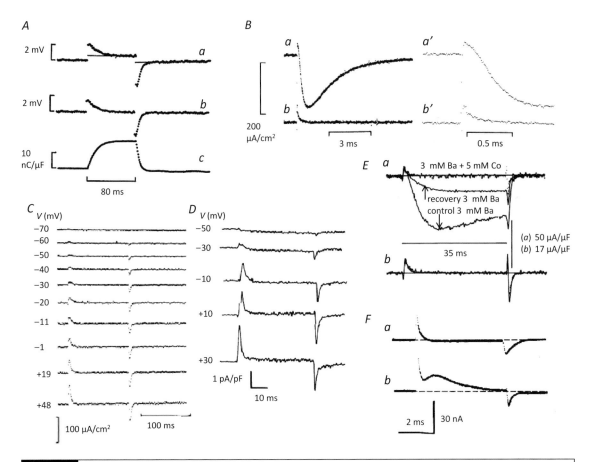

Figure 5.4 Charge movements in a range of excitable membranes following linear capacitative current corrections. (A) Charge movements in amphibian fast-twitch skeletal muscle: (a) Currents in response to voltage steps between −80 and −30 mV. Subtraction of background ionic current yields (b) the intramembrane charge movement whose (c) time integral confirms charge equality through the 'on' and 'off' of the voltage step. (B) Derivation of gating current in cephalopod giant axon in response to step from −100 to 0 mV from current (a) before and (b) following I_{Na} block by 2 μM TTX, under conditions of I_K block by intracellular Cs^+. (C, D) Charge movements in (C) rat fast-twitch muscle fibres and (D) neonatal rat ventricular myocytes produced by steps to different test voltages from a −100 mV holding potential, under conditions minimising ionic currents. (E) Current measurements using Ba^{2+} as charge carrier across Ca^{2+} channels (a) before and (b) charge movement following Co^{2+}-induced Ca^{2+} channel block in arthropod (scorpion, *Centuroides sculpturatus*) pedipalp-closer muscle. (F) Charge movements in response to test voltage steps from −63 mV to −12 mV (a) and +13 mV (b) in mollusc (*Aplysia*) neuron. ((A) from Figure 5 of Chandler et al., 1976; (B) from Figure 1 of Keynes and Kimura, 1983; (C) from Figure 1 of Hollingworth and Marshall, 1981; (D) from Figure 3 of Field et al., 1988; (E) from Figure 2 of Scheuer and Gilly, 1986; (F) from Figure 3 of Adams and Gage, 1979).

(Figure 5.4*D*). Amongst invertebrate excitable cells, they were demonstrated in barnacle muscle, scorpion long-closer muscle (Figure 5.4*E*) and molluscan neurons (Figure 5.4*F*). These various systems showed differing steady-state Q_{max}, V^* and steepness factors and kinetic properties. This is as expected from their differing expressed ionic channels. Each channel species present would contribute its own specific charge movement component with its own particular kinetic and steady-state properties. The likely presence of multiple ion channels known to exist

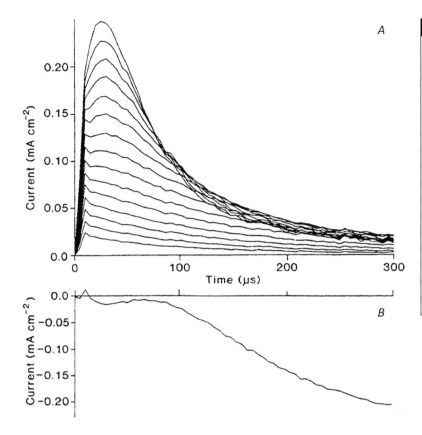

Figure 5.5 (A) Superimposed family of Na$^+$ channel gating currents recorded from a squid axon dialysed with 350 mM tetramethylammonium fluoride (TMA-F) and bathed in an artificial sea water with Na$^+$ replaced by *tris*-hydroxyaminomethane (Tris) containing 1 μM TTX. Test pulses −57 to +83 mV in steps of 10 mV. Holding potential −80 mV. Temperature 10 °C. Number of sweeps averaged was 32. (B) Initial rise of I_{Na} for a pulse to −23 mV after subtraction of gating currents in another axon bathed in an artificial sea water in which 4/5 of the [Na$^+$] was replaced by [*tris*-hydroxyaminomethane$^+$]. (From Keynes and Elinder, 1998.)

in any given excitable cell preparation may be reflected in the complex kinetics shown by molluscan neurons attributed respectively to Na$^+$ and Ca^{2+} channel contributions (Figure 5.4F*a,b*). In addition, any given component charge should include particular transitions whose voltage sensitivity is compatible with a mechanism regulating its corresponding ionic current. For example, the charge movements in the scorpion long-closer muscle showed a voltage sensitivity comparable to that of the Ca^{2+} currents they may be involved in gating that were observed in the same membrane (Figure 5.4*E*).

Meaningful studies of any observed charge movement component therefore first require its identification with its originating ion channel type. In squid axon, procedures reversibly blocking I_{Na} such as internal perfusion with Zn^{2+}, prolonged membrane depolarisation and I_{Na} inactivation using a short positive prepulse correspondingly blocked gating current. This suggested their possible identification with transitions in the Na$^+$ channel (Armstrong and Bezanilla, 1974). The existence and properties of such currents provided a molecular basis for the macroscopic Na$^+$ channel properties described by Hodgkin and Huxley.

Gating current properties fulfilled one prediction by Hodgkin and Huxley of voltage-dependent channel activation. The gating current decays approximated the sum of two exponential functions, fast and slow, extrapolated to the beginning of the step. However, gating current features appeared to exclude a discrete, similarly voltage-dependent, parallel inactivation process also suggested in their original hypothesis ('h' process: Section 4.7). Intracellular application of the proteolytic enzyme pronase reduced I_{Na} inactivation following an applied voltage step. Yet, both fast and slow gating current components remained intact. Na^+ channel inactivation instead correlated with immobilisation of a large fraction of the entire gating charge. Thus, progressively longer depolarising steps were followed by progressively reduced returning 'off' charge movement with termination of such steps. Nevertheless the timecourses of both development of, and recovery from, such immobilisation closely agreed with those of I_{Na} inactivation (Armstrong and Bezanilla, 1977).

Comparisons between gating current and I_{Na} also permitted closer analyses of Na^+ channel activation. They suggested a complex set of component reactions, each potentially generating one or more corresponding contributions to overall charge movement. Each would likely possess specific kinetic and voltage-dependence properties. Smaller test steps to successively more depolarised voltages initially elicited fast early currents (Figure 5.5A) rising quickly, before appreciable I_{Na} activation (Figure 5.5B), then decaying exponentially. However, larger depolarisations, for example the pulse to −17 mV, additionally elicited small delayed rising phases just before the start of the decay. With still larger pulses, the slowly rising phase became increasingly prominent, reaching its peak with a delay of ~30 μs.

The latter separable, q_i, component was selectively inactivated by previous depolarisation in parallel with I_{Na} inactivation. The voltage sensitivities of its kinetics and steady-state properties also more closely resembled I_{Na} than did either the total or the charge q_n remaining following such inactivation (Greeff $et\ al.$, 1982). Closer analyses of this kind have been facilitated with the availability of further techniques selectively over-expressing Na^+ channels in $Xenopus\ laevis$ oocytes for patch or voltage clamp studies. This allowed investigations of the effects of modifying channel sequences using site-directed mutagenesis or fluorescence labelling, potentially yielding more specific structural correlates of gating currents and their relationship to molecular conformational changes (Bezanilla, 2018).

5.5 | The Sodium Channel Gating Mechanism

Na^+ channel activation therefore involves detection of a membrane voltage change, for which structural evidence implicates configurational changes in the VSM (Section 5.2), detectable and quantifiable through their generation of charge movement (Sections 5.3 and 5.4). The configurational change is transmitted electromechanically to the PM, through causing allosteric rearrangement in its S5 and S6 helices. The latter likely involves the interleaved arrangement between the PM and VSM (Plate 1A) that appears conserved in all known eukaryotic voltage-dependent ion channels.

The selective transmembrane view illustrates a model of the voltage-sensing region, based on studies of the ancestral bacterial Na$^+$ channel NaChBac (Plate 2A). Its configurational changes occur about a widely conserved constriction site (HCS) containing hydrophobic amino acid residues. This likely forms a hydrophobic seal across which most of the membrane voltage is expressed, preventing transmembrane ion movement. The HCS lines the narrowest part of the pathway through which gating charge movement takes place.

The charge movement is mediated by transmembrane helix S4 of the VSM with its four to six positively charged arginine and lysine residues. These occur every three residues. Their positive charges occur along one face of the S4 helix and are stabilised within the resting membrane by forming ion pairs with negatively charged aspartate or glutamate residues in fixed positions within the surrounding S1–S3 helices (Plate 2Aa). They are held in place by the relatively negative intracellular membrane potential.

Depolarisation permits the S4 helix to move outward. In so doing, it also rotates, and tilts as it passes through the HCS (Plate 2Ab). The arginine 1 to arginine 3 region of the S4 segment may now carry gating charges through a narrow groove formed by the S1, S2 and S3 segments, possibly permitted by conformational change within the S4 segment. This screw-like rotation involves three outward 60° and 0.45 nm movements. Each such movement permits a movable positive charge to pair up with and become stabilised by the next fixed negative charge. This results in the appearance of an unpaired positive charge at the outer surface of the membrane, and creates an unpaired negative charge at the inner surface. There is an overall outward transfer of approximately one electronic charge. Over three such movements, each S4 segment would transfer the equivalent of three electronic charges (Figure 5.6).

The movement of S4 increases the size of the cleft between the S1–S2 and S3–S4 loops on the extracellular side of the activated channel (Plate 2Ab). It also drives a gating movement of the S4–S5 linker in a direction parallel to the plane of the inner surface of the membrane (Plate 2Ba–c). This may cause the entire voltage-sensing domain to rotate about the axis of the pore and exerts a torque on the S4–S5 linker, pulling the lower end of the S5 helix outward, with the S6 helix twisting in a counterclockwise manner relative to the intracellular face (Plate 2Ca,b). This shift in the positions of the PM helices opens the pore through this association of each VSM with the PM of its neighbour. These loosen the packing interactions around the activation gate and allow it to dilate and permit Na$^+$ permeation (Plate 2Da,b).

Each Na$^+$ channel contains four VSMs, one for each of the four domains DI–DIV. Different eukaryote Na$^+$ channel subtypes likely activate with differing kinetics, with the more rapid configurational changes in the VSMs of DI–DIII sufficient to cause channel opening. The VSM of DIV likely activates with the slowest kinetics. In so doing, it frees the *inactivation gate*. This intracellular linker connects the S6 helix of DIII to the S1 helix of DIV. The inactivation gate includes

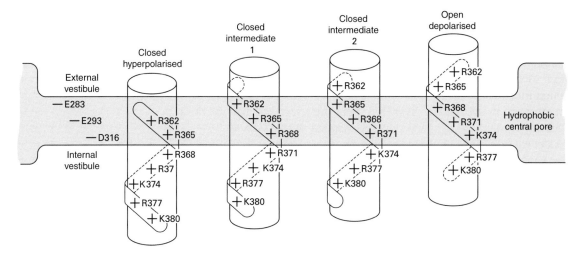

Figure 5.6 Diagrammatic representation of the screw-helical outward movement of positive charges carried by S4 in *Shaker* K⁺ channel. Each outward step transfers one electronic charge from its position on the α-helix inside the cell to the external solution. The three negative charges shown on the left occupy fixed positions on S2 and S3, and salt bridges are formed with two of the positive charges in the closed hyperpolarised state and three in the intermediate and open states. (From Keynes and Elinder, 1999.)

a cluster of hydrophobic residues containing the amino acid sequence IFMT. The inactivation gate binds to an inactivation particle receptor occurring within the S4–S5 linkers of DII, DIII and DIV. This occludes the pore, causing channel inactivation (review: Zaleska *et al.*, 2018).

5.6 | The Ionic Selectivity of Voltage-Gated Channels

The further essential feature for any given cation channel type is its ability specifically to discriminate in favour of Na^+, K^+ or Ca^{2+} permeation. Permeability studies of the node of Ranvier Na^+ channel to certain small organic cations demonstrated that such selection requires a good fit between the dimensions of the penetrating ion and those of the mouth of the channel (Hille, 1971). Only molecules measuring less than ~0.3 × 0.5 nm in cross-section could pass such a filter (Figure 5.7). However, there were striking differences in permeability between some cations of similar sizes. Hydroxylamine ($OH–NH_3^+$) and hydrazine ($NH_2–NH_3^+$), but not methylamine ($CH_3–NH_3^+$), readily permeated the channel. This was compatible with the Na^+ channel being lined at its narrowest point with carbonyl oxygen atoms. Positively charged ions containing hydroxyl (–OH) or amino (–NH₂) groups, but not those containing methyl (–CH₃) groups, could pass through the channel by hydrogen bonding with the oxygens. This suggested a geometry in which Na^+ ions could shed all but one of their hydration shell of water molecules by interacting with the strategically placed oxygen atoms, over a relatively low energy barrier. The same is true for Li^+. However, the somewhat larger K^+ ions cannot shed their hydration shell as easily. Consequently the Na$^+$ channel has a P_K only one twelfth as great as its P_{Na}.

Cryo-electronmicroscope reconstructions of the PM of the Na$^+$ channel (Plate 1A) demonstrate a central pore cavity wall lined by S5 and S6 helices from each PM (Plate 2D). The S6 helices from each domain come close at the intracellular face forming an intracellular cavity containing hydrophobic amino acids. This constitutes the activation gate, which is constricted in the channel closed state. The S5 helix is connected by an extracellular linker to a membrane-descending P-helix (P1), followed by an ascending P-helix (P2) and a further short extracellular loop connecting to helix S6. The extracellular loops from each domain form the turret-like structure of the outer mouth. These extend above the pore and form a preselection vestibule filter, whose structure is likely stabilised by disulfide bonds. The narrowest point in the vestibular region is close to where the P1 and P2 helices reverse direction within the membrane. The presence of charged aspartate (in DI), glutamate (in DII), lysine (in DIII) or alanine (in DIV) residues result in a high field strength in the constriction region. Together these may form the basis for the selectivity filter favouring permeation of Na$^+$ over other positively charged K$^+$ or Ca^{2+}.

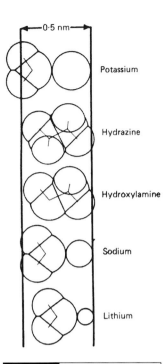

Figure 5.7 Scale drawings showing effective sizes of Li$^+$, Na$^+$ and K$^+$ ions, each with one molecule of water, and of unhydrated hydroxylamine and hydrazine ions. The vertical lines 0.5 nm apart represent the postulated space between oxygen atoms available for cations able to pass through the Na$^+$ channel. Methylamine would appear to be like hydrazine in this kind of picture, but is nevertheless unable to enter the channel. (After Hille, 1971.)

5.7 | Effect of Ion Occupancy on Channel Permeation: The Independence Principle

The above structural features may also result in the Na$^+$ channel mediating Na$^+$ permeation independent of the presence of other extracellular or intracellular ions. Other ion channels can show deviations from this *independence principle*. Ca^{2+} channels are normally permeable to Sr^{2+} and Ba^{2+}, but are completely impermeable to monovalent cations. However, with reduced [Ca^{2+}]$_o$, the channels become permeable to both K$^+$ and Na$^+$. This may reflect the presence of two high-affinity Ca^{2+} binding sites at the channel mouth. If neither is occupied, monovalent ions readily enter. If one is occupied, Na$^+$ and K$^+$ ions are electrostatically repelled. When both sites are occupied, the permitted flow of both divalent and monovalent cations again increases.

Deviations from independence were also demonstrated for K$^+$ channels in experiments measuring dependences of the ratios between K$^+$ influx and efflux, upon the driving force ($E_m - E_K$), where E_m is the membrane potential and E_K is the equilibrium potential for K$^+$, on K$^+$ permeation. These were performed in *Sepia* giant axons poisoned with DNP to abolish the Na$^+$ efflux (Hodgkin and Keynes, 1955b; Figure 5.8).

A situation in which K$^+$ does traverse the membrane in a wholly independent fashion would predict a 10-fold change in flux ratio for

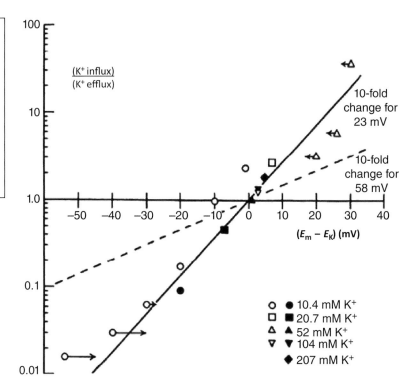

Figure 5.8 The effect of the electrical driving force $E_m - E_K$, on potassium flux ratios in *Sepia* axons. Filled-in symbols denote flux measurements using 17–29 mm lengths of fibre in the absence of applied current. The other points were obtained with axons mounted for application of current. The arrows show approximate corrections for cable effects. (From Hodgkin and Keynes, 1955b.)

a 58 mV change in the value of $E_m - E_K$. However, what was observed experimentally was a significantly greater 10-fold change for only 23 mV. This large departure from the independence relation was explained by suggesting that K^+ ions move through the membrane in narrow channels constraining them to move in single file. There would be, on average, several ions in a channel at any moment.

The latter suggestion was illustrated by a simple mechanical model (Figure 5.9*A*; Hodgkin and Keynes, 1955b) consisting of two circular chambers, each containing a number of steel balls, connected by a pore just wide enough to allow a ball to pass through it, whose length could be increased to contain three balls at a time. 100 balls blued by heating were placed in the left-hand chamber and 50 balls, identical except in being silver in colour, were placed in the right-hand chamber. When shaken vigorously by a motor for 15 s, the balls rattled about with a random 'Brownian' movement, and a certain number passed through the pore. The number of blue balls that moved from left to right was counted and compared with the number of silver balls that had moved in the opposite direction during the same time interval. The experiments were repeated a number of times to check that there were

A

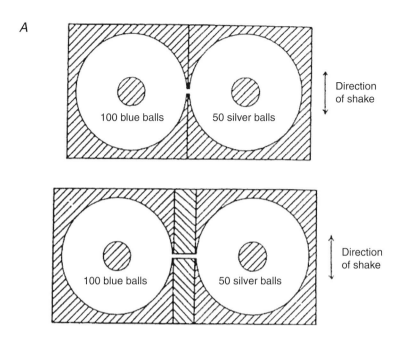

Direction of shake

Direction of shake

Figure 5.9 (A) Mechanical model mimicking K^+ flux data. Single-filing of exchange of ball bearings between the two compartments took place only when the chambers were connected by a pore long enough to contain three balls at a time, and too narrow for the balls to bypass one another. (B) The similar structure of the filter in the pore of bacterial potassium channels ((A) from Hodgkin and Keynes, 1955b, (B) from Doyle *et al.*, 1998.)

B

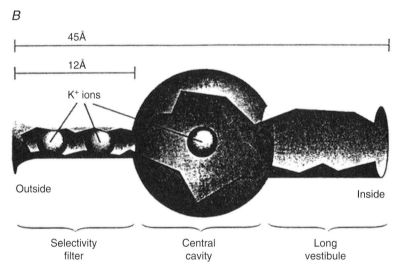

no slight asymmetries or tilt of the apparaus, and no differences between the two sets of balls, and the results were averaged.

The properties of single file K^+ permeation are consistent with X-ray crystallographic characterisations of the structure of the K^+ filter (Figure 5.9*B*). These were performed initially in bacterial K^+ channels and subsequently through a wide range of K^+ channel subtypes (Gulbis and Doyle, 2004). cDNA sequencing suggests identically structured selectivity filters, in which the four S5–S6 links create an inverted cone cradling the narrow selectivity filter of the pore, 1.2 nm in length, at its outer end. Here the main-chain atoms create a stack of sequential carbonyl oxygen rings, providing sites energetically suitable to be substituted for the hydration shell of a K^+ ion. Above and below this narrow section of the K^+ filter, the pore is somewhat wider so that the hydrated K^+ ions are free to move on, while just two unhydrated K^+ ions can be fitted into the filter itself, although within it they cannot bypass one another.

Cable Theory and Saltatory Conduction

6.1 | The Spread of Voltage Changes in a Cable System

Nerve impulse propagation following the initial excitation depends not only on the membrane electrical excitability of the nerve, but also its cable properties. These determine the resulting current flow exciting adjacent cellular regions, altering their membrane potential V from resting to threshold values. Figure 6.1A illustrates this process for a one-dimensional, cylindrical nerve fibre surrounded by extracellular fluid of resistance per unit length r_o (expressed in units of Ω/cm). The intracellular voltage V may be altered either by an imposed current through an intracellular electrode, or an inward depolarising Na^+ current, i_{Na}, (expressed in current per unit length of fibre, A/cm; Figure 6.1B) producing an action potential in the excited region. The voltage change drives an intracellular axial current i_a (expressed in A) through the intracellular resistance of unit fibre length r_a (expressed in Ω/cm). This current flows in series with the extracellular resistance r_o. Ohm's Law gives the axial current at any point along the nerve length x for the particular voltage drop along the intracellular space, (dV/dx) (Jack et $al.$, 1983):

$$i_a = \frac{1}{(r_a + r_o)} \frac{dV}{dx} \tag{6.1}$$

The current i_a alters the voltage V at distance x from the site of action potential excitation. This voltage drives a transmembrane current i_m (expressed in current per unit length of fibre, A/cm^2) at that point. Passage of i_m correspondingly reduces the remaining intracellular axial current i_a available to excite more remote regions of nerve. By Kirchoff's first law:

$$i_m = \frac{di_a}{dx} \tag{6.2}$$

Differentiating both sides of equation (6.1) and substituting the resulting expression for (di_a/dx) in equation (6.2) gives the cable

Figure 6.1 Spread of current and voltage changes in a cylindrical axon modelled by a one-dimensional cable. (A) Intracellular (i_a) and transmembrane flow of current (i_m) along the length, x, of a cylindrical axon depends upon (B) the electrical parameters of extracellular (r_o) and intracellular cytosolic resistances (r_a), the membrane resistance (r_m) and capacitance (c_m), and the driving current, exemplified here by inward Na$^+$ current (i_{Na}).

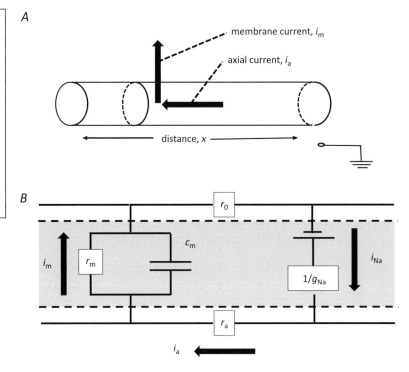

equation relating membrane current i_m to the dependence of membrane potential V with distance x for the one-dimensional cable:

$$i_m = \frac{1}{(r_a + r_o)} \frac{d^2V}{dx^2} \tag{6.3}$$

The electrical behaviour of unit length of such resting membrane reflects the properties of two major circuit components. Its capacitance c_m (expressed in μF/cm), reflects the properties of the insulating lipid bilayer and conducts a current per unit length proportional to the rate of change of membrane voltage, $c_m(dV/dt)$. In parallel with this is the membrane resistance r_m (expressed in Ω cm), which conducts a current per unit length given by Ohm's law, V/r_m (Figure 6.1B). The membrane current is the sum of the currents flowing through each circuit element. This gives the cable equation describing the *passive* properties of a biological membrane, even in the absence of voltage- or time-dependent ionic current activity:

$$\frac{1}{(r_a + r_o)} \frac{d^2V}{dx^2} = c_m \frac{dV}{dt} + \frac{V}{r_m}. \tag{6.4}$$

since:

$$i_m = c_m \frac{dV}{dt} + \frac{V}{r_m} \tag{6.5}$$

6.2 | Passive Spread of Voltage Changes Along an Unmyelinated Nerve

Equation 6.4 describes the resulting passive behaviour of the membrane organised into a cable containing multiple resistance–capacitance elements (Figure 6.2A). A full mathematical solution of Equation 6.4 results in a complicated expression, but useful insights into the expected behaviour can be derived from exploring two limiting conditions applied to a uniform cable of infinite length. First, the timecourse of the voltage change induced by a current applied at time, $t = 0$, uniformly along the entire length of the nerve, corresponds to a situation in which

$$\frac{d^2V}{dx^2} = 0 \qquad (6.6)$$

This reduces Equation (6.4) to the first order differential equation:

$$\frac{dV}{dt} = \frac{-V}{(c_m r_m)} \qquad (6.7)$$

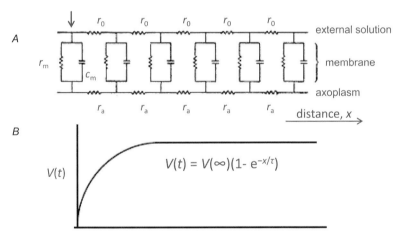

A

B

$$V(t) = V(\infty)(1 - e^{-x/\tau})$$

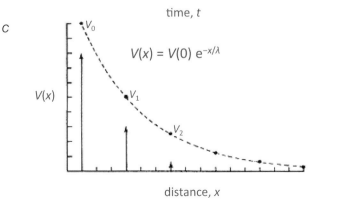

C

$$V(x) = V(0)\, e^{-x/\lambda}$$

Figure 6.2 Properties of (A) an electrical model of the passive, electrotonic, properties of a length of axon. (B) The timecourse of the voltage response to a step of current applied uniformly along the entire length of the axon. (C) The steady-state distribution of transmembrane potential when a point, $x = 0$, is connected to a constant current source.

The solution to this equation predicts an exponential approach of the voltage $V(t)$ to the final steady state voltage change $V(\infty)$,

$$V(t) = V(\infty)[1 - \exp(-t/\tau)] \tag{6.8}$$

The time constant $\tau = r_m c_m$ defines the time required for the voltage to attain ~63.2% of its asymptotic value (Figure 6.2B). Equation (6.8) thus describes the timecourse of a voltage response to a step current applied in the absence of non-uniformities over the geometry of the excitable cell.

Secondly, consider the steady-state ($t = \infty$) effect of passing a constant current across the membrane at the cable midpoint. This sets up a stable potential difference V_o between the cell interior and exterior at a point given by $x = 0$. As there is no change in voltage V with time, t,

$$c_m \frac{dV}{dt} = 0 \tag{6.9}$$

Equation (6.4) reduces to:

$$\frac{d^2 V}{dx^2} = \frac{V(r_a + r_o)}{r_m} \tag{6.10}$$

Solving Equation (6.10) predicts an exponentially declining fall of membrane potential $V(x)$ with distance x from the point of application of V_0. This passive *electrotonic* spread of potential gives an indication of the maximum, steady state, extent of current and voltage spread (Figure 6.2C):

$$V(x) = V(0) \exp\left(\frac{-x}{\lambda(\infty)}\right) \tag{6.11}$$

The *steady-state space constant* $\lambda(\infty)$, the distance at which the voltage decay has fallen to reach 63.2% of its asymptotic value, is given by:

$$\lambda(\infty)^2 = \frac{r_m}{r_o + r_a} \tag{6.12}$$

Here, the properties of the cable are dominated by transmembrane current flow through the membrane resistance, r_m, as opposed to capacitance, c_m, and the respective intracellular and extracellular axial resistances r_a, and r_o. This situation is exemplified in postsynaptic (Section 7.3) or skeletal muscle transverse tubule excitation (Section 10.5) which involve relatively slow voltage changes ($(dV/dt) \approx 0$); the voltage spread $V(x, t)$ is dominated by the r_m rather than the c_m term.

6.3 | Spread of Excitation in an Unmyelinated Nerve

In addition to these passive properties, unmyelinated nerve fibre membranes can be stimulated to produce propagating action potentials mediated by Na$^+$ channel activation. Their propagation is

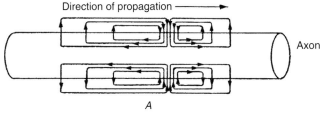

Figure 6.3 (A) Local circuit currents flowing during a propagated action potential. (B) The local circuit currents set up by a battery inserted in the core-conductor model.

similarly determined by fibre cable properties. These predict the effects of initiating an action potential of amplitude V_o over a short length of unmyelinated axon by Na$^+$ and K$^+$ current (i_{Na} and i_K, as referred to unit length of nerve) activation (Section 4.5). The initial inward i_{Na} initiates the action potential rising phase. The voltage change at this leading edge of the region of excitation drives axial current i_a through intracellular resistance r_a (Figure 6.3A; Figure 6.4A, B). These flow both ahead of and behind the active region along the length of the nerve (Figure 6.3B).

These local circuit currents reach the electrical c_m and r_m components making up the surface membrane of initially quiescent adjacent membrane regions of nerve. Here, the action potential upstroke produces a rapid membrane potential change, $(dV/dt) >> 0$. The local circuit current therefore charges c_m in preference to driving current flow across r_m (Figure 6.4C, D). This causes a time-dependent membrane voltage change $V(x, t)$, whose magnitude declines with distance x along the nerve from the point $x = 0$ of the initial excitation (Figure 6.4B). Comparison of Equations (6.4) and (6.10) indicates that $V(x)$ declines with distance with a dynamic effective length constant λ shorter than its limiting steady state value $\lambda(\infty)$.

Nevertheless, at the point where $V(x, t)$ attains the Na$^+$ channel excitation threshold, this initiates an action potential in the quiescent region. Continuation of this process results in action potential propagation along the length of the nerve. Provided that the axon has a uniform diameter and membrane properties, the consequent all-or-none action potential propagates with constant amplitude and velocity, θ. As the resulting action potential conduction velocity is determined by the spread of the initial voltage change, it similarly depends more strongly upon the membrane capacitance, c_m, than the membrane resistance, r_m, as well as varying with the axial resistances r_a, and r_o.

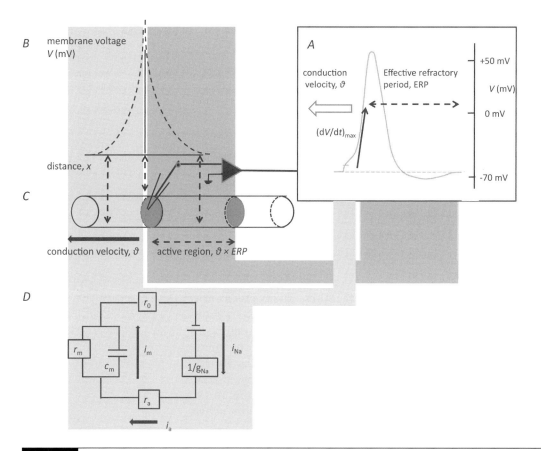

Figure 6.4 Action potential propagation along the length of the axon. (A) Action potential upstroke with slope (dV/dt) followed by recovery including effective refractory period (ERP). (B) Spreading voltage change in front and behind the excited region along the axon. (C) Propagating action potential with conduction velocity θ and active region of wavelength $\Lambda = \theta \times$ ERP results if (D) the intracellular current i_a through intracellular resistance r_a depolarises membrane ahead of the excitation wavefront to its action potential threshold. Refractoriness prevents Na^+ channel re-excitation of membrane behind the excitation wavefront ensuring unidirectional propagation.

6.4 | Action Potential Conduction Velocity and Direction

The consequent action potential conduction is normally unidirectional, even though the local current i_a flows both ahead of and behind the activated region (Figure 6.3A). It also spreads into regions that have already experienced and recovered from excitation. However, the action potential deflection is followed, not only by repolarisation, but also periods of absolute and relatively refractoriness (Section 2.4) producing an effective refractory period, ERP (Figure 6.4B). This transiently precludes re-excitation of membrane regions that have recently performed an action potential.

The process of propagating excitation followed by refractoriness maps onto an *active region* of excitation along the length of the nerve that includes not only the leading edge of the propagating action potential but also by a succeeding refractory region (Figure 6.4C). The resulting travelling wavelength Λ depends upon both conduction velocity θ and ERP: $\Lambda = (\theta \times \text{ERP})$. Any ectopic excitation applied within a distance less than $\theta \times ERP$ behind the leading edge of the action potential encounters refractory tissue and fails to elicit fresh excitation. This ensures unidirectional action potential propagation from active into initially quiescent tissue. The region behind the wave of excitation is in a refractory state. Over the ERP duration, it cannot initiate or conduct an action potential (Matthews *et al.*, 2013; Huang, 2017).

In vivo action potentials normally originate at one end of a nerve, and conduct with constant velocity θ unidirectionally away from that end. Exceptionally, experimental stimulation at the middle of an intact stretch of nerve permits excitation of initially quiescent membrane on either side of the stimulating electrode, causing action potential propagation in both directions. In clinical situations, abnormalities in wavelength characteristics or generation are fundamental to generation of cardiac arrhythmias (Sections 14.2 and 14.3)

The propagation velocity θ of the active region depends upon both c_m and i_a. The c_m term determines the quantity of charge needed to achieve the required change in membrane potential. The i_a term defines the current available to discharge c_m and in turn varies with r_a and the size of the action potential deflection. The r_a term defines the resistance of unit length of axoplasm. It depends on the cytoplasmic resistivity R_a, and the cross-sectional area πa^2 of the nerve fibre of radius, a, through the inverse relationship $r_a = R_a/\pi a^2$. As a result, large unmyelinated axons conduct impulses faster than small ones, typically with a velocity proportional to the square root of their radii (Jack *et al.*, 1983).

Finally, action potential generation and propagation are reduced with reductions in i_{Na}. This can follow compromised Na$^+$ channel expression or function produced by either pharmacological manipulation or genetic abnormality. Nevertheless, the action potential amplitude usually substantially exceeds its stimulation threshold. Conduction accordingly takes place with a large safety factor. An all-or-none conduction failure only takes place with major changes in the parameters influencing action potential initiation, amplitude or conduction.

6.5 | Saltatory Conduction in Myelinated Nerves

Another factor greatly affecting θ is nerve myelination. This increases effective membrane thickness, d, in turn decreasing c_m: the capacitance of unit membrane area depends on both its dielectric

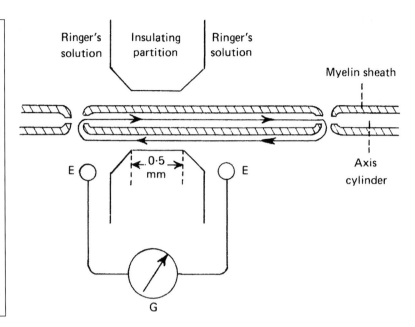

Figure 6.5 Method used to investigate saltatory conduction in myelinated nerve. The nerve fibre is drawn through a fine hole of ~40 μm diameter made in an insulator by a micromanipulator. Current flowing along the axis cylinder out of one node and into the other, as indicated by the arrows, causes a voltage drop outside the myelin sheath. The resistance of the fluid in the gap between the two pools of Ringer's solution is ~0.5 MΩ. The potential difference between them can be measured by oscilloscope G connected to electrodes E on either side. The internodal distance in a frog's myelinated nerve fibre is ~2 mm. (From Huxley and Stämpfli, 1949.)

properties as quantified by its permittivity, ε, and its thickness d: $C_m = \varepsilon/d$. Myelination also greatly increases the membrane resistance, r_m. Lillie suggested that the myelin sheath in vertebrate nerve fibres might thereby restrict inward and outward passage of local circuit current to the nodes of Ranvier, causing the nerve impulse to be propagated from node to node in a series of discrete jumps, producing *saltatory conduction* (Lillie, 1925).

Physiological tests for such a hypothesis required methods for dissecting isolated fibres from myelinated nerve trunks; these were first accomplished by Kato and his school in Japan (Kato, 1936). Ten years later Tasaki produced strong support for the saltatory theory by showing that the threshold for electrical stimulation in a single myelinated fibre was lower at the nodes than along the internodal stretches. Furthermore, conduction block by anodal polarisation and by local anaesthetics was more effective at the nodes than elsewhere (Tasaki, 1953). In collaboration with Takeuchi, Tasaki also introduced a technique for directly measuring local circuit current flow at different positions, further refining this approach.

Huxley and Stämpfli drew a myelinated fibre isolated from a frog nerve through a short glass capillary mounted in a partition between two Ringer-solution-filled compartments (Figure 6.5). The narrowness of the fluid-filled space around the nerve inside the capillary gave this a total resistance of ~0.5 MΩ. Longitudinal current flowing between neighbouring nodes outside the myelin sheath produced a measurable potential difference between the two sides of the partition. The resulting longitudinal current records showed that at all points outside any one internode the current flow was similar both in

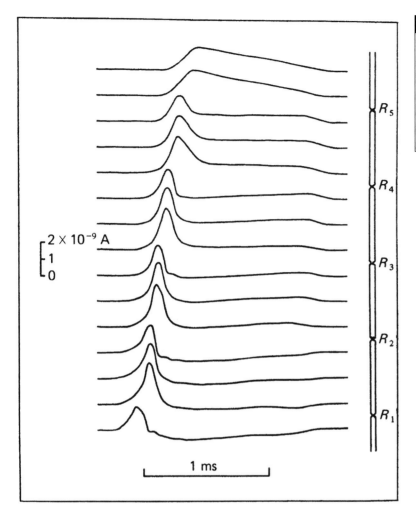

Figure 6.6 Currents flowing longitudinally at different positions along an isolated frog myelinated nerve fibre. The diagram of the fibre on the right-hand side shows the position where each record was taken. The distance between nodes was 2 mm. (From Huxley and Stämpfli, 1949.)

magnitude and timing (Figure 6.6). However, as successive nodes were traversed, the peaks of current flow were displaced stepwise in time by about one tenth of a millisecond. To determine the amount of current flowing radially into or out of the fibre, neighbouring pairs of records were subtracted from one another. The difference between the longitudinal currents at any two points could only have arisen from current entering or leaving the axis cylinder between those points (Huxley and Stämpfli, 1949).

The results of this procedure demonstrated that over the internodes there was merely a slight leakage of outward current (Figure 6.7). However, at each node there was a brief pulse of outward current followed by a much larger pulse of inward current. The current flowing transversely across the myelin sheath is exactly what would be expected for a passive leak, while the restriction of inward current to the nodes proved conclusively that the Na$^+$ channels operate only where the excitable membrane is accessible to the outside.

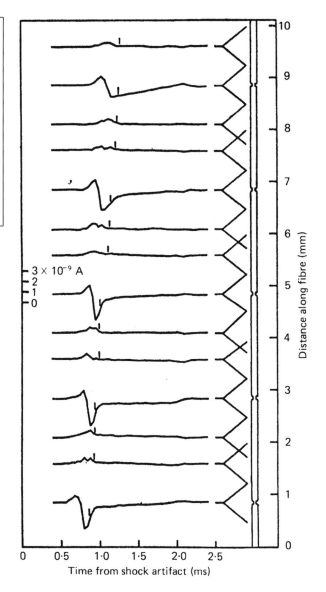

Figure 6.7 Transverse currents flowing at different positions along an isolated frog myelinated nerve fibre. Each trace shows the difference between the longitudinal currents, recorded as in Figure 6.5, at the two points 0.75 mm apart indicated to the right. The vertical mark above each trace shows the time when the change in membrane potential reached its peak at that position along the fibre. Outward current is plotted upwards. (From Huxley and Stämpfli, 1949.)

The term 'saltatory' literally means a discontinuous process. However, it would nevertheless be wrong to suppose that only one node is active at a time in a myelinated nerve fibre. The conduction velocity in Huxley and Stämpfli's experiments was 23 mm/ms, and the duration of the action potential was about 1.5 ms. The length of nerve occupied by the action potential at any moment was therefore about 34 mm. This corresponds to a group of 17 neighbouring nodes. In the resistance network equivalent to a myelinated fibre (Figure 6.8), the values of r'_n and r'_i are such that the electrotonic potential would decrement passively to 0.4 between one node and the next. The size of the fully developed action potential is of the order of 120 mV, and the

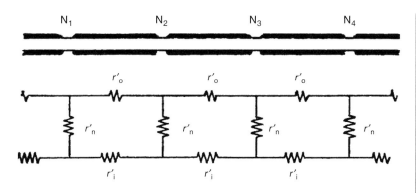

Figure 6.8 Equivalent circuit for resistive elements of a myelinated nerve fibre. For a toad fibre of outside diameter 12 μm and nodal spacing 2 mm, the internal longitudinal resistance r'_i is just under 20 MΩ and the resistance r'_n across each node is just over 20 MΩ. In a large volume of fluid the external resistance r'_o is negligibly small. (From Tasaki, 1953.)

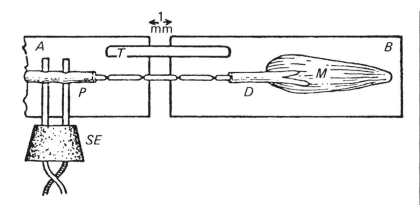

Figure 6.9 Method used to demonstrate the role of the external current pathway in a myelinated nerve fibre. A, B: insulated microscope slides; SE: stimulating electrodes; P: proximal end of frog's sciatic nerve; D: distal end of nerve; M: gastrocnemius muscle; T: moist thread providing an electrical connection between the pools of Ringer's solution on the slides. (From Huxley and Stämpfli, 1949.)

threshold depolarisation needed to excite the membrane is only about 15 mV. It follows that the conduction mechanism works with an appreciable safety factor. The impulse should be able to encounter one or two inactive nodes without being blocked. Tasaki showed that two but not three nodes that had been treated with the local anaesthetic cocaine could indeed be skipped.

A simple experiment demonstrated the importance of the external current pathway in action potential propagation along a myelinated nerve fibre (Huxley and Stämpfli, 1949). The nerve of a frog's sciatic–gastrocnemius preparation was pared down to leave only one fibre (Figure 6.9). Stimulation of the nerve at P caused a visible contraction of a motor unit M in the muscle. The preparation was now laid in two pools of Ringer's solution on microscope slides A and B, electrically insulated from each another. Its position was adjusted so that part of an internode, but not a node, lay across the 1 mm air gap separating the pools. At first, stimulation at P continued to cause a muscle twitch, but soon the layer of fluid outside the myelin sheath in the air gap dried up by evaporation, and the muscle ceased to contract.

Conduction across the gap could, however, be restored by placing a wet, conducting, thread T between the two pools. This demonstrated that an action potential arriving at the node just to the left of the air gap could trigger the node on the far side of the gap only when there was an electrical connection between the pools whose resistance was fairly low. Figure 6.8 indicates that if r'_o becomes at all large, the potential change at N_2 produced by a spike at N_1 will fall below the threshold for excitation.

As pointed out by Tasaki, there is a reservation about this experiment. Unless special precautions are taken, the stray electrical capacity between the pools may provide, for a brief pulse of current, an alternative pathway outside the dried-up myelin of the internode, whose impedance may be low enough for excitation to occur at the further node if its threshold is low. Even with such precautions, Tasaki found that impulses were still able to jump the gap if the fibre had a really low threshold, probably because simple evaporation could not make the external resistance high enough. Nevertheless, the fact that the experiment works in a clear-cut way only if the threshold is somewhat higher than it is in vivo does not prevent it from proving satisfactorily that there must be a low-impedance pathway between neighbouring nodes *outside* the myelin sheath if the nerve impulse is to be conducted along the fibre.

6.6 | Factors Affecting Conduction Velocity

Since the passive electrotonic spread of potential along a nerve fibre is an almost instantaneous process, it may be asked why the nerve impulse is not propagated more rapidly than it actually is. In myelinated fibres the explanation is that there is a definite conduction delay of about 0.1 ms at each node (Figure 6.7). This represents the time required for Na^+ to move through the membrane at the node to an extent sufficient to discharge the membrane capacity and reverse the membrane potential. Conduction in an unmyelinated fibre is slower than in a myelinated fibre of the same diameter because the membrane capacity per unit length is much greater, and the delay in reversing the potential across it arises everywhere and not just at the nodes (Section 6.1). Because the time constant for an alteration of membrane potential depends both on the magnitude of the membrane capacity and on the amount of current that flows into it (Section 6.3), conduction velocity is affected by the values of the resistances in the equivalent electrical circuit, and also by the closeness of packing of Na^+ channels in the membrane, which determines the Na^+ current density.

The effects of changing r_o and r_a are best illustrated in isolated axons. Experimentally increasing r_o by raising the axon out of a large volume of sea water into a layer of liquid paraffin reduced conduction velocity by ~20% in a 30 μm diameter crab nerve fibre and 50% in a 500 μm squid axon. Conversely, after mounting the axon in a moist chamber lying across a series of metal bars, short-circuiting the bars by a trough of mercury, increased velocity by 20% (Hodgkin, 1939). Similarly, reducing r_a by inserting a silver wire down the centre of a squid axon greatly speeded up conduction (del Castillo and Moore, 1959).

One reason why large unmyelinated fibres conduct faster than small ones is the decrease in r_a with increased fibre diameter. Assuming identical membrane properties through fibres of all sizes, it can be predicted that conduction velocity should be proportional to the square root of diameter (Section 6.3). Experimentally this does not always seem to hold. This could reflect one of the adaptations in giant axons enhancing rapid conduction being an increased density of membrane Na^+ channels. Measurements of labelled TTX binding showed that the smallest fibres of all, in garfish olfactory nerve, have the fewest channels, with a site density of $35/\mu m^2$ compared to the 90 and $100/\mu m^2$ in lobster leg nerve and rabbit vagus nerve, respectively (Ritchie and Rogart, 1977). However, in the squid giant axon there are about 290 TTX binding sites/μm^2 (Keynes and Ritchie, 1984).

6.7 | Factors Affecting the Threshold for Excitation

Excitation of a nerve fibre involves rapid membrane depolarisation to a critical level, normally ~15 mV less negative than the resting potential. The critical level for excitation is the membrane potential at which the net rate of Na^+ entry exactly equals the net rate of K^+ exit, plus a small contribution from Cl^- entry. Greater depolarisation than this tips the balance in favour of Na^+ entry. The regenerative process described in Section 4.3 then takes over and causes a rapidly accelerating entry of Na^+. After just-subthreshold depolarisation, when g_{Na} will have been raised over an appreciable area of membrane, the return of the resting potential will be somewhat slow at first, and a non-propagated *local response* may be observed.

At the end of the spike the membrane is left with its Na^+ permeability mechanism inactivated and its K^+ permeability appreciably greater than normal, reducing its excitability. Both changes raise the threshold for re-excitation. The partial Na^+ channel inactivation means that raising inward Na^+ current even to the normal critical value now requires an increased imposed depolarisation. Furthermore, the raised K^+ permeability increases the magnitude of this critical Na^+ current. Until the permeabilities for both ions have returned to their resting levels, and the Na^+ permeability system is fully reactivated, the shock necessary to trigger a second spike consequently remains above the normal threshold in size.

It has long been known that nerves are not readily stimulated by slowly rising currents: they tend to *accommodate* to this type of stimulus. Accommodation arises partly because sustained depolarisation causes a long-lasting rise in K^+ permeability, and partly because at the same time it inactivates the Na^+ permeability mechanism. Both changes occur with an appreciable lag after the onset of the imposed depolarisation. They are consequently not effective when a constant current is first applied, but become important after an elapsed time. They also persist for some time after the end of a stimulus, and so are responsible for the appearance of *post-cathodal depression*, which is a reduction of excitability after prolonged application of a weak cathodal current. As a result of accommodation, cathodal currents that rise more slowly than a certain limiting value do not produce any stimulation, since the rise in threshold then keeps pace with the depolarisation.

Another familiar phenomenon is the occurrence of excitation when an imposed anodal current is switched off. This *anode break excitation* is readily demonstrated in isolated squid axons or frog nerves. It is not seen in freshly dissected frog muscle or nerves stimulated in situ in living animals. Anode break excitation is only observed under conditions in which the resting potential becomes significantly positive to E_K because of a steady leakage of K^+. The nerve is then effectively in a state of mild cathodal depression, with g_{Na} partially inactivated and g_K well above normal. Anodal polarisation of the membrane reactivates the Na^+ permeability and reduces the K^+ permeability. The latter changes persist for a short interval after the current is switched off. The critical potential at which inward Na^+ current exceeds outward K^+ current may then become temporarily negative to the membrane potential resulting when the anodal current is withdrawn. An action potential is therefore initiated.

Divalent ions like Ca^{2+} and Mg^{2+} also strongly affect threshold behaviour in excitable membranes. In squid axons, even a slight reduction in $[Ca^{2+}]_o$ may set up sinusoidal membrane potential oscillations, while a more drastic $[Ca^{2+}]_o$ reduction results in a spontaneous repetitive discharge of impulses. Conversely, increased $[Ca^{2+}]_o$ tends to stabilise the membrane and raise the excitation threshold. Changes in $[Mg^{2+}]_o$ have similar effects on peripheral nerves, Mg^{2+} being about half as effective as Ca^{2+} in this stabilising influence.

Voltage clamp studies (Frankenhaeuser and Hodgkin, 1957) showed that the curve relating peak Na^+ conductance to membrane potential is shifted in a positive direction along the voltage axis by increasing $[Ca^{2+}]_o$, and is shifted in the negative direction by reducing $[Ca^{2+}]_o$. However, the resting potential itself is relatively insensitive to altered $[Ca^{2+}]_o$. This readily explains the relationship between $[Ca^{2+}]_o$ and threshold, since a rise in $[Ca^{2+}]_o$ moves the critical triggering level away from the resting potential, while a fall in $[Ca^{2+}]_o$ moves the critical level towards it. A moderate reduction in $[Ca^{2+}]_o$ may bring the critical level so close to the resting potential that the membrane behaves in an unstable and oscillatory fashion, and a further reduction will increase the amplitude of the oscillations to the point where they cause a spontaneous discharge of spikes. Although Ca^{2+} and Mg^{2+} have similar actions on the excitability of nerve and muscle fibres, they have antagonistic actions at the neuromuscular junction and at some synaptic junctions between neurons, because Ca^{2+} increases the amount of acetylcholine released by a motor nerve ending, and Mg^{2+} reduces it (Section 7.3). Changes in the plasma Ca^{2+} level in living animals may therefore give rise to tetany.

6.8 | After-potentials

In many types of nerve and muscle fibre the membrane potential does not return immediately to the baseline at the foot of the action potential, but undergoes further slow variations known as after-potentials. The nomenclature of after-potentials dates from the period before the invention of intracellular recording techniques, when external electrodes were used. Hence an alteration of potential in the same direction as the spike itself is termed a *negative after-potential*, while a variation in the opposite direction corresponding to a membrane hyperpolarisation is termed a *positive after-potential* (Figure 2.3). Isolated squid axons display action potentials with a characteristic *positive phase* (Figure 2.4B) almost completely absent in living animals (Figure 2.4A), while frog muscle fibre action potentials have a prolonged negative after-potential (Figure 2.4H), discussed in greater detail in Section 10.2. In some mammalian nerves, both myelinated and unmyelinated, there is first a negative and then a positive after-potential. A related phenomenon, which is most marked in the smallest fibres, is the occurrence after a period of repetitive activity of a prolonged hyperpolarisation of the membrane known as the post-tetanic hyperpolarisation.

There is no doubt that after-potentials are always connected with changes in membrane permeability towards specific ions, but there is more than one way in which the membrane potential can be displaced either upwards or downwards. In the isolated squid axon, for example, the positive phase arises because the K^+ conductance is still relatively high at the end of the spike, whereas the Na^+ conductance is inactivated and below normal. The membrane potential consequently transiently comes closer to E_K, and drops back as g_K and g_{Na} regain their usual resting values.

Different mechanisms are responsible for production of the positive after-potential and the post-tetanic hyperpolarisation in vertebrate nerves. This has been attributed to an enhanced rate of Na^+ extrusion by the Na^+ pump operating in an electrogenic mode. In other cases, a change in the relative permeability of the membrane to Cl^- and K^+ ions may play a part. There is also evidence that the presence of Schwann cells partially or wholly enveloping certain types of nerve fibre has important effects on the after-potential by restricting rates of ion diffusion in the immediate vicinity of the nerve membrane.

Neuromuscular Transmission

7.1 | The Motor Unit

A major function of peripheral nerve is to innervate skeletal muscle. Muscle receives both sensory and motor nerve fibres. The sensory nerves convey information about the state of the muscle to the nervous system. The fibres innervating muscle spindles provide information about muscle length and Golgi tendon organs detect muscle tension. Some of the variety of free nerve endings in muscle tissue convey pain sensation. Of motor fibres, γ-motor neurons provide a separate motor nerve supply for intrafusal muscle fibres within muscle spindles. However, α-motor neurons innervate the bulk of the remaining, extrafusal, muscle fibres.

Action potentials are generated in the initial segment region of the innervating α-motor neuron (Section 8.1) and propagate down its myelinated efferent axon. This divides into branches that eventually lose their myelin sheaths. Each branch terminates in a motor end-plate on an individual mammalian muscle fibre forming a *synapse*, defined as a functional point of contact at which transfer of electrical information takes place between two excitable cells. Transfer of excitation from nerve to muscle, *neuromuscular transmission,* represents a particular form of synaptic transmission; other variants are explored in Chapter 8.

Most mammalian twitch muscle fibres each make contact with a single nerve terminal, although sometimes there are two terminals originating from the same nerve axon. Frogs and other lower vertebrates show a further, tonic, muscle fibre type in which a large number of nerve terminals innervate each muscle fibre which generates slow and maintained contractions. Tonic fibres also occur in mammalian extraocular, laryngeal and middle ear muscles.

Each axon together with its individual innervated muscle fibres constitutes a functional unit of contraction, the *motor unit*, comprising typically parallel-oriented muscle fibres of a specific histological fast- or slow-twitch myocyte type (Figure 7.1A). All component muscle fibres within a single motor unit contract at the same time as all are

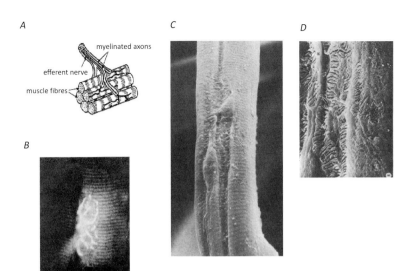

A

myelinated axons

efferent nerve

muscle fibres

B

50 μm

C

D

Figure 7.1 Structure of the neuromuscular junction. (*A*) Innervation pattern shown by motor unit. (*B*) Nerve terminals on murine sternomastoid muscle vitally stained by the fluorescent dye 4-(4-diethylaminostyryl)-methylpyridinium iodide. (*C*, *D*) Scanning electronmicroscopy of postsynaptic membrane of frog neuromuscular junction visualising membrane folds. (*C*) Isolated single muscle fibres showing imprint where the terminal arborisation of the innervating motor axon has been removed (approx. × 1000). (*D*) The junctional grooves traversed by folds of the postsynaptic membrane (approx. ×5000). ((B) from Figure 2 of Lichtman *et al.*, 1987, (C, D) From Figure 1 and 4 of (Shotton *et al.*, 1979.)

activated by the same nerve cell. In contrast, muscle fibres belonging to different motor units may well contract at different times. The contractile activity shown by the whole muscle represents a summation of the tensions generated by each component motor unit. The number of activated motor units determines the overall force of muscle contraction. Larger powerful muscles such as quadriceps femoris performing gross movements contain fewer but larger motor units, each containing ~1000 myocytes. Muscles performing finely controlled movements typically possess a proportionally greater number of smaller motor units containing ~10 myocytes.

7.2 | Presynaptic Transmitter Release

The terminal region of each axon, typically around 1–2 μm in diameter, runs longitudinally along the surface of the muscle fibre it innervates within a shallow depression, before making contact with the muscle fibre at its end-plate (Figure 7.1*B,C*). Electronmicroscopy of thin sections provides structural details of this *presynaptic region* (Figure 7.2). Here, the nerve terminal is separated from the *postsynaptic membrane* of the muscle cell by a 50–70 nm wide synaptic cleft continuous at its edges with the remaining extracellular space. The synaptic cleft is filled with extracellular mucopolysaccharide material visible to electronmicroscopy as the basal lamina. Pre- and postsynaptic membranes are anatomically distinct. This excludes direct electrical signalling between them. However, the axon terminal contains a large number of small membrane-bound *synaptic vesicles*, and

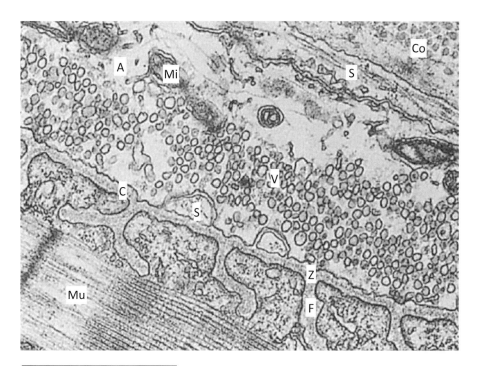

Figure 7.2 Electronmicrograph showing structure of frog neuromuscular junction. The axon terminal (A) runs diagonally across the middle of the section, covered by a Schwann cell (S) and collagen fibres (Co), overlying a muscle cell (Mu). Between the axon and the muscle cell is the synaptic cleft (C). The AChRs are concentrated at the top of a series of folds (F) in the subsynaptic membrane. The terminal contains mitochondria (Mi) and large numbers of synaptic vesicles (V). Vesicle release probably occurs at presynaptic active zones (Z). Magnification 27 000×. (Photograph supplied by Professor J. E. Heuser.)

a considerable number of mitochondria. At intervals of around 1–2 μm, the corresponding *postsynaptic* muscle membranes show transverse invaginations, or synaptic folds (Figure 7.1D). Synaptic vesicles and the synaptic cleft play a vital role in the transmission at this and other synapses.

These structural features directly reflect the mechanism by which presynaptic excitation activates postsynaptic activity. A first suggestion made in the nineteenth century implicated a presynaptically released chemical substance acting as a signalling molecule between the two cells. This idea was tested in the innervation of frog heart (Loewi, 1921). The normally spontaneous heart beat (Section 13.2) is inhibited by vagal nerve stimulation. Perfusion fluid obtained from a heart inhibited by vagal stimulation reduced the amplitude of normal beats studied in the absence of such vagal stimulation. In contrast, perfusion fluid from a normally beating heart did not have this effect. These findings suggested that vagal stimulation releases a physiologically active chemical substance, presumably from the nerve endings. Synaptic transmission is typically a chemically mediated process involving transmitter release from presynaptic terminals. The chemical substance involved in Loewi's experiments was identified as the small organic cationic ester acetylcholine (ACh) (Plate 3A). Electrical stimulation of skeletal muscle motor nerve endings similarly elicits ACh release (Dale *et al.*, 1936). Subsequent work implicated a wide range of compounds acting as synaptic transmitter substances (Section 8.1).

Histochemical studies independently demonstrated that presynaptic terminals contain ACh. This is synthesised from choline and acetyl coenzyme A by the enzyme choline acetyltransferase. Conversely, the basal lamina within the synaptic cleft contains the enzyme acetylcholinesterase which can hydrolyse the released ACh to choline and acetic acid. This limits the duration over which released ACh exerts its effects. Histochemical staining showed that this enzyme is concentrated in the synaptic cleft, especially in the subsynaptic membrane folds.

7.3 | Graded and Regenerative Components of the Postsynaptic Response

Stimulation of the motor nerve produces electrical responses in the muscle fibre. Much of our understanding of these events was derived from intracellular recording experiments (Fatt and Katz, 1951). A glass micropipette electrode, filled with a concentrated KCl solution was inserted into the muscle fibre in the end-plate region. A suitable amplifier measured the voltage between the tip of that electrode and another electrode in the external solution, giving the electrical potential difference across the membrane (Figure 7.3A).

Figure 7.3B illustrates typical findings obtained when end-plate function was empirically reduced by the paralysing South American arrow poison curare, or conditions of reduced $[Ca^{2+}]_o$, which also appeared to partly block transmission. Nerve stimulation by a pair of silver wire electrodes elicited an end-plate potential (EPP) consisting of a rapid depolarisation of a few millivolts, followed by a rather slower return to the resting membrane potential. This response was not observed if the microelectrode was inserted at some distance from the end-plate. Increasing the curare concentration or reducing $[Ca^{2+}]_o$ reduced the EPP amplitude.

In the absence of inhibition by curare or under conditions of normal $[Ca^{2+}]_o$, the EPP assumes its normal, larger, size. The EPP becomes large enough to cross the threshold for electrical excitation of the muscle cell membrane leading to an all-or-nothing propagated action potential, just as in the nerve axon (Section 10.2). The presynaptic stimulation triggering a presynaptic action potential thereby leads to a postsynaptic action potential. In contrast to the unpropagated EPP, this is conducted along the length of the muscle fibre. Serial microelectrode recordings along the length of the muscle demonstrated that action potential waveforms at the end-plate were more complicated than those recorded elsewhere along the muscle fibre (Figure 7.4). Note that the record obtained at the end-plate site indicated by positions 5 and 6 of the inset in

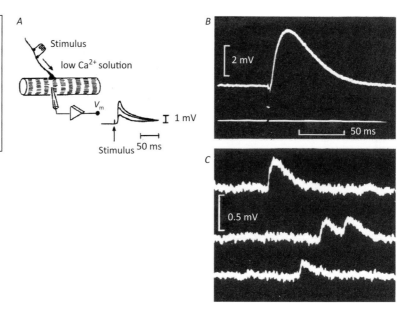

Figure 7.3 End-plate activity, frog neuromuscular junction. (A) Nerve stimulation and recording configuration demonstrating (B) end-plate potential (EPP) following presynaptic nerve stimulation and (C) spontaneous miniature end-plate potentials (MEPPs). (From Figure 1 of Katz and Miledi, 1969.)

Figure 7.4 is a combination of EPP and action potential. In contrast, recordings at greater distances from the end-plate, reflected in the greater latencies of the action potential peaks, showed more gradual rising phases free from EPP contributions. The propagated action potential is therefore mechanistically distinct from, even if it is initiated by, the preceding unpropagated EPP. It is the action potential that ultimately triggers muscle contraction.

These electrophysiological events produced by electrical stimulation of nerve could be replicated by direct experimental applications of ACh to the postsynaptic membrane. This implicated ACh release in producing EPPs. The ACh was applied by ionophoresis or iontophoresis through an appropriately placed glass electrode (Figure 7.5A). ACh is a positively charged ion (Plate 3A) and migrates in an electric field. It can be ejected from a small pipette by passing a current through it. A brief, highly localised pulse of ACh can thereby be applied to the postsynaptic membrane. This produced a muscle fibre depolarisation resembling the EPP (Figure 7.5B). The pharmacological properties of the response resembled those of the EPP, including reduction or abolition by curare. Together these findings suggest that the EPP is produced by the ACh released from the motor nerve ending.

Figure 7.6 summarises these experimental results, illustrating the separation of the generation of end-plate potentials and their resulting action potentials. It illustrates results from microelectrode recordings made both near to and remote (>5 mm) from the end-plate. A presynaptic stimulus eliciting an EPP of below-threshold

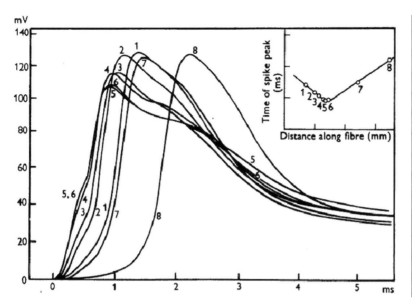

Figure 7.4 Transition between electrical activity at the end-plate to action potential firing in an amphibian muscle fibre studied at 17 °C in 9 mM $[Ca^{2+}]_o$. The microelectrode position was moved along the fibre. Records obtained at 0; 0.3; 0.45; 0.6; 0.65; 0.75; 1.75 and 2.75 mm (positions 1–8 respectively) from a reference position along the fibre. Note gradual change in the shape of the action potential and latency of the spike. Inset: plot of the timing of spike peak against distance, showing a constant propagation velocity in both directions from positions 5 and 6. (From Figure 21 of Fatt and Katz, 1951.)

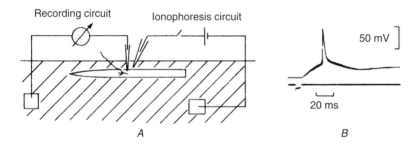

Figure 7.5 Ionophoresis technique applied to a frog muscle fibre (A) and response to an applied pulse of ACh (B). (From del Castillo and Katz, 1955.)

amplitude for action potential activation under conditions of partial AChR block by curare (Figure 7.6A) simply elicits an EPP deflection. It rises to a peak and then decays exponentially. The amplitude of this deflection declines with distance, reflecting muscle fibre cable properties, particularly its membrane resistance r_m (Section 6.3; Figure 7.6Ba). The EPP deflection is therefore restricted in extent to membrane regions immediately around the end-plate. Recordings at greater distances reveal no voltage deflection (Figure 7.6Bb). In contrast, following an EPP sufficiently large to reach action potential threshold, microelectrode recordings close to the end-plate reveal an initial deflection reflecting the EPP waveform and a subsequent superimposed action potential (Figure 7.6Ca). Finally, microelectrode recordings remote from the neuromuscular junction simply show the action potential waveform in which the initial EPP deflection has disappeared. The action potential has now become the sole mediator of postsynaptic signalling (Figure 7.6Cb; Katz, 1962).

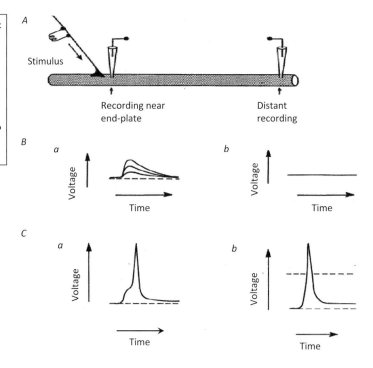

Figure 7.6 Separation of synaptic and action potential contributions to neuromuscular signalling. (A) Presynaptic stimulation of the nerve. (B, C) Recordings under conditions when EPP amplitude is subthreshold to (B) or elicits the action potential (C), at sites close to (a) and remote from (b) the neuromuscular junction.

7.4 | The Quantal Nature of Presynaptic Events

Analysis of postsynaptic events provided further important insights into presynaptic mechanisms underlying transmitter release. This is fortunate: presynaptic nerve terminals are much smaller than the postsynaptic muscle fibre. They are difficult to investigate directly by intracellular microelectrode recordings. The postsynaptic recordings demonstrated spontaneous, small (≤ 1 mV) depolarisations, *miniature end-plate potentials* (MEPPs), even in resting muscle not subject to presynaptic stimulation (Figure 7.3C). Although the normal trigger for the EPP is the arrival of the presynaptic action potential, MEPPs persist, even if the latter is blocked by tetrodotoxin (Figure 7.7A). In contrast, MEPPs were not observed in microelectrode recordings distant from the end-plate (Figure 7.7B; Fatt and Katz, 1952). MEPPs could further be replicated by ionophoretic applications of ACh. Such experiments suggested that each MEPP reflected the action of just under 10 000 ACh molecules.

MEPPs and EPPs shared physiological and pharmacological properties that were distinct from those of the action potential. MEPPs closely resembled EPPs in timecourse. They were reduced in size by curare and increased in size by anticholinesterases known to block ACh breakdown. Their frequency of occurrence was decreased

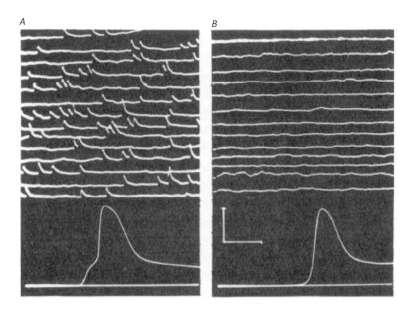

A   *B*

Figure 7.7 Spontaneous
miniature end-plate potential
recordings. (A) Results from
microelectrode recordings close to
the end-plate (upper traces). (B)
Corresponding recordings from an
electrode inserted 2 mm away in
the same amphibian muscle fibre.
Vertical and horizontal calibration
bars here denote 3.5 mV and 46 ms,
respectively. Lower traces in (A)
and (B) show corresponding action
potentials obtained in response to
presynaptic stimulation. (A) shows
the initial rise reflecting the EPP
preceding the action potential. In
(B) the action potential does not
show the initial rise, and appears
with an increased delay, reflecting
the conduction time between the
end-plate and the recording site.
The respective calibration bars here
denote 50 mV and 2 ms. (From
Figure 1 of Fatt and Katz, 1952.)

by decreasing $[Ca^{2+}]_o$ or increasing $[Mg^{2+}]_o$, and increased with increases in $[K^+]_o$ or extracellular osmotic pressure. MEPP frequency varied with $[Ca^{2+}]_o$ through a $[Ca^{2+}]_o{}^4$ relationship, consistent with specific cooperative regulatory actions of Ca^{2+} at distinct binding sites.

The findings that both $[Ca^{2+}]_o$ and presynaptic depolarisation influence MEPP production suggested a scheme in which voltage-gated Ca^{2+} channels within an axon terminal opened with depolarisation by an action potential. The resulting Ca^{2+} influx would increase presynaptic intracellular $[Ca^{2+}]_i$ and the latter might cause vesicle transmitter release. This mechanism would predict a *synaptic delay* between action potential arrival in the terminal and the ensuing postsynaptic depolarisation. A major contribution to this delay would arise from Ca^{2+}-mediated regulatory processes within the presynaptic terminal. The ACh would require an additional ~10 μs to diffuse across the synaptic cleft to combine with the AChRs. Following this, a further <100 μs would elapse before postsynaptic ionic channel opening. Frog muscle indeed shows overall mean synaptic delays of ~1 ms at 17 °C. This, however, fluctuates between release events, which show a *minimum* delay of ~500 μs.

A quantal theory of transmitter release might suggest that a single MEPP reflects release of a single quantum of ACh coming from a single transmitter-containing presynaptic vesicle. MEPPs might originate from spontaneous presynaptic release of 'packets' of ACh. Each vesicle would have a particular *probability* of release over unit time. Arrival of the presynaptic action potential might transiently increase this probability, thereby increasing generation of postsynaptic MEPPs. EPPs induced by ACh release might reflect a synchronised summation of thousands of small quantal MEPP events, each of similar amplitude and duration (Figure 7.8*Aa,b*).

Figure 7.8 (A) Records illustrating (a) individual (0.5–1 mV) quantal miniature end-plate potential (MEPP) events at a single amphibian neuromuscular junction, under conditions of reduced (0.9 mM) $[Ca^{2+}]_o$ and elevated (14 mM) $[Mg^{2+}]_o$. The wavy line provides a 50 Hz time base. Traces above the wavy line are from unstimulated nerve. Traces below the wavy line are recordings obtained in response to nerve stimulation following the dotted vertical line. Note the large number of negative responses, reflecting the probabilistic nature of EPP generation. These build up into (b) an EPP, demonstrated under conditions of elevated $[Mg^{2+}]_o$ but normal $[Ca^{2+}]_o$. EPPs fluctuate by the approximate amplitude of the MEPP. (B) Frequency distributions of (a) MEPP amplitudes obtained under conditions of reduced $[Ca^{2+}]_o$ and (b) EPP amplitudes compared with predictions that the responses statistically comprise units whose mean size and amplitude distribution are identical to those of (a). In the horizontal axes, the scale units correspond to the mean amplitude of the spontaneous potentials, of 0.875 mV in the present experiment. (From Figures 1, 2 and 7 of del Castillo and Katz, 1954.)

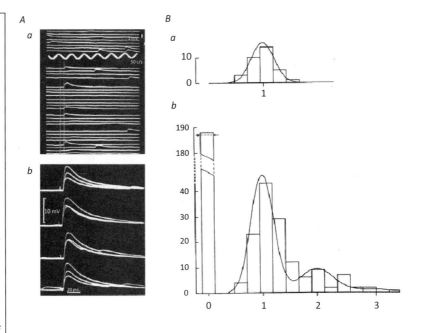

Experimental findings fulfilled the predictions of this hypothesis. Under conditions of reduced transmitter release produced by increased $[Mg^{2+}]_o$ and reduced $[Ca^{2+}]_o$, EPPs obtained in response to presynaptic stimulation fluctuated probabilistically in size. They did so in a stepwise manner. Each such step approximated the size of a MEPP (Figure 7.8Ab). It was possible to demonstrate distinct EPP peaks that could be identified with the simultaneous release of k = 1, 2, 3 or 4 MEPP quanta. Plots of the distribution of the sizes of stimulated EPPs demonstrated that the shapes and areas beneath each of these peaks fitted the *Poisson distribution*:

$$P_k = \frac{e^{-m}m^k}{k!} \tag{7.1}$$

P_k is the probability that the EPP comprises k quanta, and m is the mean number of quanta released during an EPP, also termed the *quantal content*. This condition describes a probabilistic process in which the release probability for any one vesicle is low and each individual vesicle release event is identical and independent (Figure 7.8Ba; Katz, 1962). A plot of the frequency of EPP amplitudes observed with stimulation fulfilled predictions that the responses statistically comprised units of different sizes, each generating similar distributions whose maxima were increments of a single MEPP (Figure 7.8Bb).

Independent electronmicroscope studies correspondingly demonstrated large numbers of ~50 nm diameter *synaptic vesicles* within the axon terminals (Figure 7.2). Similar vesicles occurred in terminals of other synapses thought to employ chemical transmission.

Furthermore, nearly pure synaptic vesicle fractions obtained from electric organs following homogenisation and centrifugation provided a source of large numbers of nerve endings. These indeed contained ACh (Section 7.7). Finally, studies involving rapid freezing of the muscle followed by freeze-fracture electronmicroscopy directly visualised fusion of presynaptic vesicles with the plasma membrane during neuromuscular transmission. Discharge of the contents of one vesicle into the synaptic cleft might correspond to the release of one quantum of the transmitter (Bruns and Jahn, 1995). In amphibian neuromuscular junctions, each vesicle likely contains roughly 10^4 ACh molecules.

Variations in quantal release patterns result in two phenomena altering EPP production in response to presynaptic stimulation. First, repetitive motor axon stimulation can produce EPPs of different sizes. Initially, the successive EPPs increase in size, particularly under conditions of low $[Ca^{2+}]_o$. This phenomenon is known as *facilitation*. As the larger responses comprise more quantal units, this has been attributed to enhancement of the release process by Ca^{2+} accumulation at some site within the presynaptic terminal. In contrast, *depression* resulting in succeeding responses becoming smaller and comprising fewer quantal units occurs after recent release of a large number of quanta. This can be demonstrated in the later stages of a stimulus train in the presence of an adequate $[Ca^{2+}]_o$. It has been attributed to temporary reductions in the numbers of vesicles available for release.

7.5 | Ionic Current Flows Underlying the End-Plate Potential

Fatt and Katz attributed the EPP to a general increase in ionic permeability of the postsynaptic membrane. The underlying conductance had an ion selectivity, voltage dependence and kinetic properties distinct from those shown by nerve Na^+ or K^+ channels (Section 4.6). This was apparent from ionic, excitatory postsynaptic currents (EPSCs) obtained from voltage clamped muscle postsynaptic membrane regions (Figure 7.9A). With a postsynaptic clamp at negative holding potentials close to the normal resting potential, presynaptic nerve stimulation elicited EPSCs that appeared as transient inward currents rising to a peak within a millisecond, then decaying with a ~1 ms time constant (Figure 7.9Bb) (at 25 °C, −70 mV, frog). The comparatively slower decay shown by the corresponding EPP (Figure 7.9Ba) could be attributed to the time required for recharging of the membrane capacitance that does not occur when the membrane potential is clamped (Takeuchi and Takeuchi, 1960).

Presynaptic stimulation at progressively depolarised postsynaptic holding potentials resulted in EPSCs with progressively smaller

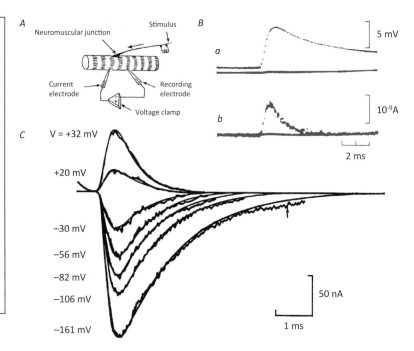

Figure 7.9 End-plate currents studied by voltage clamp. (A) Voltage clamp configuration controlling intracellularly recorded subsynaptic potential during presynaptic stimulation. (B) End-plate potential (EPP) and end-plate current (EPSC) in a curarised frog muscle fibre. (a) EPP from intracellular voltage recording. (b) Excitatory postsynaptic current (EPSC) recorded from the same end-plate with the membrane potential clamped at its resting level. (C) EPSCs elicited by a presynaptic stimulus at progressively incremented holding voltages. ((B) from Takeuchi and Takeuchi, 1959; (C) from Figure 6 of Magleby and Stevens, 1972.)

amplitudes (Figure 7.9C). The EPSCs disappeared at a ~0 mV holding potential. Further depolarisation elicited progressively larger currents that now assumed an opposite, outward direction (Magleby and Stevens, 1972). The EPSC has a *reversal potential* of close to −15 mV, at which its net ionic current is zero, which is distinct from, and falls between, the Na^+ and K^+ Nernst potentials.

This particular value of reversal potential is consistent with the EPSC being carried by more than one, Na^+ and K^+, current, each driven by its distinct Nernst potential, E_{Na} and E_K. Had only one ion species been flowing, the EPSC reversal potential would equal the Nernst potential for that particular ion. This would have been ~+50 mV if only Na^+, and ~−90 mV if only K^+ were flowing. The reversal potential, of ~−15 mV demonstrated by Takeuchi, implicates flow of both Na^+ and K^+ during the EPSC. Such a hypothesis predicted that altering the Nernst potential of any one participating ion would alter the reversal potential. This was verified by altering their respective extracellular ionic concentrations. Alterations in either $[Na^+]_o$ or $[K^+]_o$ altered the reversal potential, whereas alterations in $[Cl^-]_o$ did not. Postsynaptic activation involves simultaneous increases in membrane permeability to both Na^+ and K^+.

Finally, varying the clamped membrane potential demonstrated that EPSC amplitudes altered linearly with membrane potential, as expected from Ohm's law (Figure 7.10). This suggests that the membrane conductance at the peak of the EPSC is constant and not affected

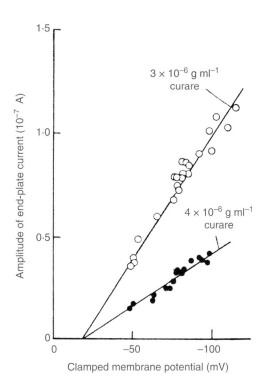

Figure 7.10 Results of a voltage-clamp experiment on a curarised frog muscle fibre, showing that the peak end-plate current varies linearly with membrane potential. (From Takeuchi and Takeuchi, 1960.)

by membrane potential. This is in marked contrast to the situation in the nerve axon, where the Na^+ and K^+ permeabilities of the membrane are strongly altered by changes in membrane potential. The EPSCs did decay more rapidly with depolarisation because of some voltage-dependence of the AChR gating process. In an already depolarised membrane the conductance switches off faster when it is no longer needed to produce depolarisation (Magleby and Stevens, 1972).

7.6 | Patch Clamp Studies

Since ACh combination with the AChR causes an increase in membrane permeability to both Na^+ and K^+, it seems likely that each AChR is closely associated with an *ion channel* through which this EPSC occurs. The channel would normally be expected to be closed and would open for a short time when ACh combines with its receptor. Experiments developing and applying the patch clamp technique for the first time directly confirmed this view (Figure 7.11; Neher and Sakmann, 1976). These studies were performed on frog muscle fibres whose motor nerve supply had been cut some time previously. Following such denervation, the entire fibre membrane surface becomes ACh-sensitive, owing to a general membrane expression of a low density of AChRs normally confined to postsynaptic end-plate membrane. The polished tip of a microelectrode can be apposed to, making a high-resistance electrical seal with, the fibre membrane providing a cell-attached patch.

Patch clamp studies can investigate current flow through *single* ion channels. These recordings revealed all-or-nothing current steps of variable durations, but constant amplitude, taking place only when the patch electrode contained ACh (Figure 7.12). This suggests that the channel can be either open or shut, and that it can only open in the presence of ACh binding. The unitary current deflections revealed a single channel conductance of ~30 pS (S = 1/Ω). The steps formed short (<1 ms) openings with random durations often clustering into bursts

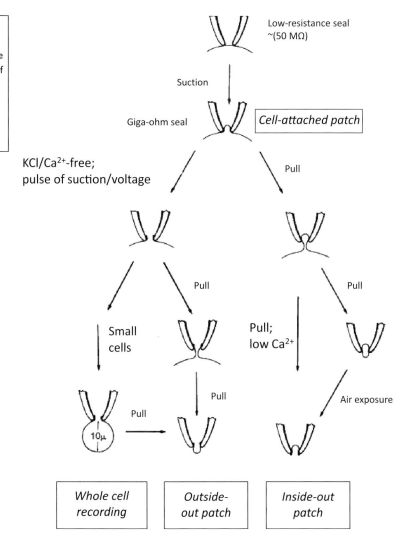

Figure 7.11 Schematic representation of procedures for forming a gigaohm seal between the tip of a micropipette and a patch of cell membrane, and of achieving 'cell-attached', 'whole-cell', 'outside-out patch' and 'inside-out patch' recording configurations. (From (Hamill *et al.*, 1981.))

ACh, 100 nM

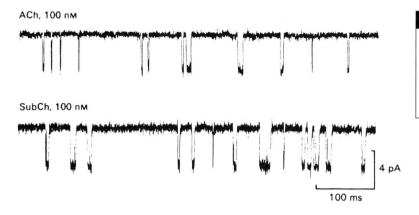

SubCh, 100 nM

4 pA

100 ms

Figure 7.12 Patch clamp records of single channel currents from frog end-plates. The upper record shows the response to ACh, the lower trace shows the response to suberyldicholine (SubCh). (From Colquhoun and Sakmann, 1985.)

of openings separated by short gaps. The bursts in turn were separated by longer time intervals.

Analysis of such recordings, and of unit channel responses to applied steps of $[ACh]_o$, suggested that ACh bound to its receptor following a reaction scheme consisting of a linear series of successive reactions between ligand (A) and receptor (R) (Plate 3B). This was consistent with a scheme in which binding of two molecules of ACh, likely to each of two ACh binding sites, was required to produce channel opening. Low ACh levels resulting from any resting leakage from the presynaptic terminal would not cause channel activation. However, increased concentrations would produce a sharply increased postsynaptic response. The EPP might result from a simultaneous opening of a large number of such channels. Finally, maintained, high levels of ACh eventually led to receptor desensitisation, suggesting a formation of further ACh-bound, but inactivated, states.

The technique of single channel recording from a very small patch of membrane has subsequently been greatly extended. Membrane currents can now be studied in a wide range of intact cells by whole-cell recording, and in cell and membrane preparations using different configurations of outside-out or inside-out patches, in which the fluid within the pipette is in contact either with the intracellular or extracellular sides of the membrane under study (Figure 7.11). This has made it possible to study the properties of ion channels in a wide range of living cells, some examples of which are explored in subsequent chapters.

7.7 | The Nicotinic Acetylcholine Receptor

Demonstration and characterisation of specific postsynaptic membrane AChRs at the neuromuscular junction was aided by the use of the polypeptide α-bungarotoxin, derived from Taiwanese banded krait snake (*Bungarus multicinctus*) venom. This causes irreversible neuro-muscular block attributable to its binding to the possible AChRs. Use of radioactively labelled toxin, made by acetylating the toxin with ^{3}H-acetic anhydride, made it possible to demonstrate autoradiographi-cally that the toxin rapidly becomes attached to the postsynaptic membrane at the end-plate regions. Counting the grains of silver produced in autoradiographs made it possible to count the number of bound toxin molecules and thereby estimate the number of AChRs. This suggested ~3×10^7 binding sites per mammalian end-plate, corresponding to an average density in that region of 10^4 sites per μm^2.

This specific α-bungarotoxin binding was utilised to isolate the AChRs in sufficient quantities for biochemical study from electric organs of the electric ray *Torpedo*. These preparations provide a rich source of AChRs. Extensive genetic, electronmicroscopic and X-ray diffraction studies suggested that the AChR is a transmembrane mol-ecule comprising four types of glycoprotein subunit, α, β, γ and δ, each with molecular weight ~55 kDa. Each AChR has two α-chains and one each of the remaining subunits, giving the stoichiometry α$_2$βγδ (Plate 3*Ca*, *D*). The ACh binding sites occur on the α-chains.

The AChR was also the first ion channel sequenced using recom-binant DNA techniques. In 1982 the Kyoto University group published the amino acid sequence of the α-subunit (Noda *et al.*, 1982). Sequences for the other subunits soon followed. The subunits varied in size from 437 amino acids (50 kDa) for the α-chain to 501 amino acids (58 kDa) for the δ-chain. The sequences for the four subunits show consider-able homology. All have four hydrophobic segments probably forming membrane-crossing helices (Plate 3*Cb*). The long section from the beginning of the chain to the first membrane-crossing helix apparently lies entirely in the extracellular side of the membrane. It contains disulfide crosslinks and sites for carbohydrate-group attachment. The α-chain contains the sites for ACh and α-bungarotoxi-n binding.

Cryo-electronmicroscope methods determined how the five sub-units are assembled to form the whole complex (Unwin, 2014). Nicotinic AChRs can be isolated from *Torpedo* electric organ as regular close-packed arrays in tubular form. These arrays can be frozen and examined in the electronmicroscope, and the images subjected to Fourier analysis to give a three-dimensional electron density map of the molecule. It was also possible to examine the effects of ACh on their structure by spraying it at the receptors just milliseconds before freezing in liquid ethane at −178 °C. These studies demonstrated that the AChR subunits assemble into a five-subunit structure with the stoichiometry α$_2$βγδ within the lipid bilayer (Plate 3*Ca*; Unwin, 2003). They possess hydrophilic, nega-tively charged amino acid side-chains that point inwards towards the 'pore', creating a selectivity filter selectively permeable to monovalent cations, whilst blocking anion permeation (Plate 3*D*).

In contrast, lipophilic side-chains point outwards towards the hydrophobic interior of the lipid bilayer, stabilising the pentamer.

The side view demonstrates that the AChR is a transmembrane protein. The large extracellular component of the two α-subunits accommodate the two ACh binding sites (Plate 3E). These ligand-binding domains (uppermost in the right panel of Plate 3E) consist of β-pleated sheets. A wide vestibule in the extracellular region narrows to a fine pore in the transmembrane region. The pore is lined by dense bent rods, probably the M2 segments of the five subunits. In the resting state the channel may be closed at its narrowest part by a large hydrophobic leucine residue. When ACh binds to the α-subunits, the M2 segments rotate to move these leucine residues out of the way so that cations can flow through the channel pore. ACh binding to the ligand-binding domain in successive α-subunits triggers rotational movements in their inner β-pleated sheets.

The nicotinic AChR channel belongs to a group of *neurotransmitter-gated channels*, themselves part of the general class of *ligand-gated channels*. It opens with binding of a particular ligand molecule. It shows important contrasts with voltage-gated channels described in Chapter 5, illustrated for the Na^+ channel in Plate 1. Unlike them, it is not gated by changes in membrane potential, and so, not surprisingly, there are no voltage sensors corresponding to the S4 segments of the voltage-gated channels.

Experiments using oocytes of the African clawed frog *Xenopus* demonstrated that assembly of these four subunits fulfilled both necessary and sufficient conditions for producing a functional ACh receptor (see also Section 5.2). Oocytes possess a normal protein translation machinery, responding to messenger RNA (mRNA) injection by synthesising its encoded protein. Two days following injection with mRNA from *Torpedo* electric organ, such oocytes responded to ionotophoretic ACh application with a rapid depolarisation (Figure 7.13; Barnard *et al.*, 1982). Furthermore, selective expression of mRNA encoding the different individual AChR subunits demonstrated such responses only with injections of mRNAs for all four of these AChR subunits. This showed that all four of the subunits were necessary to produce a functional receptor. It also showed that no extra molecular components were required.

7.8 | Specific Pharmacological Properties of the Neuromuscular Junction

Neuromuscular transmission thus depends on a complex sequence of events distinct from action potential conduction along a nerve fibre. Each event offers distinct pharmacological targets, often with

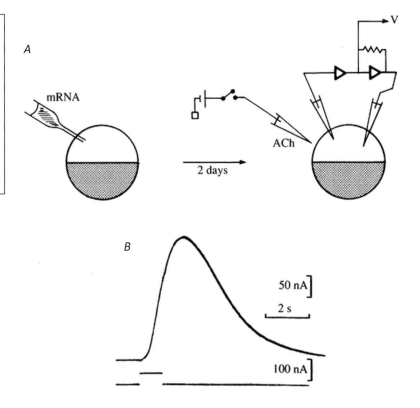

Figure 7.13 Expression of ACh receptors in *Xenopus* oocytes. Messenger RNA from *Torpedo* electric organ is injected into an oocyte; two days later the oocyte membrane potential is voltage-clamped while ACh is applied by ionophoresis (A). The record (B) shows the current response (upper trace) to ACh; the lower trace monitors the ionophoresis current. (From Barnard *et al.*, 1982.)

important clinical, therapeutic and toxicological implications. Of agents directed at *presynaptic* processes, botulinum toxin prevents ACh release. In contrast, α-latrotoxin from black widow spider venom causes a massive release of presynaptic transmitter from its available pool. This is similarly followed by a neuromuscular block.

Of antagonists to the *postsynaptic* AChR, the active constituent of curare, D-tubocurarine, exerts a competitive antagonism reversibly preventing AChR opening, thereby reducing the amplitudes of post-synaptic responses (Section 7.3). Such compounds are useful as muscle-relaxing agents in surgery. Usefully for experimental work, α-bungarotoxin produces a high-affinity, relatively irreversible AChR block. Its use facilitated AChR localisation and quantification (Section 7.7). In contrast, nicotine acts as an AChR agonist, replicating the effects of its natural ligand, ACh.

The blocker of synaptic cleft acetylcholinesterase, eserine, also called physostigmine, as well as a number of organophosphorus insecticides increase the persistence of released ACh within the synaptic cleft. This increases and prolongs EPSCs and their consequent EPPs, reflecting important physiological effects of acetylcholinesterase in shortening the duration of postsynaptic responses.

The autoimmune condition *myasthenia gravis* is characterised by fatigable muscle weakness. It is typically associated with production of immunoglobulin G auto-antibodies against the AChR. In 10% of cases it results from auto-antibodies against the muscle-specific tyrosine kinase involved in formation and maintenance of the neuromuscular junction. Auto-antibody action results in inflammatory processes that both reduce the number of AChRs and flatten the postsynaptic membrane folds. The combined result is a reduction in transmitter–receptor interaction. This compromises EPP and consequent action potential generation (Querol and Illa, 2013; Verschuuren *et al.*, 2016).

8

Synaptic Transmission in the Nervous System

Central nervous system function heavily depends on synaptic inter-actions between its constituent nerve cells. In most cases, in common with neuromuscular transmission (Chapter 7), this synaptic transmission involves chemical mechanisms. The presynaptic cell releases a chemical transmitter which produces a response in the postsynaptic cell. Different chemically transmitting synapses vary in anatomical details, but all share particular features. The presynaptic terminal contains transmitter packaged in synaptic vesicles. Pre- and postsynaptic membranes are separated by a synaptic cleft into which the vesicle contents discharge. The postsynaptic membrane contains specific receptors for the neurotransmitter concerned. Acetylcholine (ACh) is only one of a range of different central neurotransmitters (Figure 8.1). It was initially thought that any one cell would only release one neurotransmitter, but cases involving simultaneous release of two transmitters are now known. Electrically transmitting synapses also exist, which we shall consider briefly.

As with the neuromuscular junction, much of our understanding of synaptic physiology was derived from intracellular microelectrode recordings. Much such fundamental work was performed by J. C. Eccles and his colleagues on feline spinal motor neurons (Eccles, 1964), so we shall begin our account of synapses between neurons with these.

8.1 | Synaptic Excitation in Motor Neurons

α-Motor neurons, the nerve cells which directly innervate skeletal muscle fibres, provided the initial model for studies of mammalian synaptic physiology. Their cell bodies lie in the ventral horn of the spinal cord, from which their axons pass through the ventral roots to reach peripheral nerves. Their cell bodies, or somas, are ~70 μm across, and extend into a number of fine branching processes, dendrites, which are up to 1 mm long. The surfaces of both soma and dendrites are covered with small presynaptic nerve terminals. These regions of contact show typical features of chemically

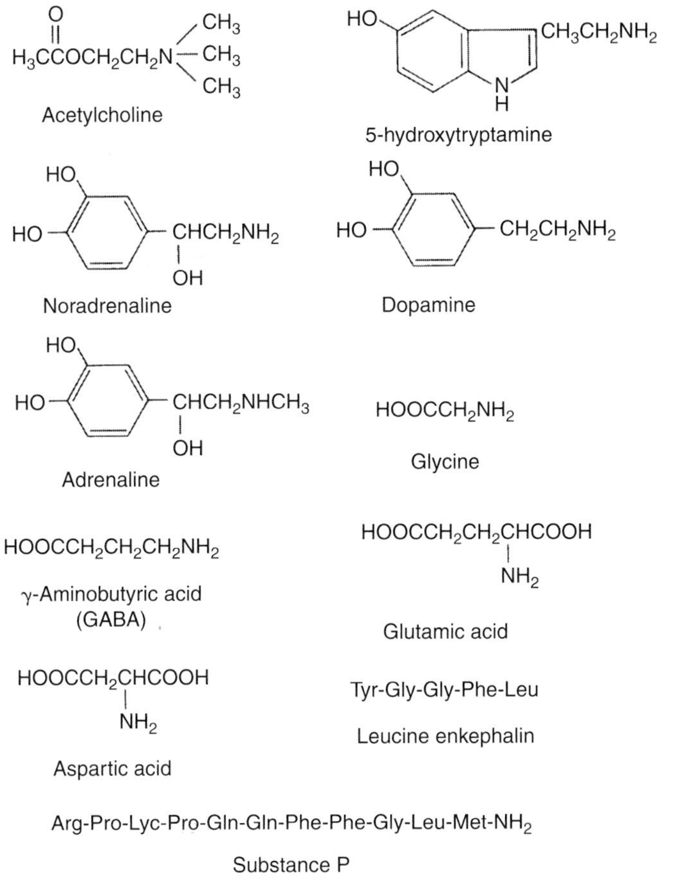

Acetylcholine

5-hydroxytryptamine

Noradrenaline

Dopamine

Adrenaline

$HOOCCH_2NH_2$

Glycine

$HOOCCH_2CH_2CH_2NH_2$

γ-Aminobutyric acid
(GABA)

$HOOCCH_2CH_2CHCOOH$
NH_2

Glutamic acid

$HOOCCH_2CHCOOH$
NH_2

Aspartic acid

Tyr-Gly-Gly-Phe-Leu

Leucine enkephalin

Arg-Pro-Lyc-Pro-Gln-Gln-Phe-Phe-Gly-Leu-Met-NH_2

Substance P

Figure 8.1 Examples of transmitter substances in the central nervous system. (From Ryall, 1979.)

transmitting synapses: a synaptic cleft and synaptic vesicles in the presynaptic terminals.

Intracellular recording demonstrates that motor neurons have a resting potential of ~−70 mV. Membrane depolarisation by ~10 mV triggers an action potential which propagates along the axon to the nerve terminals. Experiments involving injection of various ions into motor neurons indicate that the ionic basis of their resting and action potentials closely resembles those of squid axons (Sections 3.3, 3.6, 4.2). The resting potential is slightly less positive than the K^+ Nernst potential, the action potential is primarily caused by a regenerative increase in Na^+ permeability and the ionic gradients that sustain these potentials depend on active Na^+ extrusion.

8.2 | Excitatory Postsynaptic Potentials

Some of the presynaptic terminals ending on any particular motor neuron represent endings of sensory axons (known as group Ia fibres)

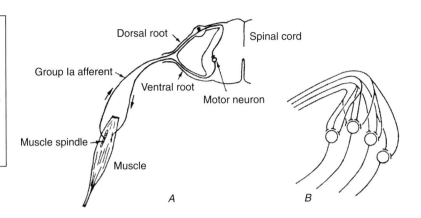

Figure 8.2 Anatomical organisation of monosynaptic stretch reflex. (A) Simplified diagram: very many stretch receptors and afferent and efferent neurons are actually associated with each muscle. (B) Branching of afferent fibres prior to synapse formation with different members of the motor neuron pool.

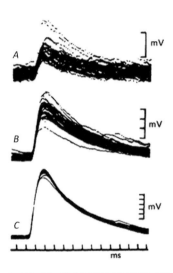

Figure 8.3 Excitatory postsynaptic potentials (EPSPs) recorded from feline spinal motor neuron in response to stimuli of increasing intensity (from A to C) applied to group Ia afferent fibres from the muscle. (From Coombs et al., 1955a.)

from muscle spindles in the muscle innervated by the particular motor neuron. Stretching the muscle excites these axons, which may excite the motor neurons supplying the muscle so that it contracts. This monosynaptic reflex (Figure 8.2) is exemplified by the knee-jerk reflex.

The corresponding postsynaptic responses of these motor neurons can be studied by microelectrode recordings in the soma. Stimulation of those group Ia fibres which synapse with the particular motor neuron studied produces brief depolarisations termed *excitatory postsynaptic potentials* (EPSPs) (Figure 8.3). Their waveforms resemble endplate potentials (EPPs) from curarised skeletal muscle fibres (Section 7.3), consisting of fairly rapid rising phases followed by slower returns to the resting potential.

Each EPSP arises from action potentials in a number of presynaptic fibres. Low-intensity stimulation applied to the nerve from the muscle excite only a few of the available group Ia fibres, giving a correspondingly small EPSP. Increased stimulus intensities excite progressively more group Ia fibres, giving a larger EPSP. Thus, responses produced by activity at different synapses on the same motor neuron add together, resulting in their *spatial summation*. A second EPSP elicited before full decay of the previous response similarly gives an enhanced net depolarisation as the second EPSP adds to the first resulting in their *temporal summation*.

Once of sufficiently large amplitude, an EPSP will cross the threshold for and result in production of an action potential. The latter then propagates along the axon of the motor neuron to the periphery where it ultimately produces contraction of the muscle fibres that it innervates (Section 7.1).

The membrane potential of a motor neuron can be altered by inserting a double-barrelled microelectrode and passing current down one barrel and recording the membrane potential through the other barrel. When the membrane potential is progressively depolarised, the EPSP decreases in size and eventually reverses in sign.

The reversal potential is ~0 mV. This suggests that the EPSP results from changes in ionic conductance independent of membrane potential, in common with the skeletal muscle EPP, likely involving Na^+ and K^+ (Section 7.5).

These electrophysiological features, together with the existence of synaptic vesicles in the presynaptic terminals, suggest that the EPSP is produced by neurotransmitter release from the group Ia terminals. Evidence particularly from cultured spinal neurons implicate glutamate as the transmitter which binds to postsynaptic membrane glutamate receptors, leading to opening of their intrinsic ion channels.

Glutamate receptors have been cloned so that some deductions can be made about their structure. The subunits have three membrane-crossing segments and one (M2) that dips into and out of the membrane from the cytoplasmic side. This structure differs from that of the nicotinic AchR, with which it shows no homology in amino acid sequence. It forms a channel made up of four or perhaps five subunits, likely surrounding a central pore.

8.3 | Inhibition in Motor Neurons

If contraction of a particular limb muscle is to be effective in producing movement, it is important that its antagonist muscles relax. In the monosynaptic stretch reflex this is brought about by *inhibition* of the motor neurons of the antagonist muscles. Figure 8.4 shows the arrangement of the neurons involved. We have seen that group Ia fibres from the stretch receptors in a particular muscle synapse with motor neurons innervating that muscle. They also synapse with small interneurons which themselves innervate the motor neurons of

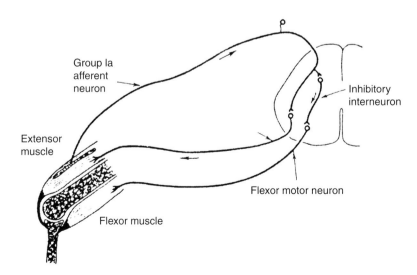

Group Ia
afferent
neuron

Inhibitory
interneuron

Extensor
muscle

Flexor motor neuron

Flexor muscle

Figure 8.4 Direct inhibitory pathway. Simplified diagram: many afferent, inhibitory and efferent neurons are involved with each inhibitory interneuron in turn innervated by several afferents, and itself innervating several motor neurons.

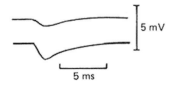

5 mV

5 ms

Figure 8.5 Inhibitory postsynaptic potentials (IPSPs) in cat spinal motor neuron, produced by stimulating group Ia fibres from the antagonistic muscle. The stimulus intensity was larger for the lower than for the upper trace, thereby exciting more group Ia fibres. (From Coombs *et al.*, 1955b.)

antagonistic muscles. It is these interneurons that exert the inhibitory action on the motor neurons.

This inhibitory action can be examined by inserting a microelectrode into a motor neuron and stimulating the group Ia fibres from an antagonistic muscle. These demonstrate small hyperpolarising potentials termed *inhibitory postsynaptic potentials* (IPSPs) (Figure 8.5).

IPSPs resemble EPSPs in form, but produce hyperpolarising rather than depolarising membrane potential deflections. Displacing the motor neuron membrane potential produces approximately linear changes in IPSP amplitude, with a reversal potential at ~ −80 mV. This is close to the Nernst potentials of both Cl^- and K^+. However, injecting Cl^- into the soma immediately reduces the reversal potential, resulting in the IPSP assuming a depolarising rather than hyperpolarising direction. This implicates increases in postsynaptic membrane Cl^- conductance in IPSP generation.

We can conclude that the mechanism producing the spinal motor neuron IPSP resembles that of the EPSP and the EPP. An action potential arriving at the presynaptic terminal causes release of transmitter, likely glycine in this case, into the synaptic cleft. The glycine combines with glycine receptors whose intrinsic ion channels open to allow Cl^- flow into the postsynaptic cell, so producing the IPSP.

In the brain itself most of the inhibitory responses are produced not by glycine, but by gamma-aminobutyric acid (GABA). GABA receptors are of two types: $GABA_A$ receptors have intrinsic ion channels, whereas $GABA_B$ receptors do not. The structure and properties of the $GABA_A$ receptors resemble those of glycine receptors. Subunits of $GABA_A$ and glycine receptors have similar sequences that show some identity with those of nicotinic ACh receptors (AChR). These three receptors form a gene family sharing a common evolutionary origin. It is likely that each receptor consists of five subunits surrounding a central pore, just as in the nicotinic AChR.

In common with EPSPs, IPSPs show spatial and temporal summation.

8.4 | Interaction of IPSPs with EPSPs

The amplitude of the EPSP-mediated depolarisation is reduced in the presence of a temporally overlapping IPSP. If the EPSP were just large enough to elicit an action potential by itself, the IPSP may reduce its amplitude below threshold for action potential initiation (Figure 8.6). A motor neuron prevented from producing an action potential cannot induce muscular contraction and so it is effectively inhibited.

The motor neuron acts as a decision-making device as to whether to initiate an action potential that propagates along the axon to its innervated muscle. If the incoming excitatory synaptic action is sufficiently in excess of the incoming inhibitory action, the resulting depolarisation crosses the threshold for action potential production and the motor neuron is excited. But reduced synaptic excitation or increased synaptic inhibition would result in a more negative membrane potential below

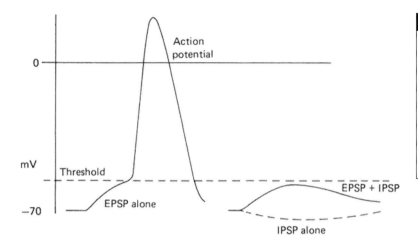

Threshold

EPSP alone

IPSP alone

EPSP + IPSP

Action potential

mV

0

−70

Figure 8.6 Interactions between excitatory and inhibitory postsynaptic potentials in the motor neuron. An EPSP just large enough to cross the threshold, thereby excites an action potential. When an IPSP occurs at the same time, the combined result is insufficient to cause excitation, and so no action potential is propagated out along the axon.

the threshold. This would prevent or terminate motor neuron excitation. The motor neuron receives excitatory and inhibitory inputs from many sources. A 'decision' based on inhibition from group Ia fibres from an antagonistic muscle may be 'over-ruled' by excitatory inputs for neurons descending from the brain.

8.5 | Presynaptic Inhibition

The inhibitory process described above involves producing hyperpolarising postsynaptic responses. Such postsynaptic inhibition accounts for most inhibitory interactions between nerve cells. In some cases, however, inhibition occurs in the absence of any postsynaptic responses to the inhibitory input alone (Figure 8.7*Aa*, *Ab*). The inhibition is attributed to synaptic inputs to the presynaptic terminal that reduce the amplitude of the presynaptic action potential, thereby reducing the number of transmitter quanta released (Figure 8.7*B*). Electronmicroscopy shows the presence of *serial synapses* (Figure 8.7*C*) in agreement with this view. This form of interaction is termed presynaptic inhibition.

8.6 | Slow Synaptic Potentials

Dale used pharmacological criteria to distinguish two types of response to ACh in peripheral tissues. *Nicotinic* responses were mimicked by nicotine and blocked by curare. *Muscarinic* responses were mimicked by muscarine and blocked by atropine. Correspondingly, there are two distinct, nicotinic and muscarinic, AChR types. Nicotinic receptors occur at skeletal muscle neuromuscular junctions. Amongst other actions, muscarinic receptors mediate responses of cardiac (Section 13.18) and smooth muscle (Section 15.9) to parasympathetic stimulation. Both receptor types occur in sympathetic ganglia, where they produce different types of responses.

Figure 8.7 Presynaptic inhibition. (A) (a) EPSP produced in a motor neuron in response to stimulation (at time E) of its innervating group Ia fibres. (b) When a suitable inhibitory nerve is stimulated just beforehand (at I), the EPSP is reduced in size, although there is no IPSP or other postsynaptic event associated with the inhibitory stimulation. (B) The probable underlying neural pathways. (C) The underlying serial synapses. ((A) from Eccles, 1964.)

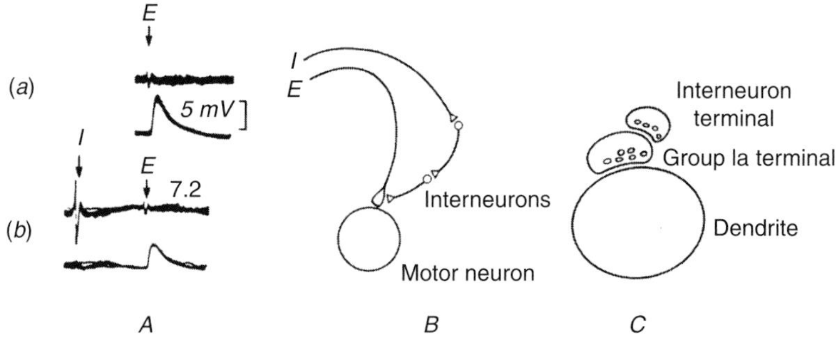

Figure 8.8 Fast and slow synaptic responses in frog sympathetic ganglion neuron. (A) Left trace: the fast EPSP produced by a single preganglionic stimulus. Right trace: a stronger stimulus excites more preganglionic fibres giving a larger EPSP sufficient to produce an action potential. (B–D) Responses observed when the fast EPSP is blocked by a curare-like compound; repetitive stimulation at various sites produces three different types of slow response. Note different time scales. (From Kuffler, 1980.)

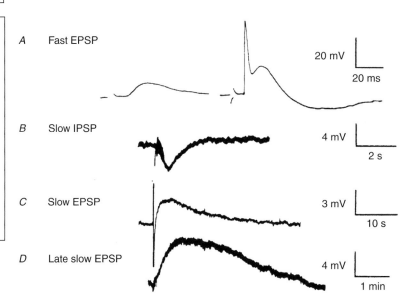

Postsynaptic cells in bullfrog sympathetic ganglia show different types of synaptic activity (Figure 8.8). A single stimulus to the preganglionic fibres produces a fast EPSP (Figure 8.8A) potentially large enough to stimulate a postganglionic action potential. The response is blocked by curare and mimicked by ACh. Thus the mechanism of production of the fast EPSP resembles that at the neuromuscular junction. It is mediated by nicotinic AChRs in which a cation-selective channel opens on ACh binding.

In some cells a slow EPSP with a much longer timecourse follows the fast EPSP. Similar responses are seen after ACh application. However, this slow EPSP is unaffected by curare, but blocked by atropine, implicating muscarinic rather than nicotinic receptor mediation (Figure 8.8C). Conductance measurements show that the slow EPSP is produced by the closure of K^+-selective ion channels.

In other cells the fast nicotinic EPSP is followed by a slow, hyper-polarising IPSP. This is also mediated by muscarinic AChRs, probably involving K^+ channel opening (Figure 8.8B). Finally, a long period of repetitive stimulation of the preganglionic fibres produces a depolar-isation which lasts for a few minutes; it is called the late slow EPSP (Figure 8.8D). The neurotransmitter which produces this is a peptide similar in structure to the luteinizing-hormone releasing hormone.

Slow potentials occur widely. Their timecourse and their long latency could be explained if channel opening or closing is mediated by an indirect process involving intermediate steps between binding at the receptor and the response of the channel, rather than the direct link which occurs in fast-acting receptors with intrinsic channels. The intermediate steps involve the activation of G-proteins and often the production of intracellular second messengers.

This second messenger concept was first introduced to clarify the role of cyclic 3', 5'-adenosine monophosphate (cAMP) in hormone action. Hormone combination with its receptor activates a G-protein. The term G-protein refers to its requiring guanosine triphosphate (GTP) binding for its activation, which in turn activates the enzyme adenylate cyclase. This produces cAMP which alters the physiological properties of the cell in some way, such as by opening or closing ion channels. Neurotransmit-ters may act in a similar fashion, or may utilise a different second messenger such as inositol *tris*phosphate, IP_3. In some cases the G-protein may act directly on the membrane channel without producing a second messenger. Figure 8.9 summarises the various mechanisms through which neurotransmitters may affect channels, and Table 8.1 outlines some of the neurotransmitter receptors involved.

Table 8.1 | *Some neurotransmitter receptors*

Transmitter	Ionotropic	Metabotropic
Acetylcholine	Nicotinic receptors	Muscarinic receptors M_1 to M_5
GABA	$GABA_A$ receptor	$GABA_B$ receptor
Glycine	Glycine receptor	–
5-hydroxytryptamine	$5-HT_3$ receptor	$5-HT_{1,2,4}$ receptors
Glutamate receptors	NMDA, AMPA and kainate receptors	$mGluR_1$ to $mGluR_5$ receptors
ATP	P_{2X} receptor	P_{2Y} receptor
Noradrenaline	–	$\alpha_1, \alpha_2, \beta_1$ and β_2 receptors
Dopamine	–	D_1-like and D_2-like receptors
Neuropeptides	–	Rhodopsin-like (e.g. substance P, enkephalin) Glucagon-receptor-like (e.g. VIP)

Abbreviations: GABA, gamma-amino butyric acid; 5-HT, 5-hydroxytryptamine; AMPA, α-amino-3-hydroxy-5-methyl-4-isoxazolepropionate; NMDA, N–methyl-d-aspartate; VIP, vasoactive intestinal polypeptide. Ionotropic receptors have their own intrinsic ion channel and mediate fast synaptic transmission. Metabotropic receptors activate ion channels indirectly via G-proteins and (usually) second messenger systems.

> **Figure 8.9** Direct and indirect neurotransmitter actions on ion channels. (A) The direct action which occurs when the ion channel is an integral part of the receptor, as in the nicotinic acetylcholine receptor (AChR). (B, C) The receptor molecule acting indirectly via through G-protein activation. The G-protein either (B) acts directly on the channel to open or close it or (C) activates an enzyme that generates a second messenger such as cyclic AMP which itself alters the state of the channel. (From Aidley, 1998.)

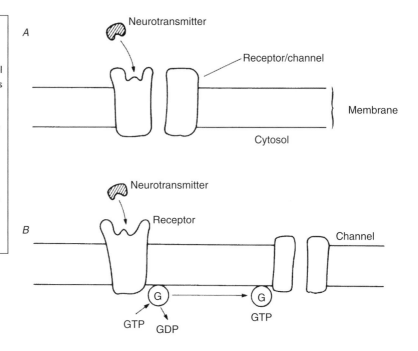

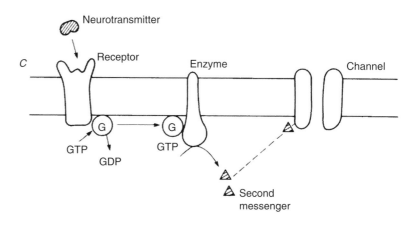

8.7 | G-Protein-Linked Receptors

The best known of the neurotransmitter receptors acting via G-proteins are the muscarinic AChRs and β-adrenergic receptors (β-ARs). There are a number of subtypes of each of these (Table 8.1). Molecular cloning techniques demonstrate that the muscarinic AChR and the β-AR (Figure 8.10) have similar structures, with identical amino acids at 30% of their residues. Their amino acid sequences are also surprisingly similar to that of the visual pigment rhodopsin, with 23% homology in each case. All three molecules are members of a large receptor superfamily and all have seven hydrophobic

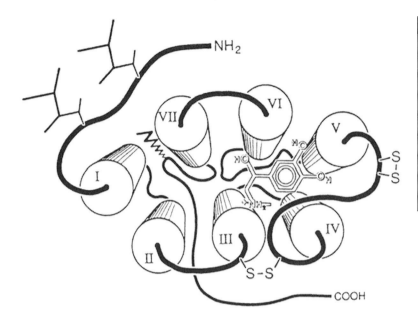

Figure 8.10 Probable arrangement of transmembrane helices in the β_2-adrenergic receptor (β-AR), with adrenaline in the binding site. The molecule is drawn as seen from the outer side of the plasma membrane. (From Ostrowski *et al.*, 1992. Reprinted, with permission, from the Annual Review of Pharmacology and Toxicology, Volume 32 © 1992 by Annual Reviews www.AnnualReviews.org.)

transmembrane segments. Different members of the superfamily respond specifically to particular neurotransmitters, neuropeptides, hormones, olfactory stimulants or, for rhodopsin and other visual pigments, the retinal isomerisations produced by light.

G-proteins consist of three subunits, α, β and γ. At rest they assume a trimeric, $\alpha\beta\gamma$, form with guanosine diphosphate, GDP, bound tightly to the α-subunit. Neurotransmitter molecule binding triggers receptor interactions with the G-protein that make it release the GDP and bind GTP in its $\beta\gamma$ subunit (Figure 8.9*B*). The α-subunit binds to and activates an effector molecule. Sometimes the $\beta\gamma$ subunit also acts as an activator. Fatty acid chains attached to the two subunits keep them in contact with the plasma membrane, so that they can shuttle between the receptor and the effector molecules.

The effector molecule is sometimes an ion channel, as occurs in the muscarinic action of ACh on the heart (Figure 8.9*B*). More commonly it is an enzyme whose activation leads to the production of a second messenger. The membrane-bound enzyme adenylyl cyclase is activated by certain G-proteins in order to synthesise cAMP from ATP. The cAMP activates the enzyme protein kinase A (PKA), which catalyses phosphorylation of some target molecule such as an ion channel (Figure 8.9*C*).

An alternative route for G-protein action is the phosphatidyl inositol signalling system. Here the G-protein activates the enzyme phospholipase C, which hydrolyses the membrane phospholipid phosphatidyl inositol to produce diacyl glycerol and inositol *tris*phosphate (IP_3). The IP_3 acts as a second messenger by activating Ca^{2+}-release channels in the endoplasmic reticular membrane, thereby raising $[Ca^{2+}]_i$ (see also Section 15.9).

These signaling systems clearly are capable of considerable amplification. One receptor molecule binding a single neurotransmitter molecule can activate a number of G-protein molecules, and each activated effector molecule will produce several second messenger molecules.

β-ARs also mediate many of the responses to noradrenaline in smooth and cardiac muscle cells (Sections 13.18 and 15.9).

8.8 | Electrical Synapses

In an *electrotonic* synapse, the presynaptic cell excites the postsynaptic cell directly through electric current flow as opposed to chemical transmission. Such synapses were first discovered in multi-cellular giant fibres involved in escape responses of crayfish and earthworms, where they directly transmit action potentials between cells. They also occur between some cells in the central nervous systems of mammals and other animals, where they can synchronise firing in adjacent cells. They are also essential for electrical transmission between cardiac (Section 13.1) and smooth muscle cells (Section 15.6).

Electronmicroscopy of electrotonic synapses demonstrates gap junction regions where the intercellular space between the two cells is much narrower than usual, being ~2 nm instead of 20 nm. These contain channels that directly connect the pre- and postsynaptic cells and permit current flow from one cell to the other. Each gap junction channel is composed of a pair of hexamers, one in each of the two apposed membranes (Figure 8.11). The hexamers are known as connexons, and their individual subunits are proteins called connexins.

Figure 8.11 Structure of gap junctions deduced from X-ray diffraction studies on material isolated from mouse liver cells. (From Makowski et al., 1984.)

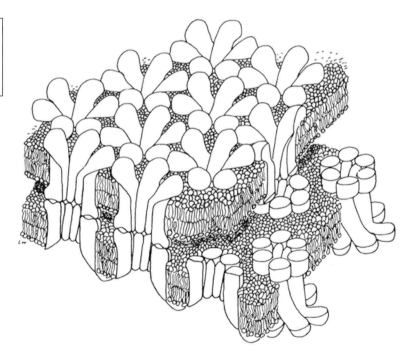

Certain excitable systems additionally show specialised sites mediating electrical, *ephaptic*, coupling across narrow extracellular clefts separating adjacent cells. These sites occur by themselves or, in cardiomyocytes, in association with connexon-mediated cell-cell coupling regions (Veeraraghavan et al., 2014). Na^+ channel opening in the membrane of an activated cell lining the cleft withdraws extracellular Na^+, depleting positive charge from the cleft. This lowers the local extracellular potential. This depolarises the membrane potential of the neighbouring quiescent cell also lining the cleft, activating its Na^+ channels, thereby triggering a fresh action potential and contributing to conduction of cell-cell excitation (Salvage *et al.*, 2020a).

8.9 | Long-Term Potentiation and Depression

Postsynaptic responses to presynaptic stimulation in central nervous system synapses also show long-lasting alterations with repeated stimulation. These are likely important in learning and long-term memory. Of these, long-term potentiation (LTP) was first demonstrated at synapses between the medial perforant path and granule cells of the dentate gyrus in hippocampi of anaesthetised rabbits (Bliss and Lomo, 1973). It followed brief high-frequency action potential trains delivered to their afferent, presynaptic, pathway. Postsynaptic responses in the population of activated granule cells revealed a lasting enhanced synaptic strength following the near-coincident pre- and postsynaptic depolarisation. The ionotropic glutamate-specific AMPA postsynaptic receptor mediates baseline chemical transmission at such excitatory synapses (Figure 8.12A). However, an additional postsynaptic, voltage-dependent, ionotropic NMDA receptor, permits Ca^{2+} entry in the simultaneous presence of: (1) the resulting membrane depolarisation which

Figure 8.12 Mechanisms of long-term potentiation. (*A*) Glutamate release activates the AMPA receptor; the NMDA receptor is blocked by Mg^{2+} at the resting potential. (*B*) AMPA activation causes net depolarising monovalent ion influx. The resulting depolarisation dissociates the blocking Mg^{2+} permitting Ca^{2+} influx through the NMDA receptor. (*C*) The latter Ca^{2+} influx increases cytosolic Ca^{2+}. This binds to Ca^{2+}/calmodulin initiating an activation of kinase and phosphatase enzymes which vary with the amplitude and waveform of the Ca^{2+}/calmodulin transient. Under certain circumstances these also result in an increase in AMPA receptor trafficking to the membrane and AMPA receptor phosphorylation, both of which result in long-term increases in synaptic efficacy.

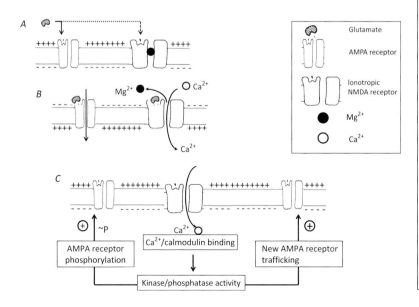

ejects bound Mg^{2+} otherwise blocking the NMDA channel at near-resting membrane potentials and (2) binding of the presynaptically released glutamate neurotransmitter. The latter elicits NMDA channel opening (Figure 8.12B; Bliss and Cooke, 2011).

Besides triggering release of intracellularly stored Ca^{2+}, the entering Ca^{2+} binds to the intracellular Ca^{2+}-sensing protein calmodulin (CaM). CaM possesses two high-affinity C-terminal and two low-affinity N-terminal Ca^{2+}-binding EF hand groups. Ca^{2+} binding converts the hydrophilic apoCaM to the hydrophobic CaM form. The latter activates a profile of different enzymic kinases and phosphatases that varies with the amplitude and kinetics of the postsynaptic Ca^{2+} transient. This can include direct actions on calmodulin kinase II (CaMKII), phosphokinase C (PKC) and calcineurin (CN), and indirect actions on phosphokinase A (PKA) and protein phosphatase 1 (PPA1) (Section 13.18). Whereas brief, steep transients induce LTP, low and prolonged Ca^{2+} transients appear to induce an, opposite, long term depression (LTD) effect.

Downstream, these mechanisms respectively enhance or reduce AMPA receptor conductance through phosphorylation or dephosphorylation (Figure 8.12C). They also increase or decrease AMPA receptor trafficking to the postsynaptic membrane. In addition, retrograde messengers from the postsynaptic membrane, possibly nitric oxide or endocannabinoids, may signal the presence of the coincident excitation to the presynaptic terminal and modify transmitter release probability (Tsien and Malinow, 1991). The relative pre- and postsynaptic contributions to the potentiation effects may vary with time after initial induction and between different classes of synapse. The resulting up- or down-regulated synaptic transmission promotes LTP or LTD.

Glutamate binding to a further, group 1, class of metabotropic glutamate receptors (mGluR) is thought also to promote LTD by initiating a phospholipase C (PLC)-mediated breakdown of phosphatidylinositol 4,5-bisphosphate (PIP_2) to inositol tris phosphate (IP_3). This triggers release of intracellularly stored Ca^{2+} and diacylglycerol (DAG) (Figure 8.9C). This leads to activation of the Ca^{2+}-sensitive PKC, whose distinct mode of AMPA receptor phosphorylation reduces AMPA conductance.

Finally, longer-term LTP and LTD effects may be mediated by brain-derived neurotrophic factor (BDNF) or synthesis of new proteins through novel gene transcription or local translation of existing transcripts. These may involve cAMP-dependent signalling cascades in turn potentially affected by catecholaminergic inputs.

8.10 | Glial Buffering of the Interstitial Space Following Neuronal Activity

Activity in central nervous system neurons also results in interactions involving their associated nonexcitable glial cells. This takes place through liberation of both K^+ and neurotransmitter, particularly

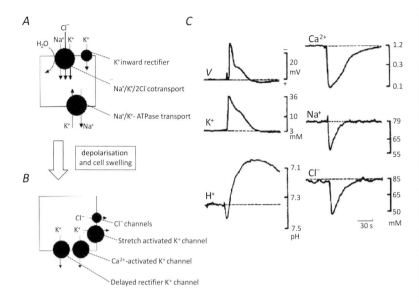

Figure 8.13 Processes producing (*A*) inward K^+ movement into glial cells and (*B*) subsequent K^+ release at remote sites with low $[K^+]_o$. (*C*) Transient extracellular direct current (d.c.) potential (marked *V*) and extracellular ionic (K^+, H^+, Ca^{2+}, Na^+, Cl^-) changes during spreading depression (SD) in rat cerebellum. Vertical scales logarithmic. Cerebellum preconditioned by reducing extracellular NaCl content and spreading depression induced by a train of local electrical stimuli. Note early increase in $[K^+]_o$, initial decrease in $[H^+]_o$ (alkalinisation) followed by acidification, and transient reductions in $[Ca^{2+}]_o$, $[Na^+]_o$ and $[Cl^-]_o$. ((A) and (B) from Figure 5 of Smith *et al.*, 2006; (C) from Figure 10 of Martins-Ferreira *et al.*, 2000.)

glutamate, with neuronal activity. Extracellular K^+ accumulation increases the extracellular $[K^+]_o$ within the restricted interstitial space of brain tissue. This would be expected to depolarise the cells that released it. Two- to threefold increases of $[K^+]_o$ would produce ~+30 mV shifts in membrane potential (Section 3.3). Increased extracellular glutamate levels could enhance these postsynaptic depolarisation events, further increasing $[K^+]_o$.

However, both extracellular K^+ and glutamate are normally cleared by astrocytes, a glial cell subtype making close contact with excitable neurons. Astrocytes mediate inward K^+ fluxes across their surface membranes through active Na^+,-K^+-ATPase transport, as well as other forms of transport process, including Na^+-K^+-$2Cl^-$ cotransport, and K^+ channel transport (Figure 8.13*A*). These ion fluxes are often osmotically accompanied by a water influx causing cell swelling (Section 3.7). Glial cell membranes facing nerve terminals and axons are also abundant in glutamate transporters.

This buffering by K^+ uptake is supplemented by movement of the intracellular K^+ to other neighbouring astrocytes through gap junctions (Section 8.8). This *spatial buffering* process may be terminated by efflux of the intracellular K^+ from cells in remote regions with a low $[K^+]_o$, through delayed rectifier, or Ca^{2+}- or stretch-activated, K^+ channels (Section 10.2). In these situations, the glial cells liberate K^+ back into the interstitial space. Such events are enhanced under conditions with abnormally high interstitial $[K^+]_o$ (Figure 8.13*B*). The latter condition depolarises the astrocytes. This also promotes K^+ release into the extracellular space through voltage-sensitive delayed rectifier K^+ channels, particularly abundant at their end feet, which are apposed to the neurons themselves. Depolarisation also results in glutamate release through voltage-sensitive organic anion channels. Cell swelling resulting from the ion fluxes (Sections 3.4 and 3.5) could also activate membrane stretch-sensitive K^+ channels and volume-sensitive organic anion channels. Finally, astrocyte cultures exhibit propagating intracellular waves of elevated $[Ca^{2+}]$ that may effect K^+ release through Ca^{2+}-activated K^+ channels. These abnormalities in glial cell mediated K^+ buffering may propagate through successive glial cells in cerebral cortical tissue. In clinical situations this can produce *cortical spreading depression* (CSD), implicated in migraine aura, or the peri-infarct depolarisation following stroke or brain injury.

Cortical spreading depression can follow K^+ release accompanying neuronal firing following pathological electrical, mechanical or chemical stimuli. Within any given cortical region, it produces transient but major ion redistributions between extracellular and intracellular, in particular glial, compartments, with $[K^+]_o$ rapidly rising to levels as high as 60–80 mmol/L (Figure 8.13C). This causes successive neuronal excitation and depolarisation followed by refractoriness resulting in a period of electrical silence. The result is a pathophysiological wave of neuronal hyperexcitability followed by quiescence propagating over the cerebral cortex (Smith *et al.*, 2006).

This pattern of fluid shifts and consequent alterations in water diffusion properties accompanying CSD were directly visualised by magnetic resonance imaging (MRI) measurements of their related apparent water diffusion coefficients. This was performed in whole rat and in feline cerebral cortex, which shares the gyral-folded anatomy of human brains (James *et al.*, 1999). Plate 4 exemplifies findings from diffusion-weighted echoplanar MRI following a localised stimulus at the centre of the suprasylvian gyrus (Smith *et al.*, 2001). Diffusion-weighted images of two-dimensional tangential cortical image slices were captured as a sequence of frames, whilst a primary spreading depression wave propagated towards the midline within the hemisphere in which it was initiated (Plate 4A). This was analysed by subtracting a single prestimulus baseline image, colour-coding demonstrable regions of reduced diffusion and superimposing this onto the reference image (Plate 4B). This revealed an area of increased signal, reflecting reduced diffusion spreading away from the point of the KCl stimulus application.

Quantitative mapping of apparent diffusion coefficient values (Plate 4C) visualised the size and propagation of the primary spreading depression wave. Superimposing the signal on an anatomical brain map (Plate 4D) localised these primary waves propagating over the suprasylvian gyrus spreading into the adjacent marginal gyrus. Each appeared as a coherent, elliptical wave of reduced diffusion propagating across the marginal and suprasylvian gyri. This initial primary wave was followed by further secondary events. These were more restricted in extent but could persist over substantial periods, even >20 min, following stimulus application. Such phenomena could be demonstrated with both sustained and transient stimuli, reflecting the regenerative and prolonged nature of both these experimental events and their clinical correlates.

9

The Mechanism of Contraction in Skeletal Muscle

Skeletal muscles are the engines of the body. They account for over a quarter of its weight and the major part of its energy expenditure. They are attached to and act upon the bones of the skeleton, thereby producing movement or exerting force. They are central to such activities as voluntary movement, maintenance of posture, breathing, eating, directing the gaze, and producing gestures and facial expressions. Skeletal muscles are activated by motor neurons (Sections 7.1 and 8.1). Their cells are elongated and multi-nucleate and the contractile material within them shows cross-striations. Skeletal muscle is a form of *striated muscle*. In contrast, cardiac and smooth muscles have cells with single nuclei. In addition, smooth muscles are not striated (Sections 13.1 and 15.1).

Of skeletal muscle fibre subtypes, type I, slow twitch ('red'), muscle, exemplified by soleus, has the densest vascular capillary supply, and the greatest mitochondrial and myoglobin content, imparting its redder colour. Its activity energetically depends upon an aerobic metabolic energy supply (Section 12.8). Its component fibres show relatively prolonged rates of increase of tension and prolonged contractions. Type II fast twitch ('white') muscle fibres, exemplified by extensor digitorum longus, show more rapid rates of rise of tension and shorter contraction timecourses. They fatigue more rapidly, reflecting their predominantly anaerobic, glycolytic, energy metabolism. Their three major subtypes, IIa, IIx and IIb, show detailed differences in contractile speed and force generation (Blaauw *et al.*, 2013).

9.1 | Anatomy

Skeletal muscle fibres are multi-nucleate cells formed by fusion of elongated uni-nucleate myoblast cells whose respective nuclei become arranged around the fibre periphery. Mature fibres may be as long as the muscle of which they form part, and are typically 10 to 100 μm in diameter (Figure 9.1). Bundles of muscle fibres are surrounded by a sheet of connective tissue, the perimysium, and the whole muscle is

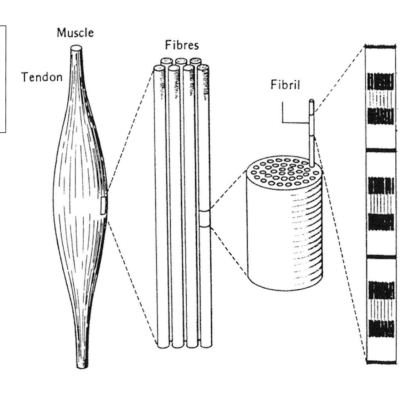

Figure 9.1 Arrangement of fibres in vertebrate striated muscle. The cross-striations on the myofibrils can be seen with light microscopy; their ultrastructural basis is shown in Figure 9.2. (After Schmidt-Nielsen, 1990.)

contained within an outer sheet of tough connective tissue, the epimysium. These connective tissue sheets are continuous with the insertions and tendons which attach the muscles to the skeleton. An extensive blood supply provides a blood capillary network between individual fibres.

The structural components of each muscle fibre reflect both its excitable and its contractile functions. The muscle fibre is bounded by its cell membrane, the sarcolemma surrounded by a thin layer of connective tissue, the endomysium. There are also extensive, specialised, internal membrane systems involved in regulating activation of mechanical activity. There is a fine system of interconnecting *transverse (T-) tubules* invaginated from the sarcolemma at regularly spaced intervals along the fibre length oriented transversely across the fibre axis (Section 10.4). In addition, a *sarcoplasmic reticulum* (SR) forms a separate intracellular membrane network of tubes and sacs acting as an intracellular Ca^{2+} storage site (Section 11.4). Muscle cells also contain organelles important in energy metabolism, cell repair and protein synthesis in common with other cells. Mitochondria, lipid and glycogen granules are particularly required in the energy metabolism that sustains contractile activity. Muscle cells also contain ribosomes and lysosomes and are abundant in a number of specific proteins such as *myoglobin*, which acts as an oxygen store, *creatinine phosphokinase*, which aids mobilisation of energy supplies, and *dystrophin*, important in preserving cell membrane integrity, abnormalities which are implicated in muscular dystrophy (Section 12.15).

9.2 | The Structure of the Myofibril

Most of the remaining fibre interior consists of the protein filaments constituting the contractile apparatus, grouped together in bundles called *myofibrils*. They have characteristic banding patterns. The bands on adjacent myofibrils are transversely aligned so that the whole fibre appears striated. In order to visualise the striations by light microscopy it is necessary to fix and stain the fibres, or to use phase-contrast, polarised light or interference microscopy.

Figure 9.2A shows the striation pattern demonstrated by these methods and by low-power electronmicroscopy. The two main bands are the dark, strongly birefringent A (anisotropic) band and the lighter, less birefringent I (isotropic) band. These bands alternate along the length of the myofibril. In the middle of each I band is a dark line, the Z line (German, *Zwischenscheibe*; *Zwischen*: spacer, *scheibe*: disk). In the middle of the A band is a lighter region, the H zone (German, *heller*: brighter), which is bisected by a darker M line (German: *Mittelscheibe; mittel:* middle). A lighter L zone in the middle of the H zone can sometimes be distinguished. The length between two Z lines is called the *sarcomere*.

Electronmicroscopy and thin sectioning techniques in the 1950s clarified the structural basis of the striation pattern (Figures 9.2B and 9.3). The myofibrils contain two sets of filaments. The thick filaments, ~11 nm in diameter, run the length of the A band. The thin filaments, ~5 nm in diameter, are attached to the Z lines and extend through the I bands into the A bands. The H zone is the region of the A band between the ends of the two sets of thin filaments. The M line is caused by cross-links between the thick filaments in the middle of the sarcomere. The thick filaments have projections from them except in a short central region which corresponds to the L zone. In the overlap region these projections may become attached to the thin filaments so as to form *cross-bridges* between the two sets of filaments (Huxley, 1963; 1976).

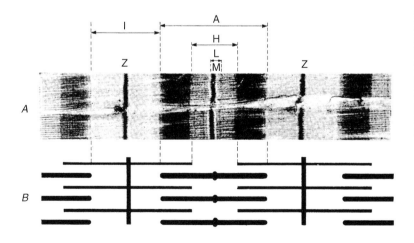

Figure 9.2 The striation pattern of a vertebrate skeletal muscle fibre as seen by electronmicroscopy of thin sections (A), and its interpretation as two sets of interdigitating filaments (B). (Photograph for (A) supplied by Dr H. E. Huxley.)

Figure 9.3 Thin longitudinal section of a glycerol-extracted rabbit psoas muscle fibre. Note cross-bridges between thick and thin filaments. (Photograph supplied by Dr H. E. Huxley.)

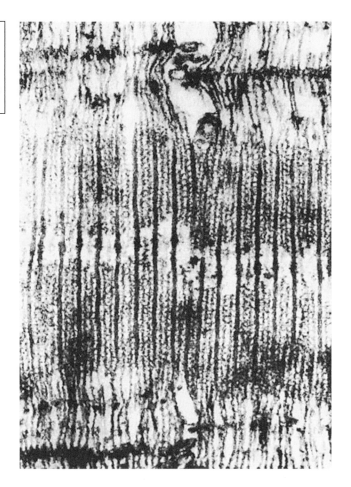

The two major proteins of the myofibrils concerned with muscle tension generation and shortening are *myosin* and *actin*. When myofibrils are washed in a solution that dissolves myosin, the A bands disappear, and further washing with a solution that dissolves actin causes the I bands to disappear. This suggests that the thick filaments are composed largely of myosin and the thin filaments largely of actin. Thin filaments also contain the proteins *tropomyosin* and *troponin*, concerned with control of contraction (Section 9.10).

9.3 | The Sliding-Filament Theory

Prior to 1954, most suggestions concerning muscular contraction mechanisms invoked the coiling and contraction of long protein molecules, akin to the shortening of a helical spring. In that year the *sliding-filament theory* was independently proposed on the basis of experiments using phase-contrast microscopy of myofibrils from glycerol-extracted muscles (Huxley and Hanson, 1954) and interference microscopy of living muscle fibres (Huxley and Niedergerke, 1954). In each case the authors showed that the A band length does not change with either muscle stretch or shortening, whether these occur actively or passively. This suggests that contraction arises from movement of the thin filaments between the thick filaments (Huxley, 1990). The sliding is thought to be caused by a series of cyclic reactions between the projections on the myosin filaments and active sites on the actin filaments. Each projection first attaches itself to the actin filament to form a cross-bridge, then pulls on it and finally releases it, moving back to attach to another site further along the actin filament. A range of subsequent studies employing a range of independent techniques converged in supporting such a theory. These are now discussed in turn.

9.4 | The Lengths of the Filaments

Electronmicroscopy of fixed and stained muscle tissue indicated that the filaments do not change in length with muscle stretch or shortening. This suggests that such muscle length changes instead involve sliding of filaments of constant length past each other. Measurements in frog muscles suggested that the thick filaments are 1.6 μm long, and that the thin filaments extend for 1.0 μm on each side of the Z line. These findings were corroborated by measurements examining diffraction patterns of an X-ray beam passing through living muscle. The resulting diffraction pattern indicated the distances between repeat units within the muscle structure. This suggested structures repeating axially at 14.3 nm and helically at 42.9 nm within the thick filaments and structures arranged on helices with pitches of 5.1, 5.9 and about 37 nm within the thin filaments. These dimensions did not change when muscle was lengthened or shortened, whether at rest or during contraction. This demonstrated that the filaments themselves did not shorten or lengthen during these changes in muscle length.

9.5 | The Relation Between Sarcomere Length and Isometric Tension

The suggestion that contraction depends on interactions between myosin and actin filaments at cross-bridges predicts that the isometric tension would be proportional to the degree of filament overlap. This idea was tested by measuring active increments of isometric tension at different known and controlled sarcomere lengths in single isolated muscle fibres, allowing for sarcomeres at the ends of a fibre assuming different lengths from those in the middle. A. F. Huxley and his colleagues developed optical servomechanism apparatus in order to maintain the contant sarcomere lengths in the middle of a fibre during an isometric contraction (Gordon et al., 1966).

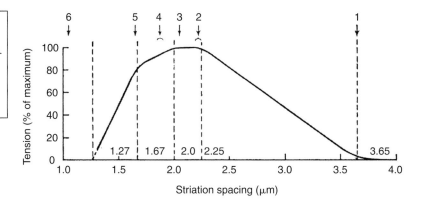

Figure 9.4 Active increment of isometric tension of frog muscle fibre at different sarcomere lengths. The numbers 1 to 6 refer to the myofilament positions shown in Figure 9.6. (From Gordon et al., 1966.)

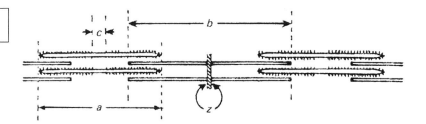

Figure 9.5 Myofilament dimensions in frog muscle.

The resulting length–tension relationship approximated a series of straight lines connected by short curved regions (Figure 9.4). There was a plateau of constant tension at sarcomere lengths between 2.05 and 2.2 μm. Above this range, tension fell linearly with increasing length, the projected line through most of the points in this region reaching zero at 3.65 μm. Below this range, tension also fell gradually with decreasing length down to about 1.65 μm, then more steeply to reaching zero at ~1.3 μm.

Matching the length–tension relationship to the dimensions of the myofilaments (Figure 9.5) supported the sliding-filament theory. The electronmicroscopy measurements indicated that the length of the myosin filaments was 1.6 μm; that of the actin filaments, including the Z line, 2.05 μm (Section 9.5). The middle region of the myosin filaments, the L-zone, which is bare of projections and cannot form cross-bridges, was 0.15 to 0.2 μm long and the thickness of the Z line was ~0.05 μm.

The length–tension diagram (Figure 9.4) could be directly related to these dimensions, beginning with long and working through to short sarcomere lengths. Above a sarcomere length of 3.65 μm (stage 1 in Figure 9.6) there should be no cross-bridges formed, and no tension development. In fact a small tension persists up to about 3.8 μm. This might reflect residual irregularities in the system. Between 3.65 μm and 2.2 to 2.25 μm, corresponding to stages 1 to 2 in both Figures 9.4 and 9.6, the number of cross-bridges would be

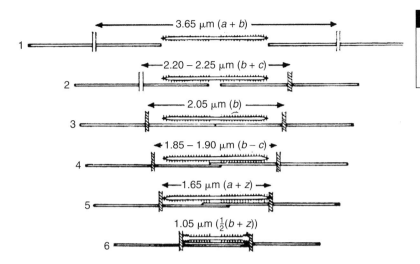

Figure 9.6 Myofilament arrangements at different lengths. The letters a, b, c and z refer to the dimensions given in Figure 9.5. (From Gordon et al., 1966.)

expected to increase linearly with decreases in sarcomere length. This correlated with the observed linear increases in isometric tension. Further shortening between stages 2 to 3 would leave the number of cross-bridges constant. This correlated with the observed plateau of constant tension. Beyond stage 3 we might expect some increase in internal resistance to shortening since the actin filaments now overlap. Beyond stage 4 the actin filaments from one half of the sarcomere might actually interfere with the cross-bridge formation in the other half of the sarcomere. Both of these would be expected to reduce isometric tension, which indeed falls at lengths below 2.0 µm. At 1.65 µm, from stage 5, the myosin filaments would hit the Z line, producing an expected considerable increase in resistance to further shortening. This correlated with a distinct kink in the tension–length curve beyond which tension fell much more sharply. The curve reaches zero tension at about 1.3 µm, before stage 6 is reached.

The experimental comparisons of myofilament structure and isometric force at different sarcomere lengths therefore provides a full quantitative confirmation of the sliding-filament theory.

9.6 | The Molecular Basis of Contraction

We have seen that the myofibrils contain a small number of different proteins. Of these myosin and actin are directly involved in the splitting of ATP and the process of contraction.

9.7 | Myosin

Myosin (molecular weight ~470 kDa) (Figure 9.7) is a complex molecule comprising six polypeptide chains: two long heavy chains and

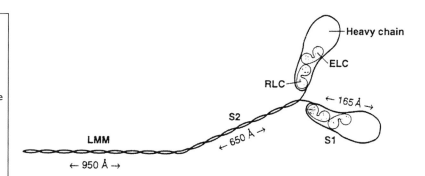

Figure 9.7 The six polypeptide chains forming the myosin molecule. The whole molecule consists of two globular heads attached to a long tail. The tail or rod is a coiled-coil formed from the α-helical regions of the two heavy chains; it is divided into a light meromyosin (LMM) and a subfragment 2 (S2) section. Each heavy chain has a globular region that combines with two light chains to form an S1 head. The light chains are of two types, essential (ELC) and regulatory (RLC). Heavy meromyosin (HMM) consists of the S1 and S2 subfragments. Enzymic activity and the molecular motor are found in the S1 heads. LMM aggregates with others to form the backbone of the myosin filament, and the S2 part of the rod connects it to the two S1 heads. (Reprinted from Rayment & Holden, 1994, Trends in Biochemical Sciences, vol. 19, The three-dimensional structure of a molecular motor, p. 129, copyright 1994, with permission from Elsevier Science.)

four short light chains. Electronmicroscopy shows that isolated myosin molecules consist of two 'heads' attached to a long 'tail'. The two heavy chains wind round each other to form the tail region, but they separate to form the two heads. The light chains, of two types called the essential (ELC) and regulatory light chains (RLC), form part of the heads.

Myosin can exert enzymic ATPase activity, hydrolysing ATP to ADP and inorganic phosphate. Treatment with the proteolytic enzyme trypsin splits myosin into two, light and heavy, meromyosin components. Electronmicroscopy revealed that heavy meromyosin has two 'heads' and a short 'tail', whereas light meromyosin is a rod-like molecule. Only the heavy meromyosin shows ATPase activity. Heavy meromyosin can be further split by papain digestion into two globular S1 subfragments forming the 'heads', and a short rod-like S2 subfragment. The ATPase activity is confined to the S1 subfragment. Light meromyosin molecules, but neither heavy meromyosin nor its two subfragments, can aggregate to form filaments under suitable conditions.

Under the right conditions, myosin molecules aggregate to form filaments with regularly spaced projections almost certainly corresponding to the projections and cross-bridges seen in thin sections of myofibrils (Figure 9.3). The middle of each filament contains a section from which these projections are absent, likely corresponding to the L zone of intact muscle fibres. Similar filaments could be isolated from homogenised myofibrils. All were 1.6 μm long, whereas the 'artificial' filaments were variable in length. Huxley suggested that the 'tails' of the myosin molecules become attached to each other to form a filament, as shown in Figure 9.8, with the 'heads' projecting from the body of the filament. This arrangement accounts for the bare region in the middle, and also suggests opposite polarities in the myosin molecules in the two halves of the filament.

X-ray diffraction studies demonstrated an axial repeat of 14.3 nm and a helical repeat of 42.9 nm on the myosin filament, as mentioned above. This suggests that a group of myosin heads emerges from the filament every 14.3 nm, and that their orientation rotates so that every third group is in line. There are probably three myosin molecules in each group, as is suggested in Figure 9.9*A*.

Figure 9.8 Asssembly of myosin molecules into a thick filament with projection-free shaft in the middle and opposite polarities of the molecules in each half of the sarcomere. (From Bagshawe, 1993, Muscle Contraction, Figure 4.7, p. 43, with kind permission from Kluwer Academic Publications.)

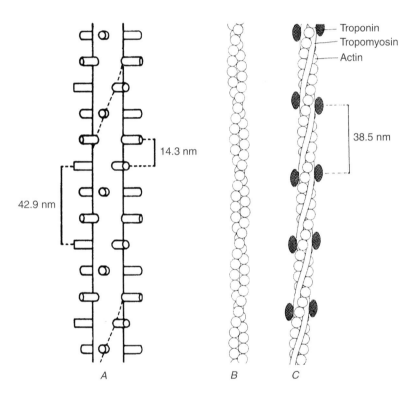

Troponin
Tropomyosin
Actin

38.5 nm

14.3 nm

42.9 nm

A   B   C

Figure 9.9 Models of the structure of the thick and thin filaments: (A) the myosin filament, (B) F-actin, (C) the thin filament. (Based on Offer, 1974 after various authors.)

9.8 | Actin

Isolated actin exists in two forms: G-actin, a more or less globular protein of molecular weight about 42 000 Da, and its fibrous polymer F-actin. Neither form shows ATPase activity, but both can combine with myosin. F-actin consists of two chains of G-actin monomers connected together in a double helix (Figure 9.9B). The thin filaments in intact muscle also contain tropomyosin and troponin, probably arranged as in Figure 9.9C.

9.9 | Interactions Between Actin, Myosin and ATP

Mixing solutions of actin and myosin results in a marked increase in viscosity reflecting formation of an *actomyosin* complex. This shows a Mg^{2+}-activated ATPase activity. 'Pure' actomyosin derived from purified actin and purified myosin splits ATP in the absence of Ca^{2+}. However 'natural' actomyosin extracted from homogenised muscle using strong salt solutions, which also contains tropomyosin and troponin, splits ATP only in the presence of a low $[Ca^{2+}]$. In the absence of Ca^{2+}, adding ATP causes natural actomyosin to decrease its viscosity, consistent with dissociation of the actin–myosin complex. These findings correlate with the absence of actin–myosin interaction and ATPase activity in the presence of an adequate $[ATP]_i$ and a very low $[Ca^{2+}]_i$, in resting muscle in vivo. Activation leading to increased $[Ca^{2+}]_i$ would lead to actin–myosin cross-bridge formation, ATPase activity, and myofilament sliding.

The amino acid sequence of the myosin heavy chain suggests that the whole of the tail section of the molecule is α-helical in structure, with the two heavy chains coiled round each other. The S1 head has a much less regular structure. X-ray diffraction studies showed that the head is divided into various functional regions. These include: an actin-binding site, an ATP-binding site and a lever arm about 10 nm long, which connects to the S2 link (see Figure 9.10). The light chains are associated with the lever arm section of the S1 head (Rayment and Holden, 1994; Rayment, 1996).

A swinging lever model suggested that the primary source of movement following cross-bridge formation is likely a change in the shape of the myosin S1 head, brought about by ATP splitting. A lever

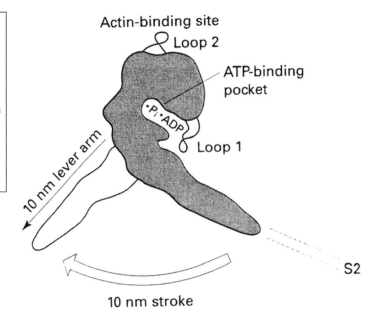

Figure 9.10 Swinging lever arm model for S1 action. A change in shape of the molecule near to the ATP-binding pocket produces a movement of about 10 nm at the end of the lever arm. This pulls on the S2 link, which is attached to the myosin filament backbone. (Redrawn after Spudich, 1994 with permission from the author and Nature, Macmillan Magazines Limited.)

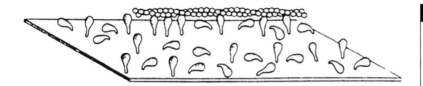

arm section swings through ~10 nm thereby pulling on the S2 link and therefore the entire myosin filament (Figure 9.10). This hypothesis was tested employing in vitro motility assays that visualised movements of single filaments under light microscopy using purified actin and myosin. One configuration of such experiments placed an actin filament labelled with a fluorescent dye on a 'lawn' of myosin S1 heads (Figure 9.11). In the absence of ATP the actin filament is bound to the S1 heads but there is no movement. On adding ATP the actin moves across the lawn at a speed comparable with the sliding of filaments in whole muscle. Use of myosin S1 mutants with either longer or shorter lever arms revealed velocities of the actin filaments that were proportional to the lengths of the lever arm (Spudich, 1994; Spudich *et al.*, 1995).

Cross-bridge action is thus a cyclical process. Each cross-bridge attaches to the adjacent actin filament and its lever arm swings thereby pulling the actin and myosin filaments past each other. It next detaches from the actin filament. The cross-bridge is then ready to attach to a new site on the actin filament and so repeat the cycle. The energy for each turn of the cycle is provided by the breakdown of one molecule of ATP to ADP and inorganic phosphate.

9.10 | The Molecular Basis of Activation

As mentioned earlier (Section 9.9), pure actin reacts with pure myosin with ATPase activity in the absence of Ca^{2+}, whereas 'natural' actomyosin containing tropomyosin and troponin, splits ATP only in the presence of Ca^{2+}. These findings implicate tropomyosin and troponin in control of muscular contraction. Tropomyosin is a fibrous protein which will bind to actin and troponin. Troponin is a globular protein with three subunits: one binds to actin, another to tropomyosin and a third combines reversibly with Ca^{2+}, undergoing a conformational change in the process.

Actin, tropomyosin and troponin occur in muscle in molecular ratios of 7:1:1. This is consistent with a model of thin filament structure (Figure 9.9C) in which tropomyosin molecules lie in the grooves between the two chains of actin monomers and a troponin molecule is attached with each tropomyosin molecule to every seventh actin monomer. The resulting repeat distance of $(7 \times 5.5) = 38.5$ nm for the troponin and tropomyosin agrees with the observation of X-ray diffractions at this distance (Section 9.4).

Figure 9.12 One of the models proposed to show how movement of tropomyosin molecules may affect actin–myosin interactions. Thin filament shown in cross-section with actin (A) and tropomyosin (TM) molecules. Two myosin S1 subunits are shown attached to the thin filaments. Tropomyosin positions are shown for the muscle at rest (dotted circle) and when active (contours). Other models give somewhat different shapes and positions for the various protein molecules, but all agree that the tropomyosin molecule moves into the 'groove' between the actin monomers on activation. It is thought that the S1 heads are unable to attach to the actin filament until this movement takes place. (From Huxley, 1976.)

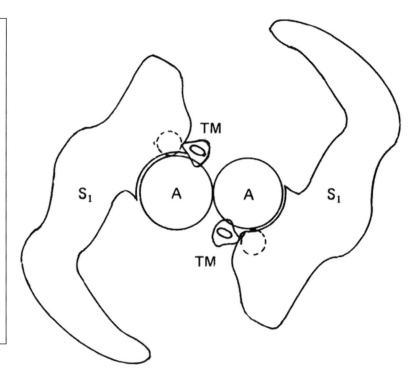

A comparison of X-ray diffraction measurements of thin filament structure under different conditions and computer simulations of such optical diffraction using electronmicrographs offer insights into the interactions between these units. In resting muscle, the tropomyosin molecules may occlude the myosin binding sites on the actin monomers, preventing actin–myosin interaction. Ca^{2+} binding by troponin causes a change in its shape. This draws its attached tropomyosin further into the groove between the two actin monomer chains. The tropomyosin molecule consequently uncovers the myosin-binding sites on seven actin monomers. This permits the myosin heads to combine with the actin, and so the muscle contracts (Figure 9.12).

9.11 | Maintenance of Structural Integrity in Contracting Sarcomeres

Additional molecules maintain anatomical integrity through these major structural changes associated with contraction in the sarcomere. The Z line contains the protein α-actinin, which binds to the actin filaments. Two large filamentous proteins also occur in the myofibril. *Nebulin* runs parallel to the thin filaments. *Titin* runs from the Z line to near the middle of the sarcomere and is bound to the thick filaments in the A band (Figure 9.13).

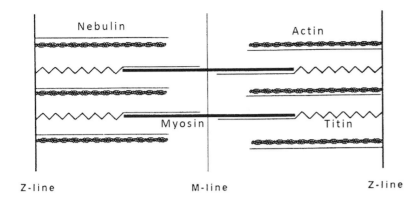

Nebulin
Actin
Myosin
Titin

Z-line   M-line   Z-line

Figure 9.13 Locations of titin and nebulin within the sarcomere. Nebulin is associated with the thin filaments. Each titin molecule is attached at one end to the Z line and at its other end is bound to a thick filament. The section in the I band is highly extensible. (From Bagshawe, 1993 Muscle Contraction, Figure 4.16, p. 56, with kind permission from Kluwer Academic Publishers.)

Titin is the largest known protein (MW ~3–4 MDa) and third most abundant muscle protein following myosin and actin. A single titin polypeptide spans the entire distance between the Z-disk and the M-line region (Plate 5A). One titin molecule exists for each actin filament. As actin filaments surround myosin in a hexagonal array in vertebrate skeletal muscles, this corresponds to about six titin filaments per half myosin.

Titin's M-line region contains its C-terminus and may have structural and kinase roles. Its N-terminal region contains binding sites for Z-disk proteins, and may have roles in assembly of the Z-disk. In the A-band region, titin is rigidly attached to myosin where it may regulate thick filament assembly and length. However, its I band region is free and elastic (Plate 5B), with its 50 nm segment just prior to insertion into the Z-band directly bound to actin. Titin thereby provides a permanent link between actin and myosin in parallel with attached cross-bridges and in series with the myosin filament.

Skeletal muscle titin contains two, tandemly arranged immunoglobulin-like (Ig) and PEVK sequences rich in proline (P), glutamate (E), valine (V), and lysine (K) domains. Cardiac muscle titin, additionally contains a N2B domain. The tandem Ig and PEVK elements predominantly extend in short and long sarcomeres respectively. In cardiac titin, extension of the N2B element dominates at the upper physiological sarcomere lengths. These three elements can be progressively recruited as a three-element molecular spring with progressive stretch of the sarcomere (Granzier and Labeit, 2007).

Apart from collagen filaments within various extracellular connective tissue layers, titin is the primary contributor to *passive* force and stiffness in cardiac and skeletal muscles. The titin segment along the I-band region is elastic, generating a passive force in response to stretch. This positions the A-band in the middle of the sarcomere, maintaining homogeneity in length between successive sarcomeres. It also allows for large elongations and passive force production that vary with titin variants associated with skeletal and cardiac muscle respectively. In cardiac muscle titin underlies the myocardial passive tension that determines diastolic filling.

Titin may also influence *active* force, particularly in cardiac muscle. Its elastic region joining the thin and thick filaments runs obliquely. This results in development of both longitudinal and radial forces, with the latter compressing the myofilament lattice (Plate 5*Ca,b*). At long sarcomere lengths, the longitudinal force slightly increases the thick filament strain. This may increase myosin head mobility. Radial compression forces may alter the distance between myosin heads and actin in contracting myocardium, thereby altering the probability of actomyosin interaction. These effects could explain the length-dependent effect of titin on maximal active tension in rat cardiac trabeculae and its Ca^{2+} sensitivity.

The Activation of Skeletal Muscle

10.1 | Background Conductances in Skeletal Muscle Membranes

Skeletal muscle resembles nerve in its membrane excitation processes, whilst including particular important further adaptations linking excitation to its contractile function. In common with nerve, a baseline Na^+, K^+-ATPase activity drives ATP-dependent efflux of $3Na^+$ in exchange for influx of $2K^+$ (Section 3.6). This results in a high extracellular $[Na^+]_o$ relative to intracellular $[Na^+]_i$ and conversely a high intracellular $[K^+]_i$ relative to extracellular $[K^+]_o$ (Sections 3.3 and 3.4). Classic microelectrode recording studies in quiescent amphibian twitch muscle fibres demonstrated resting membrane potentials E_m of typically -90 mV, close to the K^+ Nernst potential (Equation 3.1), in fast-twitch muscle studied in normal Ringer's solution (Figure 10.1A). E_m depended logarithmically upon $[K^+]_o$ (Figure 10.1B) and was unaffected by alterations in either $[Na^+]_o$ or $[Cl^-]_o$ (Figure 10.1C). These features suggest a large resting membrane permeability to K^+ relative to that of the other cations (Section 3.3; Adrian, 1956).

The latter K^+ permeability arises from a particular inward rectifying K^+ channel subtype, Kir. This was first demonstrated in skeletal muscle, but subsequently proved to exist and contribute to E_m in a wide range of excitable cell types. Kir channels lack voltage-dependent gating properties, but show nonlinear, 'anomalously', rectifying current–voltage dependences (Figure 10.1D). These properties were demonstrated through the larger voltage deflections (Figure 10.1Da, top trace) resulting from applying depolarising (Figure 10.1Da, bottom trace), compared to hyperpolarising currents across the membrane (Figure 10.1Db). They contrast with the linear current–voltage relationships shown by open voltage-gated K^+ channels in squid giant axon (Section 4.6). This rectification has been attributed to the differential block of outward compared to inward K^+ fluxes by intracellular Mg^{2+} and polyamines. Kir conductances are consequently larger at potentials close to and negative to E_K.

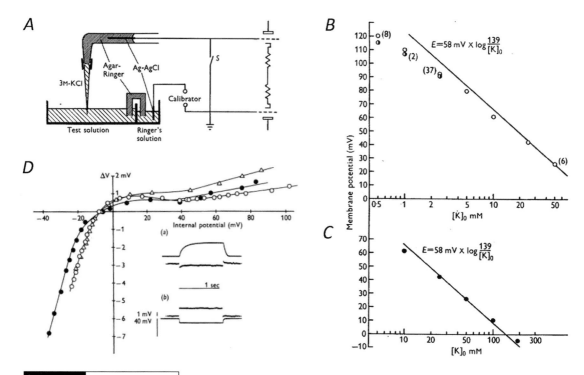

Figure 10.1 Skeletal muscle resting potentials. (A) Intracellular recording of potential differences between intracellular and extracellular compartments. (B, C) Effect of external K^+ concentration, $[K^+]_o$ (abscissa; logarithmic scale) on resting membrane potentials, E_m, (ordinates) in amphibian sartorius muscle. Readings are in (B) Cl^--containing (open symbols) or (C) Cl^--deficient (filled symbols) solutions, or in Cl^--containing solutions with half the normal $[Na^+]_o$. Solid lines: Nernst equation for K^+. (D) Current–voltage relationships using a three-microelectrode voltage clamp (see Figure 10.2) in which the ordinate ΔV approximates membrane current. Muscles in extracellular solution in which Cl^- is replaced by SO_4^{2-}. Positive ΔV denotes outward current. ((A–C) from Figures 1 and 5 of Adrian, 1956, (D) from Figure 8 of Adrian, 1964.)

They thereby stabilise E_m by increasing the current required to alter E_m. However, the slope of the current–voltage plot becomes reduced at potentials positive to E_K. This reduces the applied inward current needed to drive the membrane to the excitation threshold.

In contrast to nerve, skeletal, cardiac and smooth muscle membranes also show appreciable anionic conductances (Section 15.4). Skeletal muscle has a ClC-1 Cl^- channel that accounts for 70–80% of the resting membrane conductance. However, Cl^- is close to passively distributed across the cell membrane. Its concentrations therefore follow the Cl^- Nernst equation, with the E_{Cl} term close to the established E_m. This tends to stabilise the membrane potential E_m against effects of imposed stimuli. This stabilisation becomes apparent in experimental manipulations of $[K^+]_i$ that would otherwise negatively shift E_m. Furthermore, it is modified by pharmacological block of secondary active, Na^+-K^+-$2Cl^-$ (NKCC)-mediated inward Cl^- transport (Ferenczi et al., 2004). The Cl^- conductance also enhances recovery from action potential excitation (Sections 10.7 and 10.8). Consequently, experimental and clinical conditions such as myotonia congenita, that reduce Cl^- conductance, cause membrane instability and abnormal membrane hyperexcitability (Adrian and Bryant, 1974; Section 10.9).

10.2 | Ionic Currents Mediating Skeletal Muscle Membrane Activation

The ionic currents underlying amphibian skeletal muscle activation can be studied by a voltage clamp involving multiple, simultaneous penetrations by three microelectrodes, V_1, V_2 and I_0, inserted at respective fixed distances, l, $2l$ and $5l/2$ from the fibre end (Figure 10.2Aa). The current electrode I_0 delivers the current $I_0(t)$ needed to control the voltage at V_1. The voltage drop $(V_1 - V_2)$ is proportional to the axial intracellular current $i_a(t)$ flowing between electrodes V_1 and V_2. This approximately equals and thereby permits determination of the membrane current $i_m(t)$ in the fibre segment extending a distance $3l/2$ from the fibre end (shaded area, Figure 10.2Ab) in response to imposed test V_1 steps. The current flowing across unit surface membrane area $I_m(t)$ can then be determined.

Application of successively larger depolarising voltage steps elicited early inward followed by later outward $I_m(t)$, resembling currents previously demonstrated in nerve (Figure 10.2B). These were similarly attributable to surface membrane voltage-dependent Na^+ and K^+ channels (Adrian *et al.*, 1970). In common with the Na^+ currents, I_{Na}, in nerve, the inward currents in muscle could be described by the product of a third order of an activation (m) and an inactivation (h) variable, where m and h were themselves solutions of first-order rate equations (Figure 10.2C; Section 4.7). The Na^+ conductance is abolished by tetrodotoxin (TTX) at concentrations similar to those effective in axons.

However, skeletal muscle Na^+ channels represent a distinct Nav1.4 as opposed to the nerve Nav1.1 channel subtype. Muscle membranes also contain at least two rather than a single voltage-activated K^+ channel subtype. The first activates with depolarisation over a time-course close to that found in nerve membranes. The second activates over hundreds of milliseconds. Both are blocked by tetraethylammonium ions and have been described respectively in terms of fourth (n^4; Section 4.7) and second-order processes. In contrast to nerve, skeletal muscle I_K is inactivated by maintained depolarisations if these are extended over hundreds of ms.

The inward rectification and voltage-dependent K^+ channels discussed here are structurally distinct protein types with their own specific properties (Plate 6). The molecular organisation of voltage-dependent K^+ channels resembles that of voltage-dependent Na^+ and Ca^{2+} channels (Plate 6A) apart from their four separate domains forming single monomers of an otherwise similarly organised homotetrameric structure (Plate 6B). Each domain similarly contains six transmembrane regions including one voltage-sensing segment, S4 (Plate 6A, B). Kir channels comprise monomers containing two rather than six transmembrane helices with cytoplasmic $-NH_2$ and $-COOH$ termini (Plate 6C). An extracellular loop folds back forming the

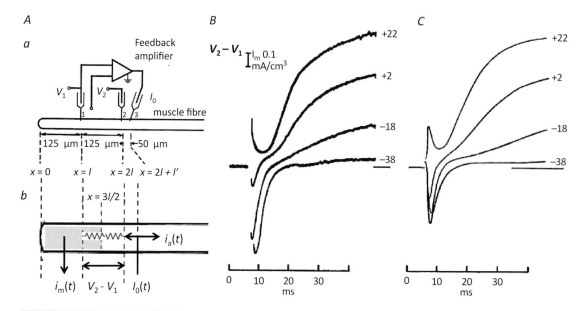

Figure 10.2 Skeletal muscle Na$^+$ and K$^+$ currents. (A) Voltage clamp characterisation of skeletal muscle surface membrane ion currents. (a) A current electrode at position 3 at distance $3l/2$ from the fibre end delivers current I_0 required to alter the voltage V_1 in the muscle end-segment monitored by the voltage electrode at position 1. (b) The voltage drop $V_1 - V_2$ between positions 1 and 2 at $x = l$ and $2l$, respectively, approximates the membrane current across the fibre end-segment over a distance $3l/2$ from the fibre end. (B) Experimental records and (C) theoretical reconstructions of inward Na$^+$ followed by outward K$^+$ currents in response to progressively larger depolarising test voltage steps. (From Figure 1 and Figure 5 of Adrian et al., 1970.)

pore-lining ion selectivity filter of the functional Kir channel, which is made up of four such subunits. Of the four known Kir channel subgroups, Kir2.x channels are persistently active. Kir3.x channels are regulated by G-protein-coupled receptors. Kir6.x channels are closed by ATP and sensitive to metabolic changes, thereby modulating E_m during exercise or other energetic demands (Section 10.7). There are also Kir1.x, Kir4.x, Kir5.x, and Kir7.x K$^+$ channel types (Hibino et al., 2010). Further background conductance and Ca^{2+}-activated K$^+$ channel species (Plate 6D, E) are considered in the chapters discussing cardiac (Section 13.6) and smooth muscle function (Section 15.4).

Skeletal muscles also contain Cav1.1 Ca^{2+} channels (Section 5.2) in their transverse (T-) tubular membranes. Their currents were demonstrated by voltage clamp methods similar to those illustrated in Figure 10.2A under conditions minimising the remaining current species. I_{Na} was eliminated by replacing [Na$^+$]$_o$ by tetraethylammonium ions and adding the Na$^+$ channel blocker TTX, I_K was minimised by the K$^+$ channel blocker tetraethylammonium and I_{Cl} current minimised by replacing [Cl$^-$]$_o$ by impermeant SO$_4^{2-}$. The observed inward Ca^{2+} currents, I_{Ca} (Figure 10.3A), had a threshold of around -40 mV, increased in amplitude with further depolarisation to a maximum value near 0 mV, then declined again as the test voltage approached the $+40$ mV Ca^{2+} Nernst potential (Figure 10.3B).

However, I_{Ca} timecourses were much slower than those of I_{Na} and I_K with times to peak of several seconds at 2–6 °C. At -24 mV, I_{Ca} reached a peak at 220 ms and decayed exponentially with a half time of 200 ms. The Ca^{2+} channel inactivation may partly result from inhibitory effects of the consequent increase in [Ca^{2+}]$_i$ (Section 5.2).

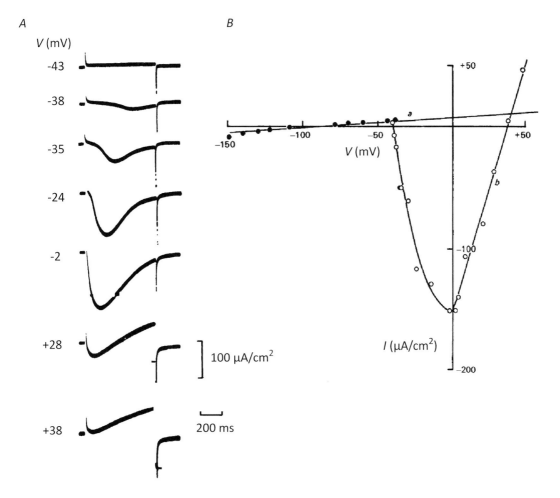

A

V (mV)

-43

-38

-35

-24

-2

+28

100 $\mu A/cm^2$

+38

200 ms

B

+50

-150

-100

-50

+50

V (mV)

a

-100

b

I ($\mu A/cm^2$)

-200

Figure 10.3 Ca^{2+} currents in amphibian skeletal muscle. (A) Membrane Ca^{2+} currents, I_{Ca} in response to command pulses to different depolarising voltages. (B) Current–voltage relationships of the observed current. (From Figure 1 and 4 of Sanchez and Stefani, 1978.)

Termination of the pulse produced fast inward tail currents. I_{Ca} activates too slowly to either contribute to normal electrical action potential excitation properties or initiate contraction in vertebrate skeletal muscle, although this does not exclude a voltage-sensing function of the Ca^{2+} channel regulating excitation–contraction coupling (Sections 11.7–11.9).

However, in invertebrate arthropod and crustacean twitch muscle and some forms of smooth muscle, I_{Ca} rather than I_{Na} mediates the action potential, as well as initiating contraction. These Ca^{2+} action potentials differ from Na^+ action potentials in their slower time-courses and inhibition by transition metal ions such as Mn^{2+}, Co^{2+}, Ni^{2+} and Cd^{2+}, as opposed to TTX. In contrast, other alkali earth cations such as Sr^{2+} and Ba^{2+} can substitute for Ca^{2+} in this regenerative response. Finally, I_{Ca} is important in action potential generation and contractile activation in mammalian cardiac muscle (Section 13.9).

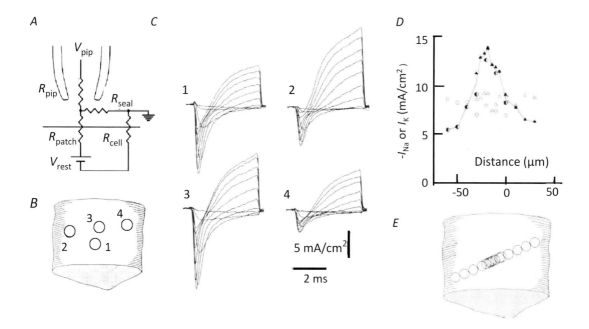

Furthermore, abnormal Ca^{2+} permeability properties may contribute
pathological changes in dystrophic muscle.

Ion channel mutations in the positively charged residues in the
voltage-sensing Nav1.4 or Cav1.1 S4 transmembrane segments
(Sections 5.2 and 5.5) are associated with inherited skeletal muscle
diseases characterised by episodes of flaccid muscle weakness. Mutations in the outermost gating charges , R1 and R2, result in cation leak
in the channel resting state, causing hypokalaemic periodic paralysis.
Mutations in the third R3 gating charge result in cation leak in
both activated and inactivated states causing normokalaemic periodic
paralysis (Jurkat-Rott et al., 2010; Cannon, 2018).

Finally, experimental studies in skeletal muscle demonstrate heterogeneous surface membrane Na^+ and K^+ channel distribution patterns. These employed a loose patch variant of voltage clamping,
applying a fire-polished single, combined, current injection and
voltage-recording pipette, making an electrical seal with a defined
area of membrane (Figure 10.4A). This technique also proved useful
in a range of other applications measuring ion currents in intact cells
(Sections 13.19, 14.1 and 14.14). The transmembrane potential was
varied by altering the controlled voltage of the intrapipette solution at
the external surface of the membrane. The resulting currents flowing
across different regions on the membrane surface to which the patch
electrode was applied were measured (Figure 10.4B, positions 1–4).
The I_{Na} and I_K showed different absolute and relative amplitudes
between different recording sites (Figure 10.4C). This is exemplified
by the differing patterns in peak inward I_{Na} (filled or half-filled

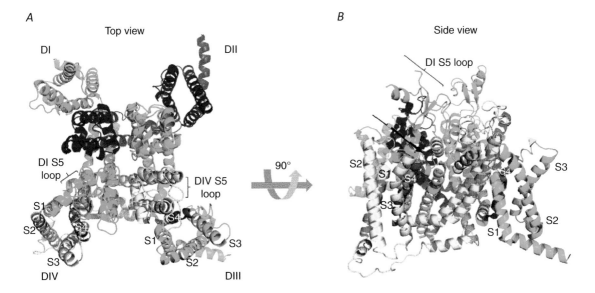

A   Top view   B   Side view

Plate 1 *Key molecules in the structure and function of nerve and muscle: the Na⁺ channel*

Model of the Na⁺ channel α-subunit, based on cryo-electronmicroscopic (cryo-EM) structure of the human Nav1.4-β_1 complex at 3.2 Å resolution (Pan et al., 2018), with the β_1 complex removed. (A) Top view, showing domains DI–DIV. (B) Side, transmembrane view. Each of the four internally homologous domains are differently coloured. Helices S1–S4 contribute the voltage-sensing modules (VSMs) as indicated for DIII and DIV. The DIII and DIV S4 helices are coloured purple for clarity. S5 extracellular loop regions for DI and DIV are indicated by brackets. (From Supplemental data in Salvage et al., 2019b.)

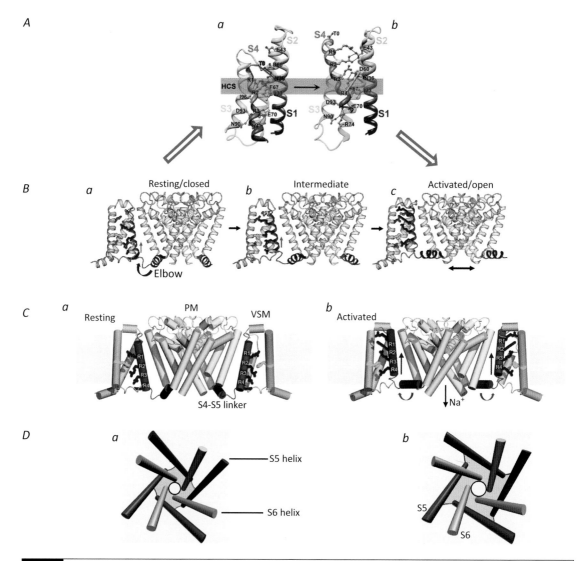

Plate 2 *Gating function in the Na⁺channel.*

(A) Model of the voltage-sensing module of the bacterial voltage-gated Na⁺ channel, NaChBac: transmembrane view. The lowest-energy models of the voltage-sensing domain in the (*a*) resting and (*b*) activated states. Side chains of the arginine (R) residues in the S4 helix mediating gating, and key residues in the S1, S2 and S3 segments labelled and shown in stick representation. (*a*) R1 forms hydrogen bonds with residues in S3. R3 makes ionic interactions with an intracellular negatively charged cluster. R4 forms an ion pair with D93 in S3. (*b*) Rearrangements in the S4 relative to the constriction site involving altered interactions involving R1–R4. (B) Molecular and (C) schematic representations in transmembrane view of coupling of voltage-sensor movement to pore opening by elbow connecting S4 to the S4–S5 linker driving the channel from its resting/closed (*Ba, Ca*) to its active/open state (*Bc, Cb*). The latter results in loosened packing interactions around the activation gate allowing it to dilate and permit Na⁺ permeation (*Da,b*). ((A) from Figure P1 of Yarov-Yarovoy et al., 2012); (B) from Figure 6D of Wisedchaisri et al., 2019; (C) from graphical abstract of Wisedchaisri et al., 2019; (D) from Figure 1C of Karbat and Reuveny, 2019.)

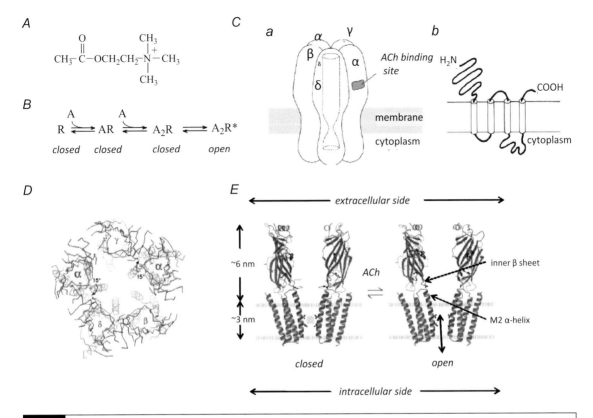

Plate 3 *Key molecules in the structure and function of nerve and muscle: the nicotinic acetylcholine receptor/channel.*

The postsynaptic membrane contains ligand-gated channels specifically binding (*A*) the transmitter molecule acetylcholine (ACh). (*B*) Linear reaction sequence shown by ACh–acetylcholine receptor (AChR) binding. (*C*) (*a*) The AChR depicted as a transmembrane molecule made up of four glycoprotein subunit types (α, β, γ and δ) assembling into a five-subunit $\alpha_2\beta\gamma\delta$ structure within the lipid bilayer, with the two α-subunits containing the ACh binding sites. (*b*) The amino-acid chain of each subunit containing four membrane-crossing segments. (*D, E*) AChR polypeptide chains fitted to electron density maps. (*D*) Plan view of α, β, γ and δ subunits viewed from the extracellular side. (*E*) Side view demonstrating the AChR as a transmembrane protein with closed and open states. ((*D*) and (*E*) from Figure 1 and 3 of Unwin, 2003.)

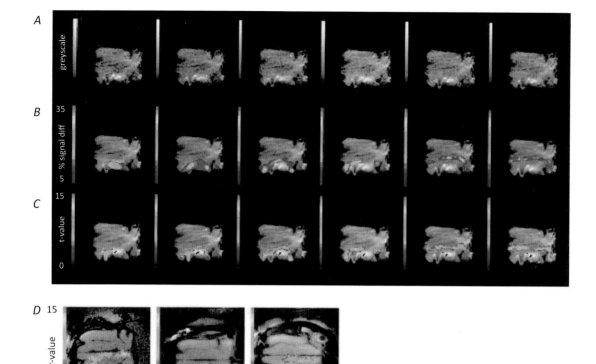

Plate 4 *Spreading depression of nerve cell activity in the central nervous system.*

(A–C) Visualisation and analysis of propagating cortical spreading depression (CSD) initiated by depolarising stimulation applied at the centre of the suprasylvian gyrus. Successive frames represent high-*b* diffusion-weighted images: field of view 5 cm; left, rostral; right, caudal; lower, left ; upper, right aspects of feline brain. (A) Single-shot, spin-echo, echoplanar high-*b* diffusion-weighted images obtained at ~1, 2, 3, 4, 5 and 6 min, respectively, after initiation of primary wave. Arbitrary grey scale designed to maximise image contrast. CSD event not apparent to the eye unless frames are animated. (B) Subtraction images derived from a reference preresponse high-*b* image, subtracted from each of the post-initiation high-*b* images. The coloured overlays represent the resulting percentage signal differences. Colour code varies from 5–35% of the reference level in each given pixel. (C) Overlays reflecting results of running *t*-tests applied to ADC map data. Colour scale represents *t* values between 0 and 15 for each individual pixel. (D) CSD events statistically mapped onto the respective anatomical gradient-echo, horizontal brain images. Field of view 5 cm. (From Figures 6 and 8 of Smith et al., 2001.)

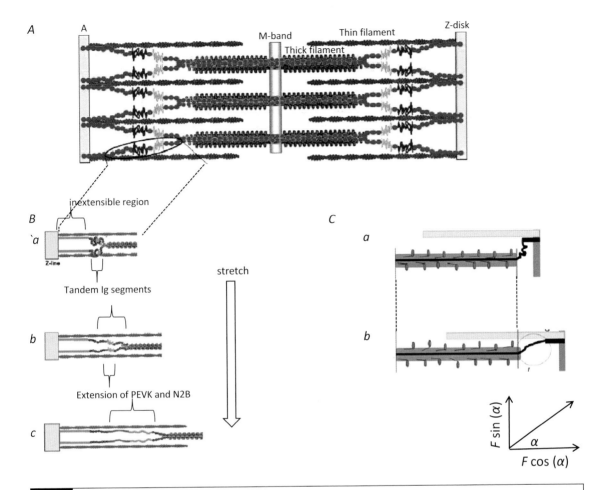

Plate 5 *Structure and functions of titin.*

(A) Structural relationships between thin-filament actin, thick-filament myosin, the M-band and Z-disk, and titin within the sarcomere. (B) Recruitment of three-element molecular spring formed by titin within the I-band of the sarcomere, with progressive stretch from (a) slack to (c) full length. (C) Titin-based modulation of myofilament lattice spacing and head mobility. Titin's force (F) can be resolved into longitudinal and radial components. The longitudinal force slightly increases thick filament strain at long sarcomere lengths. This may increase myosin head mobility. The radial force compresses the myofilament lattice and increases the probability of actomyosin interaction. ((A) from Figure 1 of Anderson and Granzier, 2012; (B) from Figure 1B of Granzier and Labeit, 2002; (C) from Figure 5C of Granzier and Labeit, 2002.)

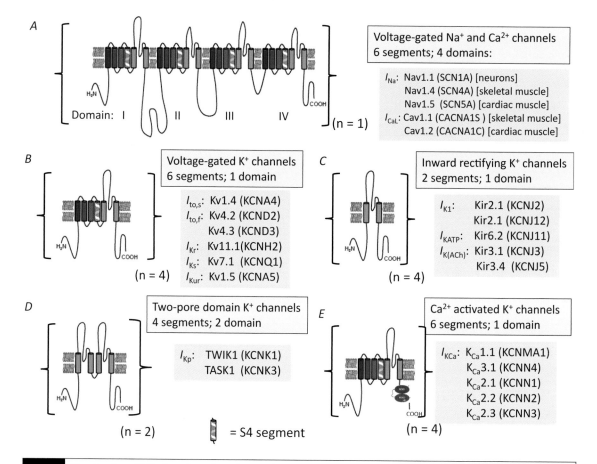

A Voltage-gated Na⁺ and Ca²⁺ channels
6 segments; 4 domains:

Domain: I   II   III   IV   (n = 1)

I_{Na}: Nav1.1 (SCN1A) [neurons]
Nav1.4 (SCN4A) [skeletal muscle]
Nav1.5 (SCN5A) [cardiac muscle]
I_{CaL}: Cav1.1 (CACNA1S) [skeletal muscle]
Cav1.2 (CACNA1C) [cardiac muscle]

B Voltage-gated K⁺ channels
6 segments; 1 domain

$I_{to,s}$: Kv1.4 (KCNA4)
$I_{to,f}$: Kv4.2 (KCND2)
Kv4.3 (KCND3)
I_{Kr}: Kv11.1(KCNH2)
I_{Ks}: Kv7.1 (KCNQ1)
(n = 4) I_{Kur}: Kv1.5 (KCNA5)

C Inward rectifying K⁺ channels
2 segments; 1 domain

I_{K1}: Kir2.1 (KCNJ2)
Kir2.1 (KCNJ12)
I_{KATP}: Kir6.2 (KCNJ11)
$I_{K(ACh)}$: Kir3.1 (KCNJ3)
(n = 4) Kir3.4 (KCNJ5)

D Two-pore domain K⁺ channels
4 segments; 2 domain

I_{Kp}: TWIK1 (KCNK1)
TASK1 (KCNK3)

(n = 2)   ▓ = S4 segment

E Ca²⁺ activated K⁺ channels
6 segments; 1 domain

I_{KCa}: K_{Ca}1.1 (KCNMA1)
K_{Ca}3.1 (KCNN4)
K_{Ca}2.1 (KCNN1)
K_{Ca}2.2 (KCNN2)
(n = 4) K_{Ca}2.3 (KCNN3)

Plate 6 *Structural motifs for ion channel α-subunits, their currents, and their encoding genes.*
Motifs for (*A*) voltage-gated Na⁺ and L-type Ca²⁺ channels carrying inward Na⁺ (I_{Na}) and Ca²⁺ current (I_{CaL}). (*B*) Voltage-gated K⁺ channels carrying slow transient outward ($I_{to,s}$), fast transient outward ($I_{to,f}$), rapidly activating (I_{Kr}), slowly activating (I_{Ks}) and ultra-rapidly activating (I_{Kur}) delayed rectifier K⁺ currents. (*C*) Inward rectifying K⁺ channels carrying K⁺ current (I_{K1}), ATP-inhibited K⁺ current (I_{KATP}), acetylcholine-gated K⁺ current ($I_{K(ACh)}$); (*D*) Two-pore-domain K⁺ channels carrying background plateau K⁺ current (I_{Kp}). (*E*) Ca²⁺-activated K⁺ channels carrying Ca²⁺-activated K⁺ current (I_{KCa}). The values of n indicate the number of represented structures making up component domains of the complete channel. S4 segments associated with voltage sensing present in (*A*) and (*B*).

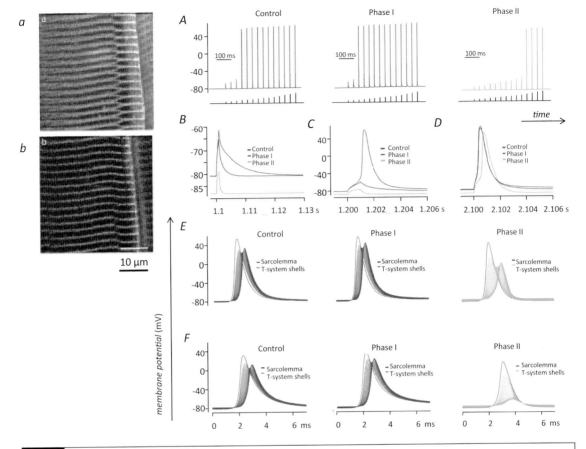

Plate 7 *Structure and function of the transverse tubular system in skeletal muscle.*

Left inset: morphology of transverse (T-) tubular system in *Rana temporaria* sartorius muscle fibres stained with extracellularly applied di-8-ANEPPS and visualised by two-photon confocal microscopy. (*a*) Maximum intensity projection of typical fibre: 15 images overlaid from serial optical sections obtained at regularly spaced imaging planes. (*b*) Single optical section from typical fibre of same preparation. Scale bar: 10 μm.

Main panels: Computational reconstructions of action potential initiation, surface action potential propagation and transverse tubular system excitation following respective 40% decreases and 500% increases in tubular Cl^- and K_{ATP} conductances known to take place with initial (Phase I) and prolonged activity (Phase II) in fast-twitch muscle.

(A) Effects during endplate excitation of sarcolemmal action potentials on membrane excitation (upper panel) of injections of 1 ms square KCl currents of progressively increasing amplitude in the central segment of a computer model representing muscle electrophysiological properties (lower panel). Current amplitudes required to excite an action potential determined under conditions where G_M was at its resting value (left panel: red traces) or altered to model conditions at the outset of (centre panel: blue traces, Phase I) or following prolonged activity (right panel: green traces, Phase II). (B) Effects on passive membrane potential responses to a current injection that was subthreshold under the three conditions. (C) Selection of a current stimulus that would trigger a surface action potential at the outset (Phase I) of activity, but not a resting fibre or following prolonged activity (Phase II). (D) A stimulating current of amplitude just sufficient to elicit an action potential in a fibre following prolonged activity (Phase II) successfully elicits activity under all the remaining situations. (E) Comparisons of surface action potential (black line) and membrane potentials in 20 concentric tubular system shells 3 mm from the point of AP excitation in resting muscle (coloured lines), at the outset of activity (Phase I) and following prolonged activity (Phase II). (F) Similar comparisons obtained when excitability is depressed with increased extracellular $[K^+]$ encountered in working muscle reproduced by a 50% reduction in maximum permeability for voltage-gated Na^+ channels. (From Figure 5 of Sheikh et al., 2001 and Figure 6 of Fraser et al., 2011.)

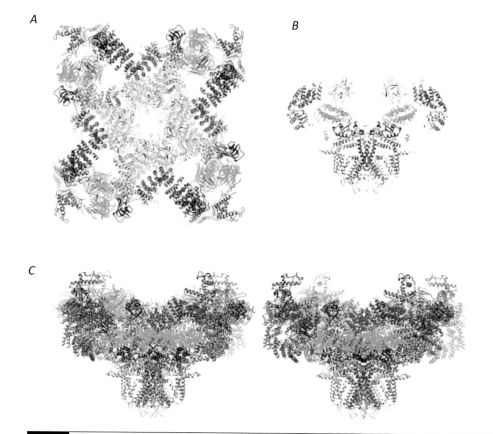

Plate 8 *Key molecules in the structure and function of nerve and muscle: the ryanodine receptor type 1 isoform.*
Cryo-electronmicrographic (EM) images of the skeletal muscle ryanodine receptor (RyR1) to close to 1.0 nm resolution. Atomic reconstructions showing (A) cytoplasmic view, omitting the transmembrane domain for clarity. (B, C) side views showing (B) a central slice and (C) a stereo view. (From Figure 2 of Samsó, 2017.)

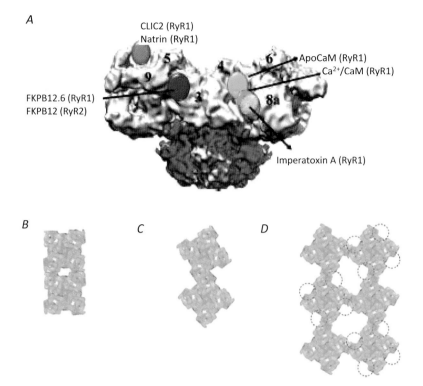

A

CLIC2 (RyR1)
Natrin (RyR1)

ApoCaM (RyR1)
Ca²⁺/CaM (RyR1)

FKPB12.6 (RyR1)
FKPB12 (RyR2)

Imperatoxin A (RyR1)

B

C

D

Plate 9 *Key molecules in the structure and function of nerve and muscle: quaternary interactions of ryanodine receptor type 1 isoform.*

(A) Localisation of binding sites for known ryanodine receptor type 1 (RyR1) ligands shown on RyR1 in side view: FKPB12 (FKBP12.6 in the case of the RyR2 cardiac isoform), ApoCaM and Ca²⁺–CaM with apo- and Ca²⁺–CaM overlap, Cl⁻ intracellular channel protein 2 (CLIC2) and imperatoxin-A. (B) Side-by-side and (C) oblique interaction of RyR1s giving rise to (D) an array of RyR1s shown with superimposed tentative footprints of the dihydropyridine receptor (DHPR1) tetrad, each represented with a dotted circumference. Every alternate RyR1 interacts with one DHPR1 tetrad. FKBP: FK506 binding protein; CaM: calmodulin; CLIC2: intracellular Cl⁻ channel protein 2. (From Figure 3 of Samsó, 2017.)

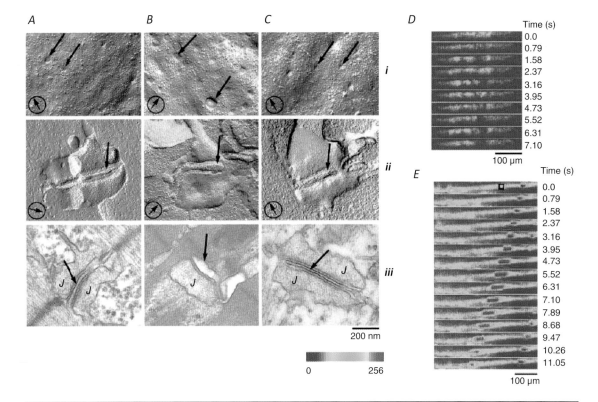

Plate 10 *Transverse tubular and junction sarcoplasmic reticular structure and their roles in release of Ca^{2+}.*

(A–C) Amphibian fibres fixed following exposure to (A) isotonic and (B) hypertonic solutions and (C) following restoration of the isotonic solutions. ((i), (ii)) Freeze-fracture electronmicrographic images demonstrating (i) diameters of T-tubular apertures (arrowed) at the surface membrane shown in P-face fractures through the plasma membrane and (ii) triad elements formed by junctional sarcoplasmic reticulum (SR) and T-tubules (arrowed). The arrow gives the direction of shadowing at the bottom left of each image. (iii) T-SR junctions in thin electronmicrograph sections with visible end-feet traversing the gap between junctional SR (J) and T-system (arrowed) in (A) and (C). Condition (B) is accompanied by (D) discrete stationary foci of increased cytosolic [Ca^{2+}] followed by (E) regenerative waves of Ca^{2+} release in Fluo-3-loaded fibres visualised under confocal microscopy. False colour scale runs from 0 to 256. (From Figures 1, 2 and 6 of Chawla et al., 2001.)

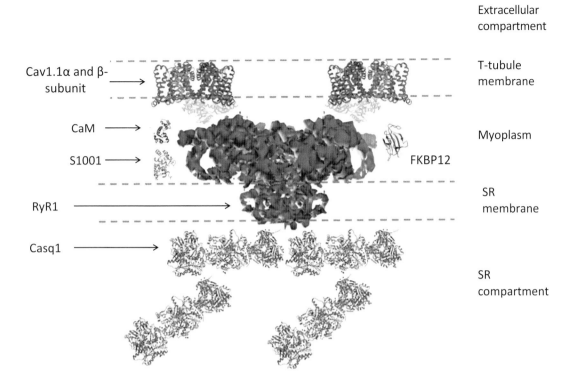

Extracellular compartment

T-tubule membrane

Cav1.1α and β-subunit →

CaM →

S1001 →

FKBP12

Myoplasm

RyR1 →

SR membrane

Casq1 →

SR compartment

Plate 11 *Molecular anatomy underlying skeletal muscle excitation-contraction coupling.*

Two of the cluster of four modified skeletal muscle Cav1.1 Ca^{2+} channels, also termed dihydropyridine receptors (DHPRIs), within the T-tubular membrane shown. Each comes into close proximity, thereby interacting with cytoplasmic components of one of the ryanodine receptor (RyR1) subunits within the sarcoplasmic reticular (SR) membrane. The SR lumen is abundant in the Ca^{2+}-sequestering protein, calsequestrin (CASQ1). The intramembrane portion of the RyR1 resides within the SR membrane and functions as an intracellular Ca^{2+} channel. The large cytoplasmic segment is the site of multiple interactions of the RyR with other regulatory molecules. FKBP: FK506 binding protein: CaM: calmodulin. (From Figure 2 of Hernandez-Ochoa et al., 2016.)

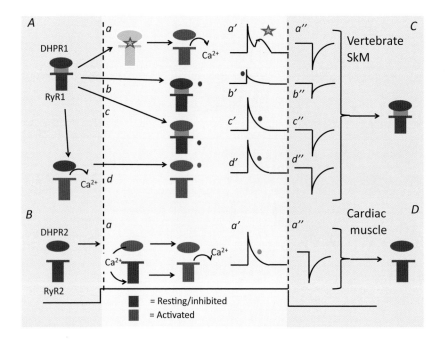

Plate 12 *Intramembrane charge movements associated with excitation-contraction coupling: DHPR-RyR coupling mechanisms in different muscle types.*

(A) Vertebrate skeletal muscle: excitation–contraction coupling involves voltage sensing by DHPR1-voltage sensors allosterically coupled to RyR1–Ca^{2+} release channels. (*a*) Configurational changes in the DHPR1 in response to a voltage step initiating reciprocal coupling in a cooperative process causing DHPR1–RyR1 dissociation (yellow figures) permitting release of intracellularly stored Ca^{2+} (green figures). (*a'*) The resulting regenerative timecourse in q_γ charge movement. Pattern accentuated by perchlorate which can reverse (*b*) charge movement block (*b'*) by DHPR1 inhibition by the DHPR blockers and q_γ charge movement inhibitors, mM-tetracaine and nifedipine, leaving q_β charge movement intact. Both (*c*) inhibition of the RyR1 configurational change by ryanodine, daunorubicin or microM tetracaine and (*d*) prior opening of the RyR1-Ca^{2+} release channel by caffeine, causing its dissociation from the DHPR1, continue to permit the DHPR configurational change. However, q_γ charge movement now shows no kinetic evidence of DHPR1–RyR1 cooperativity (*Ac', d'*). (B) Invertebrate skeletal and vertebrate cardiac muscle: excitation–contraction coupling involves voltage sensing by a DHPR2 voltage sensor mediating Ca^{2+} entry activating otherwise uncoupled RyR2–Ca^{2+} release channels. Associated charge movements (*Ba', Da''*) now do not show the cooperative kinetics illustrated in (*Aa'*). In both (A) and (B) repolarisation reverses DHPR1 or DHPR2 activation, and permits a monotonic return of the charge movement to the resting state (*Ca''-d''* and *Da'–Da''*).

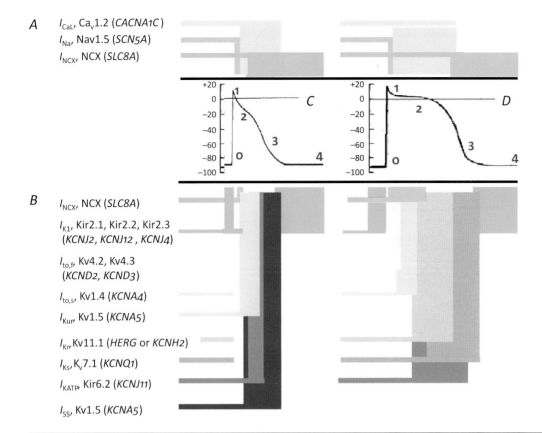

A
I_{CaL}, Ca$_v$1.2 (*CACNA1C*)
I_{Na}, Nav1.5 (*SCN5A*)
I_{NCX}, NCX (*SLC8A*)

B
I_{NCX}, NCX (*SLC8A*)

I_{K1}, Kir2.1, Kir2.2, Kir2.3
(*KCNJ2, KCNJ12 , KCNJ4*)

$I_{to,f}$, Kv4.2, Kv4.3
(*KCND2, KCND3*)

$I_{to,s}$, Kv1.4 (*KCNA4*)

I_{Kur}, Kv1.5 (*KCNA5*)

I_{Kr}, Kv11.1 (*HERG* or *KCNH2*)

I_{Ks}, K$_v$7.1 (*KCNQ1*)

I_{KATP}, Kir6.2 (*KCNJ11*)

I_{SS}, Kv1.5 (*KCNA5*)

Plate 13	*Ion channels underlying electrophysiological activity in atrial and ventricular cardiomyocytes.*

Ion channels, listed by their currents, underlying proteins and encoding genes, mediate (*A*) inward depolarising and (*B*) outward repolarising currents, driving (*C*) atrial and (*D*) ventricular action potentials. Each action potential comprises rapid depolarising (Phase 0), early repolarising (Phase 1), brief (atrial) or prolonged (ventricular) plateaus (Phase 2), Phase 3 repolarisation and Phase 4 electrical diastole. (*A*) Inward Na$^+$ or Ca^{2+} currents drive Phase 0 depolarisation; Ca^{2+} current maintains Phase 2. (*B*) A range of outward K$^+$ currents drive early Phase 1 and late Phase 3 repolarisation. Phase 4 resting potential restoration is accompanied by a refractory period required for Na$^+$ channel recovery. The resulting wave of electrical activity and refractoriness is propagated through successive SAN, atrial, AV, Purkinje, and endocardial and epicardial ventricular cardiomyocytes.

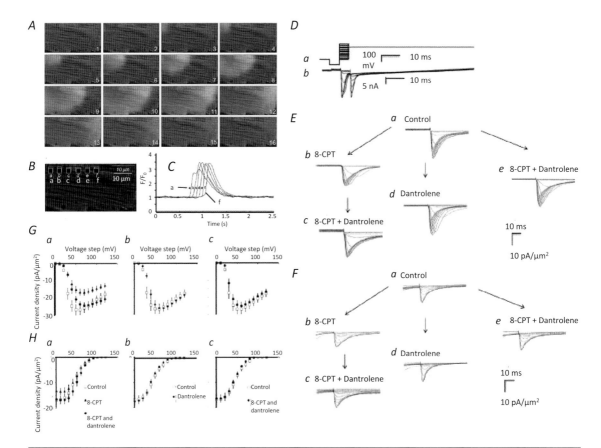

Plate 14 *Effects of Epac activation on Ca²⁺ homeostasis and feedback effects on Na⁺ channel function in murine ventricular cardiomyocytes.*

(A) Successive confocal microscope framescan images showing Ca^{2+} wave propagation along a fluo-3-loaded murine cardiomyocyte in the presence of 8-CPT. (B) Regions of interest of dimension 2×2 μm in which fluorescence emission was analysed to give (C) Ca^{2+} transients with successively delayed latencies, with distance along the long axis of the cell. (D–F) Ionic currents from an isolated murine ventricular preparation studied under loose patch-clamp. (Da) Pulse protocol imposing successively incremented depolarising steps investigating Na⁺ channel activation and inactivation. (Db) Typical recordings using a pipette (tip diameter 30 μm). Downward deflections denote inward currents in nA. (E, F) Na⁺ current (I_{Na}) densities (pA/μm²) reflecting (E) activation by a depolarising V_1 step relative to the resting potential before and following pharmacological challenge and (F) in response to the voltage step from V_1 to the final level V_2 exploring Na⁺ channel inactivation produced by the voltage step to V_1. Recordings (a) under control conditions in the absence of drug, (b) in the presence of 8-CPT (1 μmol/L), (c) in the presence of both 8-CPT and dantrolene (10 μmol/L), after initial addition of 8-CPT, (d) after addition of dantrolene alone and (e) in the presence of both 8-CPT and dantrolene. (G, H) Voltage dependences of ventricular I_{Na} activation (G) and inactivation (H) under the different pharmacological conditions. (G) I_{Na} activation (mean ±SEM) plotted against V_1 voltage step and (H) I_{Na} inactivation plotted against V_1, before and after pharmacological challenge. Results obtained (a) before (open squares) and following introduction of 8-CPT (filled triangles) and a combination of 8-CPT and dantrolene (filled circles), (b) before (open squares) and following introduction of dantrolene (filled diamonds), (c) before (open squares) and following introduction of a combination of 8-CPT and dantrolene (filled circles). ((A–C) from Figure 8 of Hothi et al., 2008); (D–F) from Figure 2 and (G, H) from Figure 4A, B of Valli et al., 2018.)

A *Intact animals:*

> a. ECG recordings: **Arrhythmic phenotype**

B *Langendorff-perfused hearts:*

> a. MAP recording: **AP waveform**

> b. Sharp electrode recording: **AP upstrokes**

> c. Multi-electrode/optical recording: **AP propagation**

C *Superfused atrium/ventricle:*

> a. Loose patch-clamp: **Ion current**

D *Isolated cardiomyocytes:*

> a. Tight patch-clamp: **Ion current**

> b. Fluo-3 confocal microscopy: **Ca²⁺ signalling**

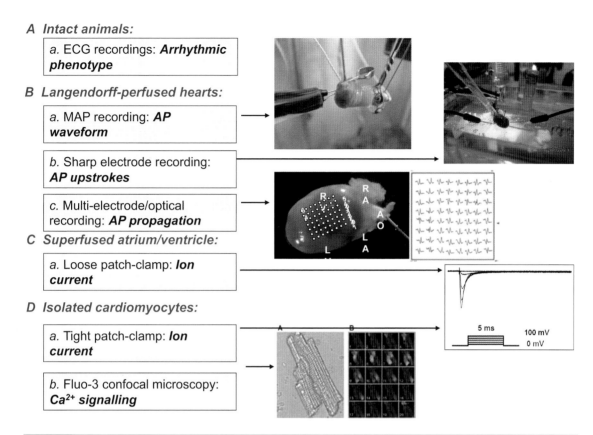

Plate 15 *Studies of arrhythmic characteristics and their underlying mechanisms from the intact animal to the single cell level*

In (A) intact animals: (a) electrocardiographic recordings in anaesthetised or ambulant animals permitting characterisation of arrhythmic phenotype. (B) Langendorff-perfused hearts: (a) monophasic action potential recordings characterising action potential waveforms and possible triggering events; (b) sharp, intracellular electrode recordings additionally providing quantification of action potential upstroke phases in single cardiomyocytes; (c) multi-electrode or optical recording mapping temporal and spatial distribution of electrical or spectrofluometric activity. (C) Superfused atrial or ventricular preparations, permitting loose patch-clamp studies of ionic currents. (D) Isolated cardiomyocytes: (a) conventional (tight) patch-clamp for characterisation of ionic current and (b) spectrofluometric studies of Ca^{2+} homeostasis in single cardiomyocytes.

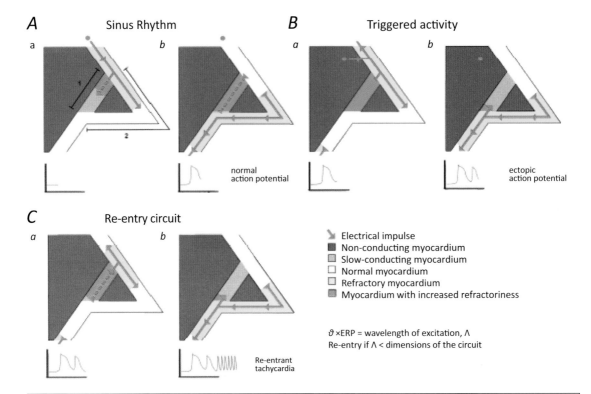

A Sinus Rhythm

a   b   normal
                                               action potential

B Triggered activity

a   b   ectopic
                                               action potential

C Re-entry circuit

a   b   Re-entrant
                                               tachycardia

Electrical impulse
Non-conducting myocardium
Slow-conducting myocardium
Normal myocardium
Refractory myocardium
Myocardium with increased refractoriness

$\vartheta \times$ ERP = wavelength of excitation, Λ
Re-entry if $\Lambda <$ dimensions of the circuit

Plate 16 *Conditions underlying generation of re-entrant and triggered arrhythmia.*

(A) Arrhythmic substrate typically involving a myocardial pathway with compromised conduction (*a* path 1; dark grey) and a second pathway with normal conduction (*b*: path 2; white). Blue arrow signifies normal action potential propagation with velocity θ and effective refractory period (ERP) of wavelength, $\Lambda = \theta \times$ ERP (yellow region) in path 2 (*b*). If the latter initiates a slow conducting action potential, this travels along path 1 (*a*). However, during normal activity, the resulting slow conducting action potential collides with refractory tissue within path 2 and cannot re-enter the circuit (*b*). (B) An abnormal triggered impulse immediately following the normal action potential (*a*) cannot enter path 2 as this remains refractory (*b*). (C) Self-perpetuating re-entrant excitation results when a retrogradely conducting action potential along path 1 (*a*) with a reduced wavelength Λ shorter than the dimensions of the propagation pathways, due to reduced θ and/or ERP, enters the beginning of path 2 (*b*). (From Figure 3 of King et al., 2013a.)

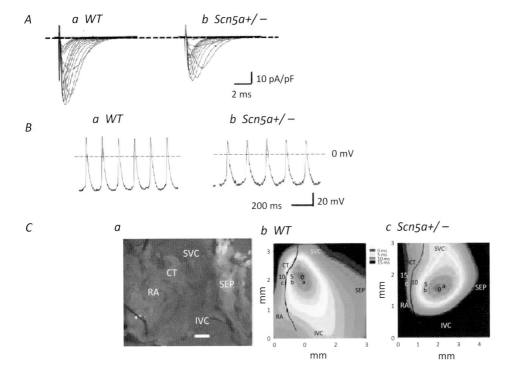

Plate 17 *The role of Na⁺ channel activity in sino-atrial pacemaking and atrial conduction illustrated in the Scn5a+/– murine model for Brugada syndrome.*

(A) Na⁺ currents in response to depolarising voltage steps in (*a*) WT and (*b*) *Scn5a+/–* cardiomyocytes. (B) Action potentials at sites near the SAN centre from intact (*a*) WT and (*b*) *Scn5a+/–* atrial preparations. (C) (*a*) Example of a SAN preparation used for electrical mapping. The mouse SAN is located in the intercaval region parallel to the crista terminalis (CT). SEP: atrial septum; SVC: superior vena cava; IVC: inferior venacava, RA: right atrium (scale bar, 200 μm). (*b, c*) Activation sequence in (*b*) WT and (*c*) *Scn5a+/–* SAN and atrium determined from extracellular potential recordings made from >200 recording sites over the SAN and surrounding atrial muscle at a 0.2–0.3 mm resolution. (From Figure 2B, 2C and Figure 4 (A–E) of Lei et al., 2005.)

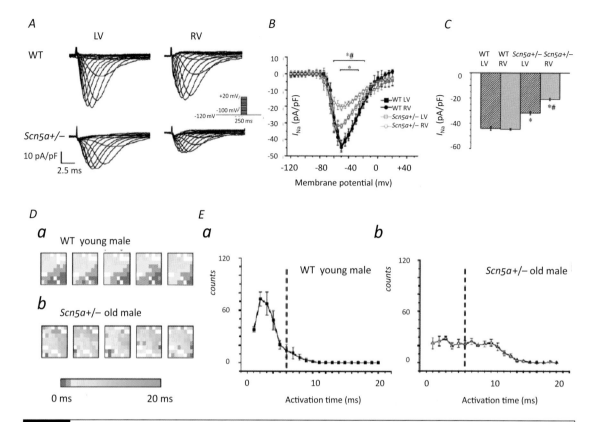

Plate 18 *Na+current and age-dependent conduction properties in Nav1.5 haploinsufficient Scn5a+/− hearts.*

(A) Na+ currents, I_{Na}, normalised to cell capacitance from myocytes from left (LV) and right ventricles (RV) of wild-type (WT) and Scn5a+/− hearts. (B, C) Corresponding (B) current–voltage relationships and (C) maximum I_{Na}. * denotes effect of genotype and # denotes effect of cardiac ventricle (LV or RV). (D) Activation maps of conduction latency from five successive cardiac cycles in (a) young male WT and (b) old male Scn5a+/− heart. (E) Frequency distributions of activation times in (a) young male WT and (b) old male Scn5a+/− . ((A–C) from Figure 5a,b,c of Martin et al., 2012; (D, E) from Figure 3a,g and 5a,g of Jeevaratnam et al., 2012.)

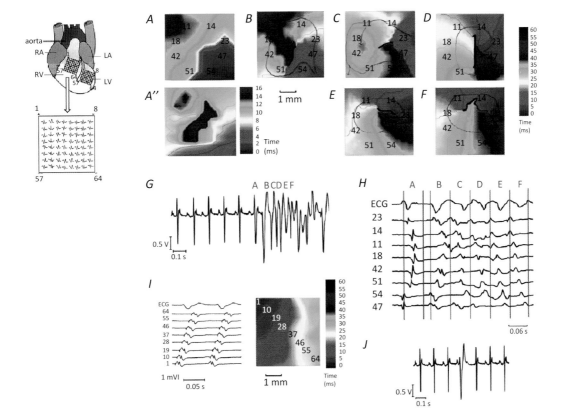

Plate 19 *Re-entrant circuit initiation of ventricular arrhythmia in murine Scn5a+/– hearts.*

Inset: positioning of multi-electrode recording arrays on right (RV) and left ventricular (LV) epicardium of *Scn5a+/–* heart.

Main Panels: (A–F) typical isochronal RV action potential propagation maps in isolated Langendorff-perfused *Scn5a+/–* heart following flecainide challenge at the onset of ventricular tachycardia (VT). Thick black lines denote lines of conduction block. Thin arrows denote lines of propagation. (A) Crowded isochronal lines in the last sinus beat, demonstrating area of conduction slowing. (A") Repolarisation map of the last sinus beat. Note repolarisation heterogeneity within the same area. (B) Superimposition of a premature ventricular beat leading to line of block; impulse propagation flows around it. (C) A second ventricular ectopic event causing formation of a re-entrant circuit. (D) The circuit continuing into the next beat to initiate VT with migration of the line of block to result in non-stationary vortex generating the polymorphic arrhythmia shown in panels (E) and (F). (G) Corresponding electrocardiogram (ECG) trace showing the ventricular ectopic event (VE) that then initiates a run of polymorphic VT. (H) Part of the same ECG trace, together with eight electrogram traces, at the initiation of VT. Electrogram numbers correspond to the channel numbers of the array shown in the maps. The point within each electrogram trace marked by each letter corresponds to the correspondingly labelled propagation maps. (I) propagation map, ECG, and electrogram traces of the VT propagating as a wave front across the LV from its onset in the RV. (J) ECG trace of a VE occurring after the T-wave. (From Figures 1 and 4 of Martin et al., 2011b.)

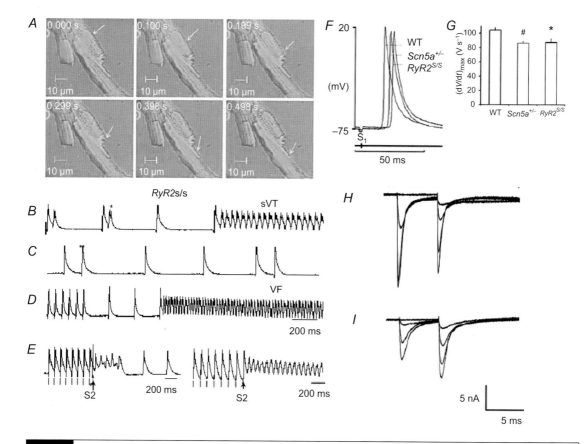

Plate 20 *Pro-arrhythmic phenotypic features of murine RyR2-P2328S hearts.*

(A) Sequential confocal microscope images of single fluo-3 loaded ventricular homozygotic RyR2–P2328S (RyR2$^{S/S}$) myocyte showing Ca^{2+} waves after isoproterenol (100 nM) challenge. Path taken by typical Ca^{2+} wave arrowed. (B–D) Ventricular epicardial monophasic action potentials in spontaneously beating RyR2$^{S/S}$ hearts. (B) Spontaneous early after-depolarisations (EADs) (*) followed by episodes of sustained monomorphic ventricular tachycardia (sVT). (C) Coupled beats. (D) Persistent ventricular fibrillation (VF) after ending of regular S1 pacing. (E) S2 extra-stimuli producing episodes of non-sustained VT before (left trace) and sustained (>30 s) VT following isoproterenol challenge (right trace). (F–G) Intracellular electrode action-potential recordings in RyR2$^{S/S}$, Scn5a$^{+/-}$, and WT atrial myocytes. (F) Left atrial intracellular action potentials showing conduction latencies. (G) Corresponding maximum action-potential upstroke rates (dV/dt)$_{max}$. (H–I) Loose patch-clamp Na$^+$ current records during a 100 mV, 50 ms duration activation step following 50 ms duration prepulses between 20 and 100 mV for (H) WT and (I) RyR2$^{S/S}$ atria. (From Figures 5, 8 and 9 of Goddard et al., 2008 and Figure 5 of King et al., 2013b.)

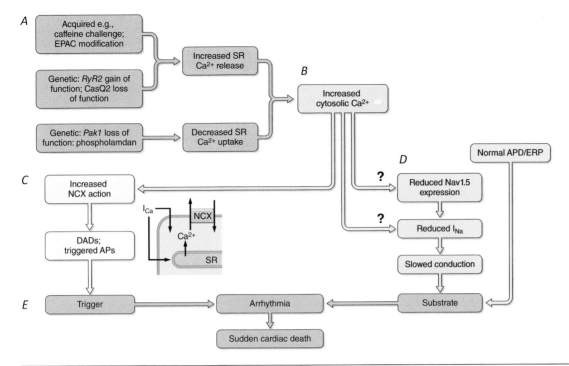

Plate 21 *Cellular mechanisms underlying Ca²⁺-induced arrhythmia.*

(A, B) Altered cytosolic [Ca²⁺] arising from increased RyR2-mediated release of sarcoplasmic reticular (SR) Ca²⁺ or decreased SR Ca²⁺-ATPase-mediated Ca²⁺ reuptake from cytosol to SR due to either acquired or genetic abnormalities. (C) Electrogenic Na⁺–Ca²⁺ exchange (NCX) activity causing transient inward current (I_{ti}) leading to pro-arrhythmic diastolic after-depolarisation (DAD). (D) Downregulated Nav1.5 expression and/or Nav1.5 function with pro-arrhythmic action-potential conduction slowing, despite unchanged action-potential recovery features reflected in (action-potential duration)/(effective refractory period) ratio, (APD/ERP). Combination of (C) and (D) culminate in (E) ventricular or atrial arrhythmias. (From Figure 18 of Huang, 2017.)

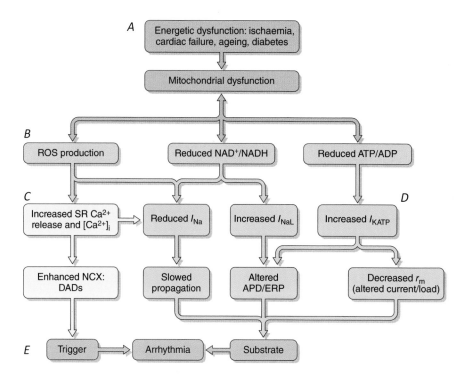

Plate 22 *Energetic dysfunction and arrhythmic phenotype.*

Simplified relationships between: (A) energetic dysfunction associated with ischaemic conditions, cardiac failure, ageing and diabetes, (B) mitochondrial dysfunction associated with ROS production, altered $NAD^+/NADH$ and ATP/ADP ratios, and their influence on (C) RyR2-mediated SR Ca^{2+} release increasing cytosolic $[Ca^{2+}]_i$, in turn increasing Na^+/Ca^{2+} exchange-mediated DAD triggering, and (D) Na^+ and K^+ channel activity affecting action potential excitation, propagation and recovery to cause (E) arrhythmias. (From Figure 20 of Huang, 2017.)

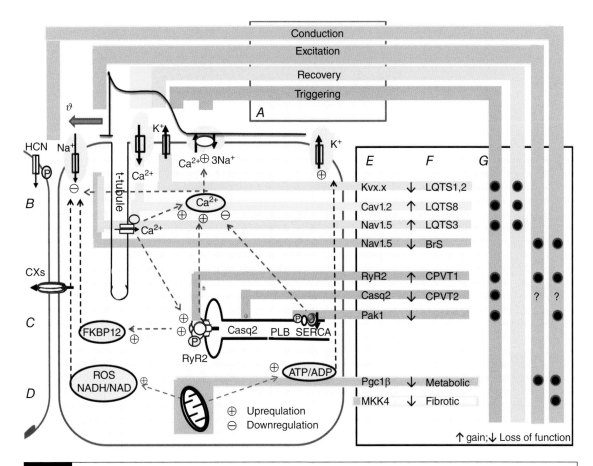

Plate 23 *Studies of arrhythmic mechanisms in genetically modified exemplars of ion channel disorder recapitulating human arrhythmic syndromes.*

(A) Arrhythmic mechanisms involving action potential excitation, conduction and recovery, and ectopic triggering. These derive from altered function in: (B) cardiomyocyte surface membrane biomolecules mediating electrophysiological activation, (C) feedforward mechanisms activating Ca^{2+} signalling in excitation–contraction coupling, and (D) further downstream effects reflecting cellular energetics and pathological remodelling processes. Mapping of (A)–(D) onto experimental study of (E) ion-channel phenotypes using genetically modified murine models involving specific biomolecules, whose gain- or loss-of-function modifications in turn recapitulate (F) known human arrhythmic syndromes. The latter accordingly showed one or more (G) forms of arrhythmic substrate or triggering as indicated by the bullet points.

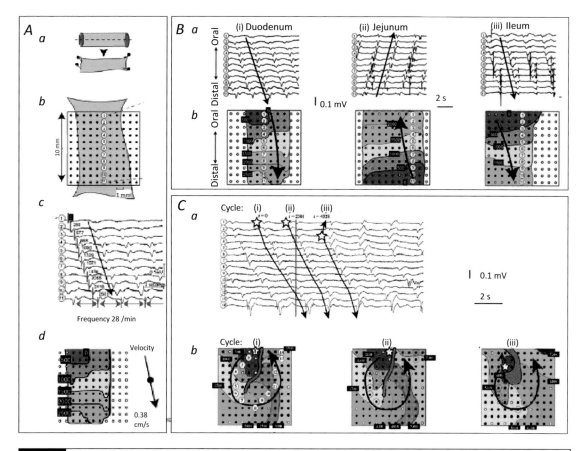

Plate 24 *Multi-electrode array electrogram studies of slow wave propagation maps from isolated rat small intestinal segments.*

(A) Recording methods illustrating (a) use of cut open isolated small intestinal segment, on which (b) a 121-electrode array is placed on its serosal surface. (c) Typical recordings obtained from an exemplified column of electrodes used to derive (d) isochronal maps representing wave latencies over the recording surface. (B) Electrograms (a) and propagation maps (b) from (i) duodenal (ii) jejunal and (iii) ileal segments. (C) Slow propagation resulting in re-entrant activity in a 10 × 10 mm tissue area. (a) Multi-electrode array electrogram recordings at sites designated in (b)(i). (b) Propagation pattern from three successive focal discharges (marked by star along the upper border)((i)–(iii)). Arrow and isochronic plot demonstrates direction of propagation. ((A) adapted from Figure 1 of Lammers et al., 2011); (B) adapted from Figure 5 of Lammers et al., 2011); (C) adapted from Figure 3 of Lammers, 2015.)

symbols) and final outward I_K (open symbols) (Figure 10.4D) during 80 mV depolarisations when plotted against distance across the fibre (Figure 10.4E).

10.3 | The Surface and Transverse Tubular Membrane Systems in Skeletal Muscle

The skeletal muscle surface membrane also contains localised areas whose stimulation preferentially results in transduction of electrical stimulation into muscle contraction. These were demonstrated by applying local stimulating currents through an external microelectrode at successive points along the length of the fibre surface. Local contractile activity only occurred with stimulation at specific points along the length of the muscle fibre surface. Viewing under polarised light microscopy related these effective stimulation sites to definite regions within the muscle striation pattern. These sites were clearly related to the regular striations formed by the myofibrillar contractile elements (Section 9.2; Huxley and Taylor, 1958). In amphibian muscle, these sites occurred opposite the Z line where the A bands adjacent to the I band opposite the electrode were drawn together (Figure 10.5).

Initially, it was thought that the inward conducting mechanism was the Z line itself. However, similar experiments in crab muscle fibres localised these 'active spots' not to the Z line, but near the boundary between the A and I bands. These findings implicated some transverse structure located at the Z lines in frog muscles and at the A–I boundary in crab muscles.

Anatomical studies identified these 'active spots' with a transverse (T-) tubular membrane system formed from surface membrane invaginations at regular intervals along the muscle fibre length. These opened into an extensively networking T-tubular system whose lumen remains continuous with the remaining extracellular space. Reconstructions from electronmicrograph preparations based on serial 0.7 μm cross-sections demonstrated that these networks occurred at regular intervals along the fibre in a plane transverse to the fibre axis (Figure 10.6; Peachey and Eisenberg, 1978). These occurred at junctions between the A and I bands in mammalian skeletal muscle and the Z line in mammalian cardiac and frog skeletal muscle. They surrounded each individual myofibril and resulted in a membrane system of total surface area six to ten times that of the sarcolemmal membrane cylinder.

Electronmicrographs of progressive stages in sarcomere development in *Xenopus laevis* skeletal muscle illustrate the geometrical relationship between these T-tubular networks, myofibrils and a further, intracellular sarcoplasmic reticular (SR) membrane system not connected to either T-tubules or extracellular fluid. The series of

Figure 10.5 Successive cine films from local stimulation experiment delivering depolarising currents by a pipette placed at different positions along the length of an amphibian sartorius muscle. Fibre viewed under polarised light. The A sarcomeric bands appear dark. Pipette applied to A (above) and I bands (below) respectively. Left image taken 'before' and right image taken 'during' stimulation. Local contraction only observed where the pipette is placed on the I-band which is related to the position of the transverse (T-) tubules. (From Figures 1–4 of Plate 1 from Huxley and Taylor, 1958.)

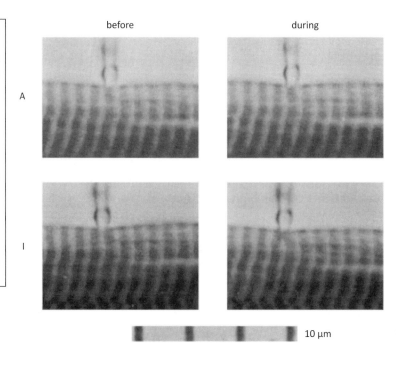

before   during

A

I

10 μm

Figure 10.6 Reconstruction of transverse (T-) tubular network over the muscle fibre cross-section derived from tracing successive portions of the network from micrographs of serial transverse slices each ~0.7 μm in thickness (×1400). (From Figure 2 of Peachey and Eisenberg, 1978.)

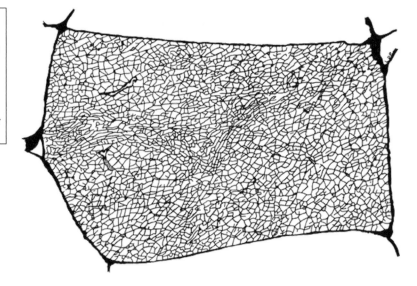

micrographs initially show that relatively few organelles have developed apart from periodic parallel lines representing successive T-tubular lumina along the length of the developing muscle (Figure 10.7A). A second SR membrane system develops in the sarcomere region intervening between successive T-tubules (Figure 10.7B). With full sarcomeric development, T-tubular cross-sections appear in close alignment with the developed myofilaments (Figure 10.7C; Huang and Hockaday, 1988).

A

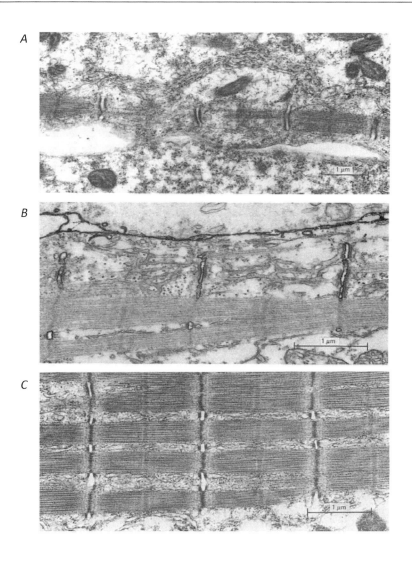

B

C

Figure 10.7 Electronmicroscope appearances in developing *Xenopus* skeletal muscle. Differential appearances of (*A*) transverse (T-) tubular, (*B*) sarcoplasmic reticular (SR) membrane systems during development. (*C*) Eventual appearance of T-tubules in close alignment with ordered thick and thin filament myofibrillar contractile components. (From Figures 3, 4 and 5 of Huang and Hockaday, 1988.)

10.4 | Surface and Transverse Tubular Components of the Muscle Action Potential

Skeletal muscle activation begins with excitation of a rapidly conducted action potential which then propagates over the muscle surface membrane from the neuromuscular junction to the ends of the fibre (Section 7.3). However, T-tubular depolarisation is required to initiate contraction. The T-tubular lumina are continuous with the remaining extracellular space permitting direct transmission of electrical changes between the surface and T-tubular membranes. This continuity, and the detailed in vivo structure of the T-tubular network can be directly visualised in viable muscle. Membrane in contact with extracellular fluid is stained with the fluorescent membrane-impermeant dye di-8-ANEPPS; this accesses the entire T-tubular network. Plate 7 (inset) illustrates two-photon confocal microscopic images of the morphology of stained amphibian sartorius muscle T-tubular system. The T-tubular lumina and regularly occurring T-tubular system bifurcations near the mouths of the T-tubules were reconstructed from a maximum-intensity projection constituted from overlays of 15 regularly spaced serial optical image sections (Panel *a*). Panel *b* shows a single optical section from a typical fibre of the same preparation (Sheikh *et al.*, 2001).

A hypothesis implicating T-tubular excitation in activating contraction makes a number of predictions. The T-tubules must effectively spread the surface electrical wave into the fibre interior to an extent that would activate even T-tubular membrane in the depths of the muscle fibre. Both experimental observations (Adrian *et al.*, 1969) and computational reconstructions (Adrian and Peachey, 1973) demonstrated that a passive T-tubular membrane excitation through cable spread of the electrical changes initiated by a surface action potential (Sections 6.1 and 6.2) would be unlikely to produce sufficient T-tubular depolarisation to successfully initiate contraction. This is illustrated in Figure 10.8A, showing the surface action potentials at the edges of the block diagram, with a computationally calculated passive spread of T-tubular excitation in traces between these representing electrical changes in T-tubular membrane in the depths of the muscle fibre.

These problems of T-tubular excitation are overcome in skeletal muscle through expression of activatable Na^+ channels within the T-tubular membranes. These should occur at a density sufficient to generate action potential activity following excitation from the surface, but that would not themselves produce membrane currents sufficient to conversely influence the surface membrane voltage. Na^+ channels indeed occur in the T-tubular membranes, but at a lower density than on the surface.

Verification of these predictions by direct measurement of T-tubular potentials was not possible until subsequent introduction of potentiometric dyes. Nevertheless, spread of depolarisation into the T-tubules could be assessed under microscopy by visually observing propagation of mechanical activity from superficial to axial myofibrils, in voltage clamped frog twitch muscle fibres (Adrian *et al.*, 1969). These findings suggested that effective T-tubular depolarisation indeed depended on I_{Na}. Under conditions of reduced $[Na^+]_o$ or action potential block by TTX, it was only possible to induce shortening over the entire fibre cross section, using voltage steps several mV depolarised relative to threshold. Otherwise, only superficial and not axial myofibrils contracted. In contrast, effective radial spread of activation took place under conditions of normal $[Na^+]_o$ in the absence of TTX. A just-threshold depolarisation successfully produced contraction of even the most axial myofibrils in some preparations. The radial spread of activation took place with a ~0.6 ms delay at 20 °C for a 100 μm diameter fibre. The delay increased with decreasing temperature with a Q_{10} of 2.0 (Costantin, 2011).

10.5 | Partial Separation of Surface and Transverse Tubular Electrophysiological Activity

Successful computational reconstruction of the observed surface action potential waveform combined with action potential generation throughout the T-tubular system following the surface excitation (Figure 10.8B) also required the existence of an access resistance, r_{ac}, restricting external electrical access to the T-tubular lumina. This might be provided by partial luminal constrictions in the mouths of the T-tubules (Adrian and Peachey, 1973).

Such a partial T-tubular isolation in turn predicted a corresponding partial separation of early components of the action potential resulting from depolarisation of the surface membrane from likely later components of the action potential attributed to T-tubular activity. Experimentally observed action potential waveforms indeed comprised early rapid deflections quickly recovering to form prolonged after-depolarisations (arrowed: Figure 10.9A, B). This contrasts with the after-hyperpolarisation phases shown by nerve action potentials (Section 2.2). The after-depolarisation phases became more prominent with increases in temperature, likely reflecting the temperature sensitivity of an active rather than a passive process. However, progressively increasing temperature eventually resulted in an all-or-none disappearance of the delayed T-tubular component, whilst sparing the surface membrane action potential component (Figure 10.9C; Padmanabhan and Huang, 1990). This fulfils expectations of a progressive increase in surface action potential conduction velocities that finally results in their transit times through any given region of surface membrane becoming insufficient to initiate T-tubular excitation.

These findings are compatible with a rapid surface membrane spread of the depolarisation wave, whilst ensuring sufficient local circuit current to trigger a separate, more slowly propagated T-tubular depolarisation. The presence of the T-tubules results in a five- to tenfold higher membrane capacitance referred to unit muscle length when compared to axonal surface membranes. Increased capacitance would tend to reduce action potential conduction velocity (Section 6.3). However, the partial isolation of the T-tubules from the surface membrane by the access resistance, r_{ac}, would reduce the slowing effect of this T-tubular capacitative load and maximise surface action potential conduction velocity.

Experiments detaching the T-tubular lumina from the extracellular space, leaving otherwise functioning surface membrane verified

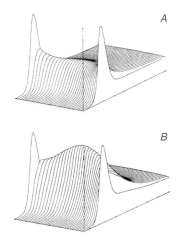

Figure 10.8 Action potential generation in surface and transverse (T-) tubular membranes of amphibian skeletal muscle. Vertical axis: membrane potential changes resulting from action potentials plotted against the time along the horizontal axis running to the right. These are represented through the surface membrane and T-tubules, along the cross section of the fibre, of diameter 50 μm, along the horizontal distance axis running to the left. In (A) regenerative activity propagating along the length of the surface membrane is permitted only to spread passively into a (T-) tubular system modelled without Na⁺ channels. In (B) the presence of T-tubular Na⁺ channels results in an inward propagation of regenerative activity, and a successful T-tubular excitation. In both (A) and (B) T-tubular and surface membranes are separated by a 150 Ω cm² access resistance. (From Adrian and Peachey, 1973.)

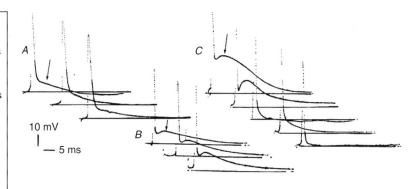

Figure 10.9 Surface and transverse (T-) tubular (arrowed) components of amphibian sartorius muscle fibre action potentials (A). A 4 °C temperature increase shortens the initial peak, but results in more prominent after-potentials with noticeable rising phases (B). However, further warming results in an all-or-nothing disappearance of the delayed T-tubular wave (C). (From Padmanabhan and Huang, 1990.)

10 mV

— 5 ms

this prediction. Such a tubular isolation was accomplished by an osmotic shock produced by introduction and withdrawal of extracellular glycerol (Fraser *et al.*, 1998). As expected, this also abolished contractile activation, leaving surface electrical activity intact. Confocal microscope sections demonstrated a normal T-tubular structure in di-8-ANEPPs-stained muscle prior to the osmotic shock (Figure 10.10*Aa*). Adding di-8ANEPPs followed by osmotic shock trapped the dye within a detached vesiculated T-tubular system, even following dye washout from the extracellular fluid (Figure 10.10*Ab*). In contrast, adding di-8ANEPPs both 15 min (Figure 10.10*Ac*) and 45 min after osmotic shock (Figure 10.10*Ad*) did not produce such internal membrane staining. These findings together demonstrate a separation of the T-tubular lumina from the extracellular space by the osmotic shock (Sheikh *et al.*, 2001).

Figure 10.10*B* shows the corresponding action potential waveforms (panel (i)) and their time derivatives giving the corresponding rates of voltage change (dV/dt) (panel (ii)). Fibres spared osmotic shock demonstrated intact early depolarisation and recovery, as well as after-depolarisation phases, marked by the thick lines below the traces, attributed to surface and T-tubular membrane excitation respectively (Figure 10.10*Ba*(i)). Osmotically shocked fibres showed a loss of the after-depolarisation phase consistent with a selective loss of the T-tubular component of the action potential (Figure 10.10*Bb*(i)). However, action potential conduction velocities, deduced from the latencies between the stimulus artefact, and time at peak action potential upstroke rate (dV/dt), (Figure 10.10*Bb*(ii)) were paradoxically unaffected by T-tubular detachment, notwithstanding this likely reducing the total, surface and tubular, membrane capacitance of such muscle fibres.

These findings suggest that at least part of the local circuit current (Section 6.3) produced by the surface action potential is directed along the membrane surface rather than into the T-tubular system. This would maximise the propagation velocity of the surface action potential. T-tubular activation would nevertheless be initiated provided sufficient local circuit current initiates a separate activation of

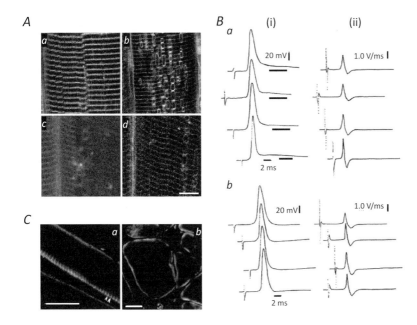

A

C

B
a (i) (ii)

20 mV| 1.0 V/ms |

2 ms

b

20 mV| 1.0 V/ms |

2 ms

Figure 10.10 Experiments on action potential waveform and conduction velocity in detubulated amphibian skeletal muscle fibres. (A) Single-photon confocal microscopic optical sections using the fluorescent membrane-impermeant dye di-8-ANEPPs added to the extracellular solution. (*a*) Fibre spared osmotic shock showing stained intact (T-) tubular system. (*b*) Osmotic shock procedure applied in the presence of extracellular dye: dye trapped in T-tubular vacuoles, even following dye washout from remaining bathing solution. (*c*, *d*) Appearances after osmotic shock followed by dye addition after 15 min (c) and 45 min (*d*). Scale bar: 10 μm. (B) Action potential waveforms (i) and corresponding time derivatives giving (dV/dt). (ii): (*a*) Traces from fibres spared osmotic shock: intact after-depolarisation phases (marked by horizontal bar beneath traces). (*b*) Osmotically shocked fibres, with loss of the after-depolarisation, but unchanged latencies as determined from the time interval from stimulus artefact to peak (dV/dt). (C) Plasmalemmal and T-tubule Na⁺ channel distributions in muscle fibres visualised by immunostaining using an antibody to a conserved epitope of voltage-gated Na⁺ channel α-subunits, using confocal microscopy. (*a*) Longitudinal section demonstrates immunostaining particularly at the mouth of the T-tubules (arrowheads; scale bar, 20 μm). (*b*) Transverse sections show preferential distribution close to the (T-) tubular system perimeter (Scale bar, 20 μm). (From Figures 4B, 4C, 6, 7a,c of Sheikh *et al.*, 2001.)

T-tubular action potentials as the propagating surface action potential passes each successive T-tubular opening. Such a notion was compatible with the selective immunological localisations of increased Na⁺ channel densities at the peripheries of the T-tubules that could mediate a T-tubular excitation process partially distinct from the surface excitation (Figure 10.10*Ca,b*).

10.6 | Electrophysiological Relationships Between Surface and Transverse Tubular Propagation of Excitation

These features of surface and T-tubular excitation ensuring both maximal surface conduction velocity and full T-tubular excitation were explicable in terms of the respective cable properties of the surface and T-tubular membranes (Sections 6.1–6.3). In common with nerve, skeletal muscle possesses a surface membrane comprising parallel resistance, r_m, and capacitance, c_m, components (Figure 10.11*A*). However, this is connected to a T-tubular component (Figure 10.11*B*). The latter was initially approximated by a simple lumped luminal resistance r_L in series with parallel, membrane resistance, r_T and capacitance, c_T components. These elements were separated from the surface membrane by access resistance, r_{ac} (Figure 10.11*B*). Subsequent, more precise, representations employed geometrically more realistic distributed T-tubular network representations (Figure 10.11*C*).

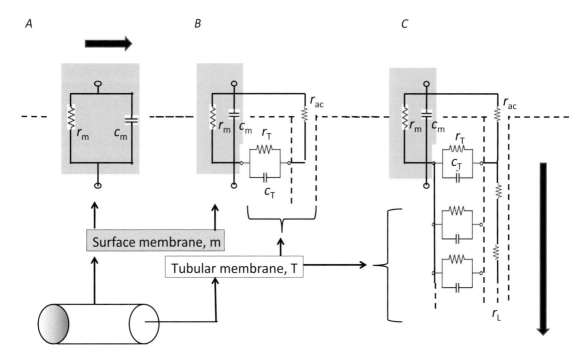

Figure 10.11 Electrical representations of (A) surface membrane with addition of a (B) lumped and (C) distributed transverse (T-) tubular membrane, showing surface and T-tubular membrane, resistances r_m, r_T and capacitances c_m, c_T, and T-tubular access resistance r_{ac}. The last model (C) provides a more realistic electrical representation of the (T-) tubular system shown in Figure 10.6.

Cable theory would predict that spread of surface membrane action-potential electrical excitation is driven by its rapid upstroke phase, with its large rate of voltage change (dV/dt). As $i_m \approx c_m$(dV/dt), most of the transmembrane current, i_m, charges the membrane capacitance, c_m, as opposed to flowing through the r_m component. The resulting conduction velocity θ is determined primarily by the axial resistance r_a, and the capacitance term c_m (Section 6.3). In contrast, the i_m associated with the more gradual T-tubular excitation mainly involves the T-tubular membrane resistance, r_T. Propagation of T-tubular depolarisation is consequently primarily affected by the resistance terms r_{ac}, r_L and r_T (Pedersen *et al.*, 2011).

This situation was examined in rat extensor digitorum longus muscle fibres from which voltage signals were obtained from three voltage-recording, intracellular electrodes spaced along the fibre length (Figure 10.12*Ba*). These were placed at different distances from a current electrode injecting subthreshold sinusoidal input currents over a range of low, direct current (d.c., 0 Hz) to high frequencies (800 Hz; Figure 10.12*Aa*). This made it possible to determine the frequency dependences of the length constants, λ, and conduction velocities, θ, using the amplitudes and phases of the sinusoidal wave-forms of the resulting passive output membrane voltage changes (Figure 10.12*Bb,c*, upper panels). These plots conformed to expectations of a distributed model (Figure 10.11*C*) for muscle membrane geometry (Figure 10.12*Bb,c*, lower panel). Such a distributed model also resulted in a greater λ and θ at the higher signal frequencies

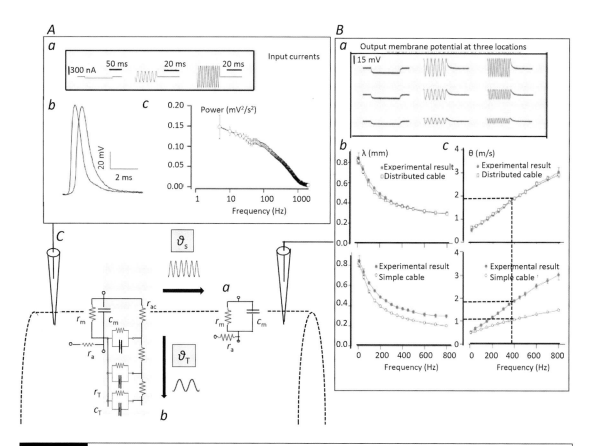

Figure 10.12 Transfer impedance analysis of surface and transverse (T-) tubular conduction in rat extensor digitorum longus muscle fibres. (A) Application of (a) steady and time-varying, subthreshold sinusoidal input currents over 0 to 800 Hz or (b) stimulation of an action potential through a intracellular microelectrode represented as a (c) spectrum of amplitudes at different frequencies. (B) Recordings of (a) outputs of the sinusoidal stimuli at three separate microelectrode recording sites spaced along the length of the fibre to derive (b) their amplitude and extent of spread, λ, as well as (c) their velocity of conduction, θ, at different frequencies. Comparison of experimental findings with those obtained from a distributed (top panels) and from a simple electrical model (bottom panels) indicated (C) preferential filtering of (a) higher frequencies into surface and (b) lower frequencies into T-tubular activation. Mathematical symbols as in Figure 10.11. (From Figures 1A, 2A, 3A, 3B of Pedersen et al., 2011.)

relative to a simple model leaving the entire T-tubular capacitance entirely accessible from the membrane surface. Thus, in skeletal muscle, higher-frequency signals are preferentially propagated along the surface as opposed to the tubular membrane. This would leave the lower-frequency signals to be admitted to and charge the T-tubular membrane.

To establish in vivo implications for these properties, these frequencies and their corresponding velocities were then compared to those displayed by stimulated action potentials (Figure 10.12Ab). These had conduction velocities around ~1.92 m/s that mapped onto a frequency of ~400 Hz in the plots of passive conduction of a sinusoidal signal (dotted lines, Figure 10.12Bc top panel). At such a frequency, the distributed model would show a substantially enhanced surface action potential conduction velocity compared to expectations from a simple model (dotted lines, Figure 10.12Bc bottom panel). Furthermore its representation as a frequency spectrum revealed that the action potential waveform was adapted to achieve such a conduction velocity. This spectrum was concentrated at 0–1 kHz and had a median frequency of ~396 Hz (Figure 10.12Ac). The latter, and higher frequencies than this, could be filtered towards contributing to surface membrane action potential conduction (Figure 10.12Ca). The spectrum also included lower frequencies that would exert little influence on the higher-frequency processes of surface action potential propagation. These instead would be filtered into initiating T-tubular excitation and the consequent slower kinetic processes involved in excitation–contraction coupling discussed in Chapter 11 (Figure 10.12Cb).

10.7 | Physiological Modulation of Surface and Transverse Tubular Membrane Function

The effects of the T-tubular membrane resistances r_T on T-tubular excitation reflect contributions made by alterations in the key background Cl^- and K^+, particularly ClC−1 and K_{ATP}, conductances. Alterations in these are implicated in functional adaptations of skeletal muscle to different physiological conditions, particularly exercise.

First, exercise markedly alters these conductances. This was replicated by episodic sustained action potential firing, recorded by a voltage electrode, in response to periodic trigger pulses delivered by a second current electrode in rat muscle (Figure 10.13Aa; Pedersen et al., 2009; Fraser et al., 2011). Membrane conductance was measured from the voltage deflection produced by a hyperpolarising current pulse between firing episodes (Figure 10.13Ab). Muscle activity initially ('Phase I') reduced the background membrane conductance to ~40% of its resting level (Figure 10.13Aa(ii)). This was attributable to a phosphokinase-C (PKC)-dependent inhibition of ClC-1 Cl^- channels (Section 10.1). It was observed in both extensor digitorum longus fast- and soleus slow-twitch muscle fibres. This reduced background conductance would correspondingly reduce the applied current required to alter the membrane voltage to the Na^+ channel activation threshold and enhance membrane excitability (Figure 10.13C).

Secondly, more prolonged firing ('Phase II') increased the membrane conductance to around five times its resting level in fast- but not slow-twitch fibres (Figure 10.13Aa,b(iii)). This was attributable to activation of both ClC-1 and ATP-sensitive K^+ channels (Section 10.1), in a scheme suggesting a simple division of the sequence of decreased and increased membrane conductances respectively into Phase I and Phase II responses (Figure 10.13B). The increased background

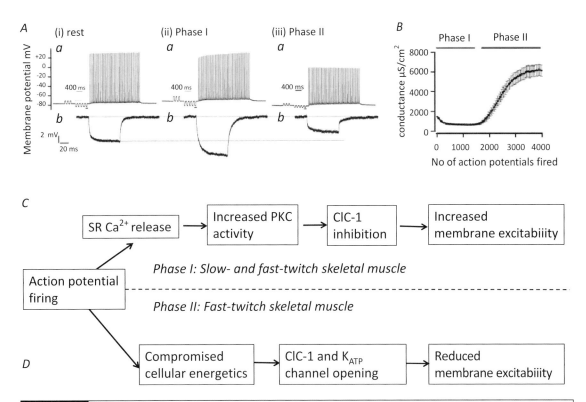

Figure 10.13 Changes in membrane conductance with the onset of, and following prolonged action potential firing, measured by recording electrodes (Pedersen et al., 2009). (A) Consecutive 15 Hz firing episodes induced by a current injection electrode. (a) 15 Hz action potential trains in a fast-twitch extensor digitorum longus fibre during the first ((i) 'rest'), the 40th ((ii) 'Phase I'), and the 80th train ((iii) 'Phase II') of 49 action potentials in cycles each lasting 7 s. (b) Conductance measurements between action potential trains from voltage responses to current steps. Note decreased and increased membrane conductance as indicated by the voltage deflections in Phase I (ii) and Phase II (iii) respectively, as well as shifts in resting potential prior to stimulus application. Thus: (B) The onset of action potential firing was associated with reduced membrane conductance in fast-twitch fibres ('Phase I'), findings also seen in slow-twitch fibres. However, prolonged firing ('Phase II') markedly increased conductance, but only in fast-twitch fibres. (C, D) Dynamics of membrane conductance changes and their underlying cellular mechanisms with sustained action potential activity in fast- (C, D) and slow-twitch (C) skeletal muscle fibres. ((A, B) from Figure 9A, C of Fraser et al., 2011, (C, D) based on Figure 9 of Pedersen et al., 2009.)

conductance would increase the applied current required to alter the membrane voltage to threshold, decreasing membrane excitability (Figure 10.13D).

Thirdly, activation of ATP-sensitive K^+ channels causes an efflux of cytosolic K^+ into the transverse T-tubular lumina. The latter are partially isolated from the remaining extracellular fluid (Section 10.4). They form a restricted extracellular compartment with which diffusion equilibration with remaining extracellular fluid is relatively slow. This results in a region within which K^+ can accumulate during prolonged action potential firing. This would result in a local sustained membrane depolarisation, which would promote Na^+ channel inactivation and refractoriness and thereby further depress membrane excitability.

10.8 | Reconstruction of Transverse Tubular Functional Changes During Exercise

Computational modelling of the physiological consequences of these T-tubular conductance changes combined these with the T-tubular cable properties discussed above (Section 10.6; Fraser *et al.*, 2011). They explored effects of 40% decreases (Plate 7 centre column, Phase I) and 500% increases (Plate 7 right column, Phase II) in T-tubular conductance relative to that in resting muscle (Plate 7, left column), known to occur at the onset and with prolonged exercise in a fast-twitch muscle.

This first tested surface membrane excitability leading to action potential generation (Plate 7*A* top row) in response to 1 ms duration step current stimuli at varying amplitudes (lower panels). This determined the stimulus amplitudes required to excite an action potential (top panels). Compared to the required stimulus strengths in resting muscle (left panel, control) or with the decreased membrane conductance at the outset of activity (centre panel, Phase I), the increased membrane conductances resulting from prolonged activity (right panel, Phase II) resulted in a requirement for markedly increased stimulus strengths for action potential generation suggesting an eventual decrease in excitability.

Secondly, subthreshold brief current injections gave passive membrane potential decays that were significantly prolonged with the decreased T-tubular conductance during Phase I (Plate 7*B*). Effective excitation thresholds were also modified (Plate 7*C*). It was possible to select a current stimulus that would only trigger a surface as opposed to a T-tubular action potential in Phase I, but not either in a resting fibre or following Phase II prolonged activity. In contrast, a current of amplitude just sufficient to elicit an action potential in a fibre following Phase II prolonged activity also successfully elicited activity at rest or in Phase I activity (Plate 7*D*).

Thirdly, exercise conditions also altered T-tubular activation during surface membrane action potential firing. Plate 7*E* demonstrates normal surface (black line) and T-tubular action potential timecourses and amplitudes shown for 20 concentric T-tubular system shells 3 mm from the point of action potential excitation during Phase I (coloured lines). The voltage deflections, particularly in the T-tubules, were significantly reduced following Phase II prolonged activity. Further depressing surface membrane excitability to produce a 50% reduction in maximum permeability for voltage-gated Na^+ channels as might occur with the increased extracellular $[K^+]_o$ encountered in working muscle further accentuated the above effects (Plate 7*F*).

A rapid reduction in membrane conductance in the T-tubules at the start (Phase I) of muscle activity enhances neuromuscular transmission and T-system excitation. However, the increased conductance after prolonged, Phase II, activity inhibits these processes, contributing to muscle fatigue. These effects are accentuated by the accompanying depression of excitability caused by membrane depolarisation arising from accumulated increases in T-tubular $[K^+]_o$ in active muscle.

10.9 | Functional and Clinical Implications of Altered Transverse Tubular Membrane Properties

These findings indicating inverse relationships between membrane conductance and muscle excitability have direct implications for normal muscle T-tubular function. Reduced membrane conductance at the onset of exercise enhances T-system excitation. It also facilitates recovery of excitability and force production in muscle previously depolarised by elevated $[K^+]_o$. It could counteract fatigue caused by electrolyte shifts in active muscle (Nielsen *et al.*, 2001; Pedersen *et al.*, 2004). However, prolonged muscle activity leading to pronounced reductions in the cellular energetic state of the fibres increases Cl^- and K^+ permeability particularly in fast- as opposed to slow-twitch muscle. Its effect in reducing fibre excitability could furnish a fibre type specific fatiguing mechanism linking cellular energetic state to muscle fibre excitability. Metabolically poisoned frog muscles show a high membrane conductance, and are completely inexcitable, despite persistence of normally polarised membrane potentials (Pedersen *et al.*, 2009).

The findings also impact upon abnormal muscle function. ClC-1 Cl^- conductance tends to stabilise the membrane potential at its resting level (Section 10.1). Conversely, loss of ClC-1 channel function would lead to membrane hyperexcitability. This may form the basis of myotonic dystrophies, autosomal dominant disorders characterised by abnormal sarcolemmal Na^+ or Cl^- channels. Of these, *myotonia congenita* is characterised by an abnormal or absent sarcolemmal ClC-1 Cl^- channel. The resulting membrane hyperexcitability leads to involuntary repetitive action potentials and sustained muscle contraction. The clinical result is a myotonia taking the form of abnormally prolonged or repetitive muscle contractions following voluntary relaxation (Adrian and Bryant, 1974). Myotonic dystrophy is also associated with generalised muscle weakness and ageing; myotonia congenita may also cause palatopharyngeal dysfunction, leading to dysphagia, or difficulty in swallowing (Thornton, 2014; Turner and Hilton-Jones, 2014).

11

Excitation–Contraction Coupling in Skeletal Muscle

11.1 | Dependence of Excitation–Contraction Coupling on Membrane Potential

Excitation–contraction coupling refers to the sequence of events connecting action potential activation to initiation of tension generation by actin–myosin interaction. This process shows physiological and pharmacological properties and kinetics distinct from those of its initiating electrical changes. Its immediate trigger is the voltage change produced by membrane excitation. The in vivo action potential involves changes in Na^+ conductance, delayed K^+ rectification and inward rectifying K^+ conductances (Sections 10.2 and 10.4). However, these processes affect contractile activation entirely through their effects on membrane potential. Imposed depolarisation itself induces contractile activation, even when action potential generation is inhibited by tetrodotoxin (TTX) challenge or replacing extracellular Na^+, $[Na^+]_o$, by choline.

The initial experimental studies directed at excitation–contraction coupling depolarised membranes of amphibian semitendinosus muscle fibres by exposing them to high $[K^+]_o$ (Section 10.1). This manoeuvre induced prolonged K^+ contractures (Figure 11.1A). For example, increasing $[K^+]_o$ to 20 mM depolarises the membrane to approximately –50 mV; this initiated a prolonged contracture developing over 2–3 s to reach a steady plateau tension. Larger depolarisations steeply increased both the rates of development and maximum amplitudes of this mechanical response. In contrast to the all-or-nothing nature of the action potential, maximum contracture tensions were graded, albeit steeply, with $[K^+]_o$ through a sigmoidal relationship (Figure 11.1B). The transition from full relaxation to maximum tension often to values comparable to full tetanic tension was achieved within a voltage increment of $<+20$ mV.

These K^+ contractures could be reversed by restoring normal $[K^+]_o$. In addition, after reaching its maximum plateau level, contracture tension declined to baseline even during the maintained depolarisation. Both the extent and rate of this inactivation also steeply increased

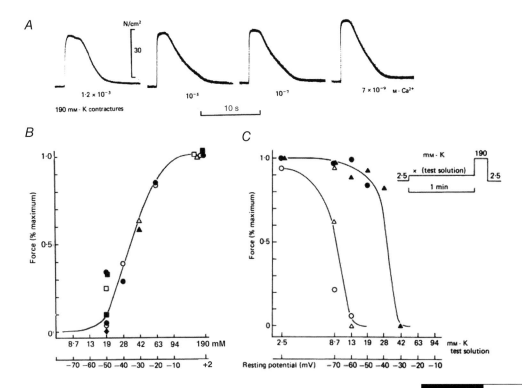

Figure 11.1 Voltage dependence of contractile activity in amphibian skeletal muscle. (A) Contractures induced by exposure to 190 mM $[K^+]_o$ at different Ca^{2+} concentrations, $[Ca^{2+}]$ (in M). (B) Dependence of force generation, normalised to maximum force, on $[K^+]_o$ (logarithmic scale) also expressed as the corresponding membrane potential. (C) Inactivation of K^+ contractures brought about by a conditioning depolarisation (inset). Filled symbols: experiments in 3-0-3.2 mM $[Ca^{2+}]$; open symbols: experiments in 3.05-3-2 mM $[Mg^{2+}]$, and 10^{-9} M $[Ca^{2+}]$. ((A) from Figure 2, (B) from Figure 5 and (C) from Figure 6 of Lüttgau and Spiecker, 1979.)

with increasing depolarisation (Figure 11.1C). Contracture plateaus in frog semitendinosus fibres at 20 °C lasted ~7 s at −40 mV, but ~2 s at 2 mV. The fibres also recovered from inactivation, regaining their ability to contract, to extents and rates varying with the subsequent repolarisation. These findings suggest that excitation–contraction coupling comprises separable activation and inactivation processes each with their own distinct voltage-dependent kinetic and steady-state properties (Costantin, 2011).

11.2 | Involvement of Intracellular Ca²⁺ in Excitation–Contraction Coupling

A second major feature of excitation–contraction coupling is its dependence on alterations in subcellular Ca^{2+} distribution and its consequences for the cytosolic Ca^{2+} concentration, $[Ca^{2+}]_i$. First, activation of the actin–myosin interaction initiating mechanical activity requires Ca^{2+} binding to troponin (Section 9.10). In isolated actomyosin preparations obtained by homogenising muscle cells followed by differential centrifugation of the resulting homogenate, increasing free $[Ca^{2+}]$ to ~1 μM induced ATPase activity and actin–myosin precipitation. Experiments microinjecting EGTA-buffered Ca^{2+}-containing solutions in crustacean muscle fibres and studies in chemically or mechanically skinned vertebrate skeletal or cardiac muscle suggested similar effective free $[Ca^{2+}]_i$ was required to achieve contraction threshold.

Secondly, in intact muscle fibres, contractile activation persists with altered extracellular Ca^{2+} concentration, $[Ca^{2+}]_o$, though this modifies the voltage dependence of its inactivation (Figure 11.1C). Contracture tension generation in response to elevated $[K^+]_o$ persisted in vertebrate twitch muscle in Ca^{2+}-free, EGTA and Mg^{2+}-containing solutions (Figure 11.1A, B). The increased $[Ca^{2+}]_i$ required for contractile activation is thus dependent upon Ca^{2+} release from an intracellular store, rather than its entry from the extracellular space.

Thirdly, biochemical experiments identified a potential subcellular fraction containing such a store from which Ca^{2+} could be released into and subsequently retrieved from the cytosol. This intracellular Ca^{2+} sequestering fraction was first isolated as a relaxing factor from the microsomal fraction by differential centrifugation of homogenised rabbit muscles. It contained 100 nm diameter microsomal vesicles appearing to correspond to the sarcoplasmic reticulum (SR) in vivo. The isolated vesicles accumulated Ca^{2+} against a concentration gradient, sequestering $[Ca^{2+}]_i$ sufficiently to dissociate Ca^{2+} from the contractile proteins, in the presence of ATP. They also had a substantial Ca^{2+} storage capacity. This was demonstrated in skinned fibres by precipitating the accumulated Ca^{2+} with intraluminally introduced oxalate. Their membranes showed an active Ca^{2+} transport process dependent upon a Ca^{2+}-ATPase action capable of generating 1000-fold transmembrane Ca^{2+} activity gradients, further discussed in Section 11.3. The resulting Ca^{2+} storage capacity was more than sufficient to account for the observed total increase in $[Ca^{2+}]_i$ associated with contractile activation (Ebashi and Endo, 2003).

11.3 | The Measurement of Intracellular Ca^{2+}

Studies exploring the role of Ca^{2+} in excitation–contraction coupling were greatly facilitated by measurements of changes in optical properties of particular molecules specifically binding to and thereby detecting intracellular alterations in ionised Ca^{2+} concentration ($[Ca^{2+}]_i$). This approach to studying Ca^{2+} homeostasis subsequently proved central to investigations in a wide range of other cell types. The synthetic probe molecules ideally selectively bind Ca^{2+} as opposed to other ions, particularly Mg^{2+} and H^+, and do so with consistent binding stoichiometry and dissociation constants permitting determinations of ~10 nM $[Ca^{2+}]_i$ levels. In addition, they should be readily introducible into the cell intracellular space, rapidly diffuse and achieve uniform concentrations throughout the cytosolic compartment without traversing

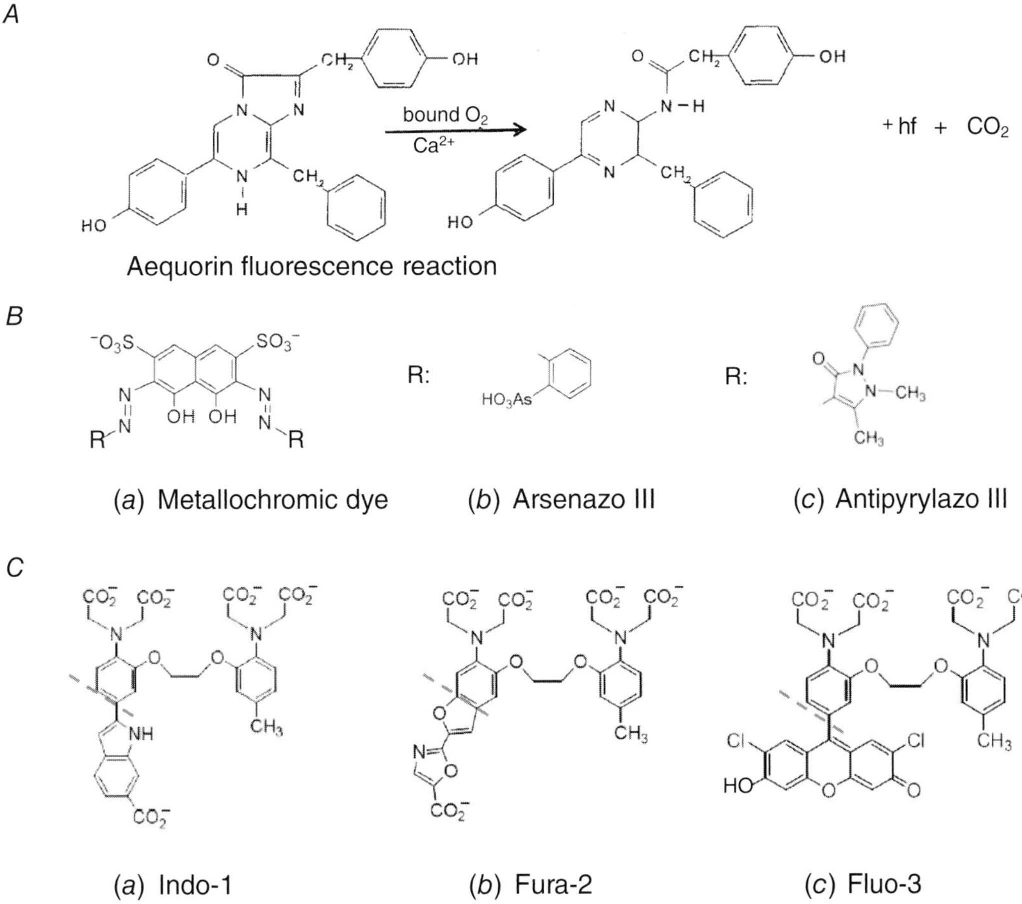

A

Aequorin fluorescence reaction

+ hf + CO$_2$

B

(a) Metallochromic dye   (b) Arsenazo III   (c) Antipyrylazo III

C

(a) Indo-1   (b) Fura-2   (c) Fluo-3

Figure 11.2 Ca^{2+}-sensitive agents used in studies of skeletal muscle excitation–contraction coupling. (A) Photoreaction underlying aequorin fluorescence. (B) Metallochromic absorbance dyes: (a) parent structure, (b, c) functional groups R, found in arsenazo III (b) and antipyrylazo III (c). (C) Fluorescent Ca^{2+}-sensitive compounds developed from an eight-coordinate tetracarboxylate Ca^{2+} chelating site with stilbene chromophores: (a) indo-1, (b) fura-2 and (c) fluo-3. ((C) adapted from Figure 3 of Huang, 2018.)

intracellular membranes to enter other compartments. In studies of skeletal muscle, these 'dyes' should respond sufficiently rapidly to signal millisecond timecourses of Ca^{2+} changes.

A classical example of a Ca^{2+}-detecting molecule is the Ca^{2+}-activated photoprotein aequorin extracted from the bioluminescent jellyfish *Aequoria forskalea*. This emits a blue light following a chemical reaction triggered by Ca^{2+} binding to one of its prosthetic groups (Figure 11.2A). However, although aequorin luminescence signals offered reproducible representations of Ca^{2+} transients, aequorin suffers from low Ca^{2+} affinities and nonlinear, 2.5th power relations between signal and [Ca^{2+}]$_i$. The latter complicates quantitative interpretation particularly in systems with heterogenous myoplasmic ion distributions. However, subsequently developed metallochromic dyes (Figure 11.2Ba) signalled absorbance changes in response to Ca^{2+} with higher and more linear Ca^{2+} affinities. Their more rapid Ca^{2+} binding (arsenazo-III, 2–3 ms: Figure 11.2Bb; antipyrylazo-III, <<~1 ms: Figure 11.2Bc) could track rapid [Ca^{2+}]$_i$ changes in skeletal muscle.

This binding was detected using absorbance measurements at light wavelengths at which such absorbances were either maximal or minimal. Simultaneous tracking of two such wavelengths could be used to correct for accompanying $[Mg^{2+}]_i$ or $[H^+]_i$ changes.

A further group of Ca^{2+}-sensitive compounds were developed from an eight-coordinate tetracarboxylate Ca^{2+} chelating site attached to stilbene chromophores (Figure 11.2C). They showed further improved Ca^{2+}-selectivity and could be used in fluorescence studies. Their intracellular access was often conveniently achieved through extracellular introduction of their membrane-permeant acetomethoxy (-AM) derivatives. In some, Ca^{2+} binding additionally produced wavelength in addition to intensity changes. Indo-1 (Figure 11.2Ca) has a dual emissions peak, with its main emission shifting from 475 to 400 nm with Ca^{2+} binding. Fura-2 (Figure 11.2Cb) emits at a single 510 nm wavelength, but has a peak excitation that shifts from ~380 to 350 nm with Ca^{2+} binding, giving emission ratios directly related to $[Ca^{2+}]$. Fluo-3 (Figure 11.2Cc) is amenable to visible, including 488 nm argon laser, excitation, increasing its fluorescence with Ca^{2+} binding, about a single-peak 525 nm emission wavelength. Its more rapid Ca^{2+} dissociation permits tracking of rapid $[Ca^{2+}]_i$ kinetics in both skeletal and cardiac muscle.

Finally, use of fluorescent proteins expressed through introduction of their encoding genes into the cell types under study involves transfection of their DNA, as opposed to direct introduction of the protein molecules. This group of agents, exemplified by green fluorescent protein (GFP), has had major impact on cell physiological studies extending to cell types other than muscle cells (Miyawaki et al., 1997).

11.4 | Voltage-Dependent Release of Intracellularly Stored Ca^{2+} in Excitation–Contraction Coupling

Classic studies of aequorin fluorescence provided direct evidence that imposed depolarisation caused elevations in $[Ca^{2+}]_i$ (Figure 11.3A; Ashley and Ridgway, 1968). Aequorin solutions injected into large muscle fibres of the barnacle, *Balanus nubilis*, produced a faint glow of light, measureable using a photomultiplier tube, suggesting an increase in free cytosolic Ca^{2+}, with electrical stimulation (Figure 11.3B). The timecourse of the Ca^{2+} transient followed that of the depolarisation, but preceded the ensuing tension change (Figure 11.3C).

Further studies quantifying $[Ca^{2+}]_i$ transients employed the absorbence dyes arsenazo III or antipyrylazo III to measure $[Ca^{2+}]_i$ in amphibian muscle (Kovács et al., 1979). Muscle fibres were stretched across an extracellular pool sealed from a surrounding compartment

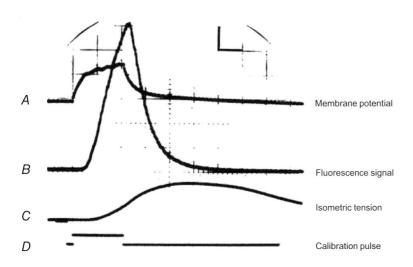

A — Membrane potential

B — Fluorescence signal

C — Isometric tension

D — Calibration pulse

Figure 11.3 Membrane depolarisation, aequorin transient and force generation in barnacle muscle fibre. (A) Membrane response; vertical scale bar denotes 20 mV, horizontal scale bar denotes 100 ms. (B) Ca^{2+}-mediated light emission; vertical scale bar denotes 1.9×10^{-9} lm/cm. (C) Isometric tension; vertical scale bar denotes 5 g/cm. (D) Stimulus mark and 1 V calibration pulse. (From Figure 3 of Ashley and Ridgway, 1970.)

made directly continuous with the fibre intracellular space by a cut in the fibre. The latter was also used to introduce the dye whose absorbence of an incident light beam was detected through an objective (Figure 11.4A). Antipyrylazo III proved popular for selective and sensitive kinetic measurements of ionised $[Ca^{2+}]_i$. Ca^{2+} binding increases antipyrylazo III absorbances particularly at wavelengths of 660 and 800 nm at which Mg^{2+} has no effect. This permitted selection of wavelength pairs (720–790, 675–690 nm), where Ca^{2+} transients could be measured without interference from Mg^{2+}. In a buffered reaction mixture at pH 7 containing 100 mM KCl, the antipyrylazo III-Ca^{2+} complex has a dissociation constant K_D of 160 µM, and a relaxation time of 180 µs. Even large antipyrylazo III concentrations did not appear to bind to cells or organelles or otherwise affect cellular function.

The antipyrylazo III signals obtained in response to voltage clamp steps (Figure 11.4B) demonstrated a distinct ~ −40 mV threshold, beyond which Ca^{2+} release increased steeply with voltage. Even small, ~1.9 mV, further depolarisations caused e-fold changes in the calculated peak $[Ca^{2+}]_i$ (Figure 11.4C; Maylie et al., 1987). This saturated at a maximum amplitude at positive voltages. The Ca^{2+} signals nevertheless persisted even with strong depolarisations to +170 mV or conditions of $[Ca^{2+}]_o$ chelation by EGTA, both of which would reduce or even reverse the inward driving force for Ca^{2+}. This finding attributes the observed $[Ca^{2+}]_i$ changes to a release of intracellularly stored Ca^{2+} that saturates at positive voltages, rather than extracellular Ca^{2+} entry. It also accounts for skeletal muscle contraction persisting in the absence of $[Ca^{2+}]_o$. This release of intracellularly stored Ca^{2+} ultimately would initiate mechanical activity through its binding to the regulatory protein troponin (Section 9.10).

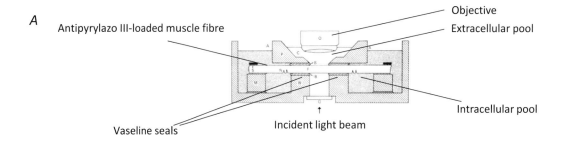

A

Antipyrylazo III-loaded muscle fibre

Objective

Extracellular pool

Intracellular pool

Vaseline seals

Incident light beam

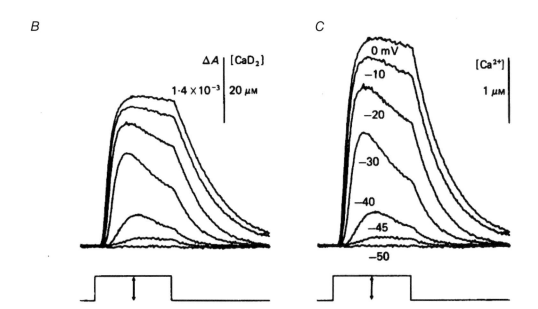

B

ΔA | [CaD$_2$]

$1 \cdot 4 \times 10^{-3}$ | 20 μM

C

0 mV
−10
−20
−30
−40
−45
−50

[Ca^{2+}]

1 μM

Figure 11.4 Ca^{2+} transients in skeletal muscle in response to voltage clamp steps to different test voltages. (*A*) Typical layout for measurement of free [Ca^{2+}]$_i$ transients in a muscle fibre in a vaseline gap voltage clamp. (*B*) Antipyrylazo III absorbance and (*C*) calculated free cytosolic [Ca^{2+}] recordings produced by 100 ms duration steps from resting potential to different test voltages (From Figures 1 and 10 of Kovacs et al., 1983.)

11.5 | Triad Complexes Between the Transverse Tubular and Sarcoplasmic Reticular Membranes

Release of the store of activating Ca^{2+} thus involves transduction of a membrane depolarisation initiated at the fibre surface. This is transmitted into the transverse (T-) tubules (Section 10.5) in turn causing a Ca^{2+} flux across the SR membranes (Section 11.2) that elevates [Ca^{2+}]$_i$. The extensive branching of the T-tubular network ensures that this has membrane elements in close anatomical proximity to both the myofibrils and the SR, even in the depths of the muscle fibre. In amphibian muscle this close proximity takes place in the region of

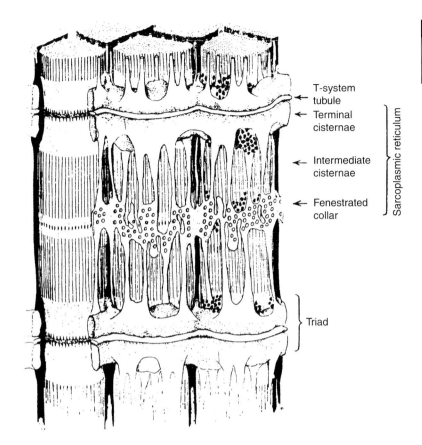

T-system
tubule

Terminal
cisternae

Intermediate
cisternae

Fenestrated
collar

Sarcoplasmic reticulum

Triad

Figure 11.5 The internal membrane systems of a frog sartorius muscle fibre. (From Peachey, 1965.)

the I-band. In most other skeletal muscles, including crab muscles, this takes place at the A–I boundaries. This directly corresponds to the sites at which local stimulation elicited muscle contraction (Section 10.3).

The SR forms a space continuous along the sarcomere length and throughout the fibre cross section. It forms flattened sacs, making up the longitudinal SR, surrounding the myofibrils from Z line to Z line. However, the SR lumina are separate from both the extracellular space and the T-tubular lumina from which they are electrically isolated. Electrical signals in the T-tubular membranes cannot be directly conducted into the SR membrane (Figure 11.5). The overall membrane surface capacitance of muscle reflects that of its surface and T-tubules, but excludes that of the SR.

However, the two membrane systems, T-tubular and SR, come into close geometrical proximity, forming triads. Here, a central T-tubular, terminal cisternal element is interposed between two slightly flat-tened terminal cisternal SR elements (Porter and Palade, 1957). There is, however, no continuity between these T-tubular and junctional SR membranes. Serial (80 nm) thin sections through the triad structures confirmed the presence of geometrically close but anatomically separated T-tubular and SR membranes. Furthermore, despite their close

Figure 11.6 Schematic representation of transverse (T-) tubular and sarcoplasmic reticular (SR) terminal cisternal structures that may be involved in excitation–contraction coupling. Terminal cisternal membranes of the SR come into close geometrical relationship with the transverse (T-) tubular membrane at the triad region. End-feet, identified with ryanodine receptor Ca^{2+} channels, occur in regular geometric array in cisternal membrane and fill the T–SR gap. Alternative feet occur in close relation to junctional tubular tetrads. Ca^{2+}-ATPase protein occurs in non-junctional SR membrane. Junctional SR lumen contains the Ca^{2+} binding protein calsequestrin. (From Block *et al.*, 1988.)

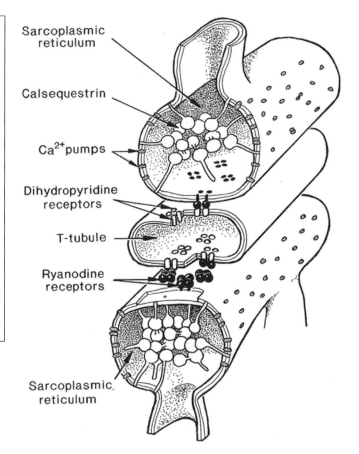

Sarcoplasmic reticulum

Calsequestrin

Ca^{2+} pumps

Dihydropyridine receptors

T-tubule

Ryanodine receptors

Sarcoplasmic reticulum

geometrical proximity, there is little evidence for any direct electrical communication between T-tubules and the SR. The T–SR complexes are nevertheless thought to be central to excitation–contraction coupling. Electronmicroscopy revealed that the T-tubular and junctional SR membranes appear connected by an array of structures resembling foot processes. These were subsequently shown to be anchored in the SR membrane (Figure 11.6; Block *et al.*, 1988), suggestive of direct coupling mechanisms by which the two membrane systems might communicate. Furthermore, both these triadic foot processes and a Ca^{2+} channel subsequently demonstrated from the SR terminal-cisternal membrane exhibited a specific binding for the plant alkaloid ryanodine.

11.6 | Triggering Molecules for the Release of Sarcoplasmic Reticular Ca^{2+}

Detailed examination of the foot processes employed freeze drying, rotary-shadowing, electronmicroscopic studies to isolate and study their underlying ryanodine-binding membrane receptors (RyR) occurring in the terminal cisternal SR. They demonstrated that each RyR consists of four subunits. Each is formed from large protein chains with over 5000 amino acid residues. Their C-terminal ends include a short 500 amino acid region resembling the transmembrane segments of the nicotinic cholinergic receptor (Section 7.7). Each such sequence assembles into an association of tetramers giving an overall molecular weight $>2 \times 10^6$ Da. The ryanodine-binding property requires the assembled tetramer.

Plate 8 illustrates recently reported, ~1.0 nm resolution, cryo-electronmicroscopy (cryo-EM) maps of the skeletal muscle, type 1, ryanodine receptor isoform RyR1. The atomic reconstructions in cytoplasmic view (Plate 8A) omit the transmembrane domain for clarity. They demonstrate a cytoplasmic region with a flat square prism $275 \times 275 \times 120$ Å shape, formed by tetrameric subunit association in square array, with an overall left-handed rotational asymmetry. Handle and clamp domains respectively form the sides and the corners of the square. The side, central slice (Plate 8B) and stereoscopic views (Plate 8C) confirm an overall mushroom shape, revealing the ~$120 \times 120 \times 60$ Å transmembrane region extending into the SR membrane. This is connected to the cytoplasmic region through four thick columns. The latter include six to eight transmembrane helices for each of the four subunits of which five to six were detected in the cryo-EM maps. Their inner helices create the pore-forming region mediating Ca^{2+} channel function in the RyR1 (Plate 8A).

Entry of the channel into the open state permits Ca^{2+} flow out of the SR lumen into the cytosol surrounding the myofibrils. The maps also demonstrate that the cytoplasmic portion extends ~10 nm beyond the cytoplasmic cisternal membrane surface. It could thus provide the foot processes spanning the gap between T-tubular and SR membranes. Images from negative staining of purified RyR1s suggested a complex pathway for Ca^{2+} flux through the RyR1 which involves a single pore in the transmembrane region and four branches outside the foot process.

The RyR C-terminal regions show consensus sequences expected for regulatory binding of Ca^{2+}, calmodulin, FK506 (tacrolimus)-binding protein (FKBP12) and nucleotides (Plate 9A). The central region of the RyR possesses two EF hand groups permitting Ca^{2+} binding, likely mediating direct regulatory actions of Ca^{2+} on RyR activity. Calmodulin (CaM) acts as a Ca^{2+}-sensing protein through its two high-affinity Ca^{2+}-binding C-terminal and two low-affinity N-terminal EF hand groups. Prior to such Ca^{2+} binding, its apoCaM form activates the channel. Following Ca^{2+} binding, Ca^{2+}–CaM becomes hydrophobic and this inhibits the RyR. The immunophilin FKBP12 shows a 1:1, nM affinity for individual RyR1 subunits with FKBP12 dissociation accompanying RyR1 activation. RyR action is also modulated by ATP and oxidants, including reactive oxygen species (ROS), and shows diverse post-translational modifications. Finally, the intracellular Cl$^-$-specific CLIC2 channel expressed in skeletal muscle, cardiac muscle and brain may exert inhibitory actions on both RyR1 and RyR2.

The skeletal muscle T-tubule membrane contains particles grouped in fours ('tetrads') positioned opposite the SR feet. These constitute the only major intramembrane particles in the T-tubular membrane. They likely correspond to groups of four voltage-gated modified Ca^{2+} (Cav1.1) channels, also known as dihydropyridine receptors (DHPRs) reflecting their binding of dihydropyridine derivatives. They exist in different, DHPR1 and DHPR2 isoforms in skeletal and cardiac muscle respectively. In toadfish swimbladder muscle, the tetrads are aligned in two equidistant rows, forming groups of four particles or tubular tetrads arranged with an off-register alignment.

DHPR1-RyR1 coupling occurs at regions of close apposition between the terminal cisternae of the SR containing two parallel rows of RyR1s, and the T-tubules, which show groups of four DHPR1s or tetrads. Every alternate RyR1 interacts with one DHPR1 tetrad. RyR1s show both side-by-side (Plate 9*B*) and oblique interactions (Plate 9*C*). In the latter case they give rise to molecular arrays of RyRs (Plate 9*D*), with the tetrads showing the same skew angle within the plane of the membrane as the RyR1s, but with distances between tetrads twice that between feet (28 nm). Superimposing the RyR arrays on a tentative footprint of the DHPR1 tetrads, each represented with a circumference (Plate 9*D*) predicts that the tetrads interact with every alternate RyR1. All the DHPR1 tetrads are closely applied to the cytoplasmic portions of a RyR1. Conversely, half of the RyR1s have associated DHPR1s. In amphibian muscle, the remainder have been identified with a further isoform, RyR3.

11.7 | Tubular Voltage Detection Mechanisms Triggering Excitation–Contraction Coupling

Events coupling tubular depolarisation with contractile activation are thought to begin with detection of this voltage change by a voltage sensor, involving changes in molecular configuration in the tubular membrane DHPR1s. These molecular transitions have been detected as charge movements under experimental conditions designed to minimise the otherwise larger, ionic, currents (Sections 5.3 and 5.4; Schneider and Chandler, 1973; Adrian and Almers, 1976*a*; Chandler *et al.*, 1976). Imposed depolarising steps from resting potential to test potentials known to initiate contraction elicited an extra outward current, indicating a slight current–voltage non-linearity. The current decayed with time to a steady-state level. A tail current whose integral equalled that of the extra outward current was recorded at the end of the test voltage step. This was compatible with a return of the intramembrane charge to its resting position (Figure 11.7). This charge movement increased with increasing depolarisation along a sigmoidal relationship that saturated with large voltage displacements. These features were consistent with a charged protein undergoing configurational changes

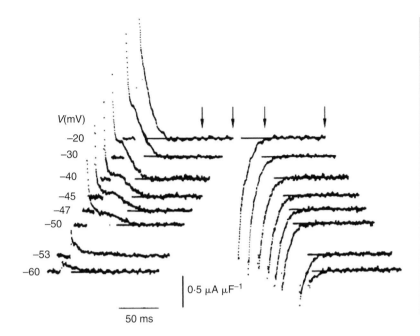

V(mV)
-20
-30
-40
-45
-47
-50
-53
-60

0·5 µA µF^{-1}

50 ms

Figure 11.7 Charge movements in response to progressively increasing depolarising voltage steps in the membrane of amphibian skeletal muscle consisting of an initial q_β decay followed by a prolonged q_γ transient progressively speeding up with further depolarisation. The recovery currents taking place with return of the membrane potential to the resting level (arrowed) are shown laterally displaced, and consist of simple decays. The integrals of the 'on' and 'off' currents indicate equal amounts of charge moved by the depolarising and repolarising parts of the voltage steps (From Figure 2 of Huang, 1994.).

within the membrane likely representing reversible movements of charged functional groups responding to voltage change, as opposed to passage of ionic current across the membrane (Section 5.3). These currents were much smaller than ionic currents and so their occurrence would not interfere with action potential propagation.

Currents of this kind reflect configurational changes in their underlying charged chemical groups. Such dipole relaxations would be expected to generate simple exponential kinetics (Section 5.3; Adrian and Almers, 1976b). However, biological membranes contain a wide variety of charged proteins underlying their different functions, and each could well contribute their own charge movement components (Section 5.4). These have proved pharmacologically and electrophysiologically separable into a number of individual components (Figure 11.7). In skeletal muscle, the most recognised are the q_α, q_β and q_γ components. Of these, the q_β component shows a relatively rapid exponential decay to baseline. However, its voltage dependence extends over a wide range of membrane potentials, reflecting a voltage sensitivity insufficiently steep to account for the corresponding voltage sensitivity of Ca^{2+} release (Huang *et al.*, 2011).

Consequently, the most interesting charge component in excitation–contraction coupling is q_γ. It contributes to delayed, as opposed to simple, exponential currents over a limited voltage range close to the contractile threshold. These can be observed independently in the absence of the earlier q_β decays under some conditions. Larger depolarisations result in these delayed components becoming larger, more rapid in timecourse and merging with the earlier q_β

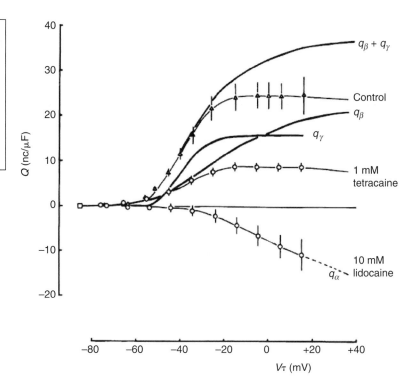

Figure 11.8 Resolution of charge–voltage curves for q_β, q_γ and q_α charge movement components, through comparison of charge–voltage curves obtained in the absence of pharmacological agents, and following addition of the q_γ blocker tetracaine, and the q_β and q_γ blocker, lidocaine. (From Huang, 1982.)

decays. The q_γ component is very steeply voltage-dependent in both its kinetic properties and the extent of its steady-state charge transfer. These parameters match the corresponding voltage dependence of the Ca^{2+} release process described above (Section 11.4). The q_γ component was also selectively inhibited by the contractile inhibitor tetracaine, strongly suggesting functional roles in excitation–contraction coupling. Such pharmacological properties proved useful in separating the steady-state voltage dependences of the different q_β, q_γ and q_α components of charge (Figure 11.8).

Charge movements persist even under conditions when SR Ca^{2+} is depleted, consistent with these steeply voltage-dependent intramembrane molecular configurational changes occurring upstream of SR Ca^{2+} release (Chawla *et al.*, 2002). Following detubulation (Section 10.5), q_γ charge movement is abolished. This left charge movements corresponding to those produced by q_β. The application of tetracaine, which abolishes q_γ charge movement leaving q_β alone, now had no further impact. These findings localised q_γ charge movement to the T-tubules, whilst indicating a more uniform distribution of q_β through both the surface and T-tubular membranes (Huang and Peachey, 1989). The localisation of both the excitation–contraction coupling process and q_γ to the T-tubules suggests that the electrical signature of the voltage sensor involved in triggering excitation–contraction coupling is likely to be q_γ (Huang, 1993).

Pharmacological studies using Ca^{2+} channel antagonists additionally associated the q_γ charge movement with configurational changes in the DHPR1, abundant within the T-tubules (Schwartz et al., 1985). These studies also reported parallel inhibition of sarcoplasmic reticular Ca^{2+} release (Rios and Brum, 1987). The dihydropyridine effects additionally showed a voltage dependence that paralleled their specific DHPR binding. Finally, in direct parallel with these effects, nifedipine reduced the steeply voltage-dependent q_γ charge movement, leaving the q_β component intact (Huang, 1990). The q_γ charge movement thus likely provides the electrical signature for DHPR1 action initiating excitation–contraction coupling, through which insights into the underlying configurational transitions may be derived. Finally, the q_γ charge appeared to occur at membrane densities similar to those of the DHPR1, of around $200-230/\mu m^2$ of T-tubular membrane (Huang and Peachey, 1989).

These findings specifically associate the q_γ charge both with the DHPR1 and with sensing of the T-tubular depolarisation that ultimately triggers excitation–contraction coupling. This sensing process may involve functional groups similar to those involved in ion-channel gating, for which the S4 voltage-sensing segment, which is rich in positive arginine residues, is of potential importance (Section 5.2). In fact, phenylglyoxal, known to neutralise charged groups on arginine residues, left subthreshold q_β charge intact, whilst reducing the q_γ component. This further associates the q_γ charge with the S4 segment and, given its localisation, strongly implicates the q_γ DHPR1 charge as the excitation–contraction coupling voltage sensor.

11.8 | Sarcoplasmic Reticular Ca²⁺ Release Through the Ryanodine Receptor

The available evidence implicates the RyR as both a transducer of the voltage-dependent configurational change in the DHPR1 and the mediator of the resulting Ca^{2+} release. This is consistent with its location in the terminal cisternal SR membrane close to its junction with the T-tubular membrane. Mammalian muscle mainly expresses a single, type I, skeletal muscle isoform of the ryanodine receptor (RyR1). Many non-mammalian vertebrate, including amphibian, skeletal muscle types equally express two isoforms, skeletal muscle RyR1 and neuronal RyR3. Both subtypes can act as SR Ca^{2+}-release channels. RyR1 can be gated by either T-tubular depolarisation- or Ca^{2+}-induced Ca^{2+} release (Section 13.14). The latter may involve RyR1 A- and I1 sarcoplasmic Ca^{2+} binding sites respectively promoting and inhibiting RyR1 opening, resulting in its bell-shaped dependence on cytoplasmic $[Ca^{2+}]$. However, bilayer studies indicate that the latter are only effective at μM and mM $[Ca^{2+}]$. SR levels of divalent ions may also influence skeletal muscle RyR1 channel opening, likely through a further luminal, L-binding site. Increases in SR luminal $[Ca^{2+}]$ from 0 to 1 mM increased RyR open probabilities and decreased RyR sensitivity to inhibition by Mg^{2+}. RyR3 primarily shows a Ca^{2+}-induced Ca^{2+}

release (Laver, 2018). Genetic RyR1 knockout results in a lethal dyspedia, or loss of foot processes connecting T-tubular and SR membranes, and a loss of excitation–contraction coupling.

RyR1 protein isolated from rabbit skeletal muscle and studied in lipid bilayers gives rise to Ca^{2+} channels blocked by ryanodine itself, the contractile inhibitor tetracaine at μM concentrations, the anthraquinone daunorubicin and the hexavalent inorganic dye ruthenium red. The channel properties were identical to the native channels in heavy vesicles formed from SR terminal cisternae. These in vitro gating properties were consistent with a role for RyR1 in excitation–contraction coupling in intact skeletal muscle. Both addition of ryanodine or tetracaine to the extracellular solution and ruthenium red injected into intact muscle fibres both depress mechanical activity.

RyR-mediated gating was shown to be modifiable in viable isolated muscle preparations, by experiments manipulating the membrane morphology of the triad structures. These examined membrane morphological appearances from amphibian muscle fibres fixed following exposure in Ringer's solutions under isotonic control (Plate 10A), hypertonic extracellular conditions containing 350 mM sucrose (Plate 10B) and following restoration of the original isotonic conditions (Plate 10C). Exposure to hypertonic conditions caused cell volume changes, in turn increasing intracellular ionic strength owing to the consequently altered intracellular ion concentrations. These changes together could modify T-tubular and SR anatomy, and cause protein dissociation. The latter would potentially affect RyR1–DHPR1 interactions at the triad structures.

Increased extracellular tonicities did indeed elicit altered freeze–fracture electronmicroscopic appearances of the following. (i) Significant but reversible increases were observed in the diameters of the T-tubular apertures (arrowed) at the surface membrane visible in P-face fracture planes through the plasma membrane. (ii) In the triad elements formed by junctional SR and T-tubules (the arrow giving the direction of shadowing at bottom left of each image), T-tubular diameters reversibly increased from ~25 to ~85 nm, and volume fractions of muscle occupied by JSR reversibly decreased from ~2.7 to ~1.7. (iii) Thin electronmicrograph sections showed T-SR junctions with visible foot processes traversing the gap between junctional SR and T-system (arrowed). Hypertonicity reversibly reduced the gap between T and SR membranes from ~9.90 to ~6.6 nm.

These changes accompanied an alteration in RyR gating properties resulting in an extreme form of Ca^{2+}-induced Ca^{2+} release (Sections 13.14, 13.15, 14.13, 14.14 and 15.8) not normally seen in skeletal muscle. These manifest initially as discrete stationary foci of increased cytosolic Ca^{2+} in fluo-3-loaded fibres studied under confocal microscopy. This is illustrated by the colour representation keyed to the false colour map in the inset (Plate 10D). These were followed by regenerative waves of Ca^{2+} release visualised from serially timed images (Plate 10E). The waves persisted despite sequestration of

$[Ca^{2+}]_o$ and DHPR1-voltage sensor inactivation by nifedipine, but were inhibited by the known RyR blocking agents, tetracaine and ryanodine (Chawla *et al.*, 2001).

11.9 | Structural Evidence for DHPR-RyR Coupling

These electrophysiological and anatomical findings together suggest a hypothesis in which initial detection of T-tubular membrane depolarisation by DHPR1-voltage sensors through their voltage sensitivity causes conformational transitions identifiable as q_Y charge movements. These transitions are directly communicated to the RyRs within adjoining SR membrane through the allosteric contact between them (Figure 11.9). This orthograde interaction between the DHPR1s and RyR1s causes complementary configurational changes in the RyR1s resulting in channel opening and release of SR Ca^{2+} into the cytosol.

A structural layout for such a coupling mechanism would place T-tubular DHPR1s in close proximity to cytoplasmic components of the RyR1. Plate 11 illustrates two of the cluster of four DHPR1s, each interacting with one of the RyR1 subunits schematised in Plate 9D. Coupling of T-tubular depolarisation to SR Ca^{2+} release would first involve detection of T-tubular membrane depolarisation by the DHPR1 voltage sensors, transitions producing q_Y charge movement. Each of

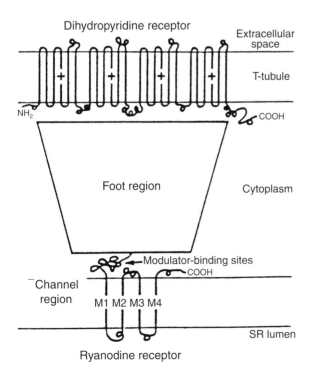

Dihydropyridine receptor

Extracellular space

T-tubule

NH₂

COOH

Foot region

Cytoplasm

Modulator-binding sites

COOH

Channel region

M1 M2 M3 M4

SR lumen

Ryanodine receptor

Figure 11.9 Schematic diagram to show the membrane topology of the ryanodine receptor (RyR) in skeletal muscle. The receptor is a Ca^{2+}-release channel anchored in the sarcoplasmic reticulum (SR) membrane. Only one of its four identical subunits is shown. Each subunit is closely associated with a voltage-gated Ca^{2+} channel (the dihydropyridine receptor, DHPR) in the T-tubule membrane, which probably acts primarily as a voltage sensor. (From Takeshima *et al.*, 1989. Reprinted with permission from Nature 339, copyright 1989 Macmillan Magazines Limited.)

the tetradic cluster of four voltage-sensing DHPR1s would orthogradely interact with one of the four cytoplasmic components of the RyR1, whose intramembrane portion within SR membrane functions as an intracellular SR Ca^{2+} release channel. Finally, the large cytoplasmic segment of the RyR1 could additionally permit multiple interactions with other regulatory molecules, potentially influencing either the DHPR1–RyR1 interaction and/or Ca^{2+} transit through the RyR1-Ca^{2+} release pathway (Plate 9A; Hernandez-Ochoa et al., 2016).

Structural data additionally suggests reciprocal, retrograde actions of the RyR1s on the DHPR1s, predicted by such a direct allosteric interaction. Ryanodine at blocking concentrations (100–200 μM) expected to drive RyR1s to a closed state caused large conformational changes in the RyR1 cytoplasmic domain. These were accompanied by significant (~2 nm) shifts in distances between DHPR1s, suggesting their involvement as a complex of DHPR1s coupled to RyR1. Finally, an allosteric coupling could permit cooperative properties in such DHPR1-RyR1 coupling that might account for the complex kinetics shown by q_γ charge movement.

11.10 | Physiological Evidence for DHPR-RyR Configurational Coupling

Physiological experiments exploring such reciprocal DHPR1-RyR1 interactions investigated the effects of a range of DHPR1- or RyR1-specific agents (Figure 11.10), applied alone or in combination (Figure 11.11), on steady-state and kinetic properties of q_γ charge movement (Huang et al., 2011). DHPR1 antagonists would be expected directly to reduce the total quantity of q_γ charge movement, as well as the subsequent orthograde downstream RyR1 activation. In contrast, whilst influencing SR Ca^{2+} release, RyR1 modification would not alter total available steady-state q_γ charge. However, it could influence both its steady-state voltage dependence and the kinetics of q_γ charge movements, were RyR1 to reciprocally and allosterically interact with the DHPR1.

First, challenges using the RyR antagonists, ryanodine and daunorubicin, at concentrations known to drive RyRs to a closed state, fulfilled the latter predictions (Figure 11.11A). These manoeuvres specifically preserved total available voltage-dependent charge movement (Figure 11.12C), in contrast to a direct DHPR1 antagonist action that reduced steady-state q_γ charge (Section 11.7). However, q_γ charge movements no longer showed distinguishable delayed kinetics, instead becoming rapid decays (Figure 11.12A) with positively shifted charge–voltage curves, despite this unchanged maximum saturating charge (Figure 11.12C).

Secondly, subsequent addition of the RyR agonists perchlorate (8 mM) and caffeine (~5 mM) rescued these antagonist actions (Figure 11.11B). Total available voltage-dependent charge movement was unchanged, but the typical pattern of delayed q_γ kinetics was

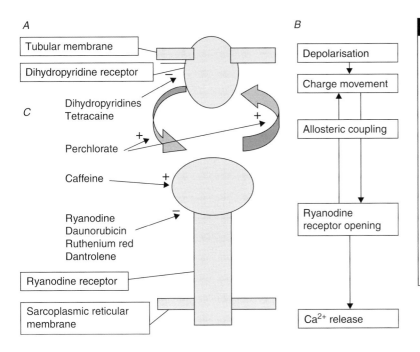

A

Tubular membrane

Dihydropyridine receptor

C

Dihydropyridines
Tetracaine

Perchlorate

Caffeine

Ryanodine
Daunorubicin
Ruthenium red
Dantrolene

Ryanodine receptor

Sarcoplasmic reticular
membrane

B

Depolarisation

Charge movement

Allosteric coupling

Ryanodine
receptor opening

Ca^{2+} release

Figure 11.10 Schematic representation of the functional relationship between voltage-sensing transverse (T-) tubular dihydropyridine receptors (DHPRs) and Ca^{2+}-releasing sarcoplasmic reticular ryanodine receptors (RyRs) (*A*), summarising the underlying physiological events and interactions between them in (*B*) triggering of excitation–contraction coupling. (*C*) summarises pharmacological agents directed at either the DHPRs or the RyRs used in clarifying these relationships. The symbol + denotes agonist, and − denotes antagonist, inhibitory effects.

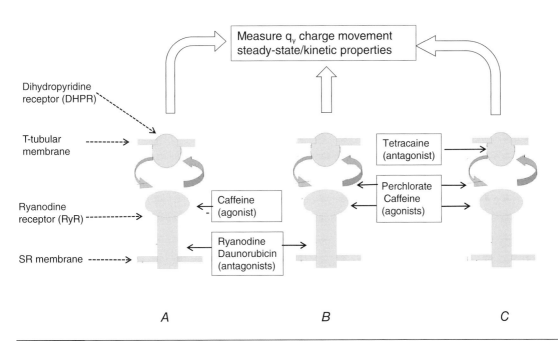

Measure q_γ charge movement
steady-state/kinetic properties

Dihydropyridine
receptor (DHPR)

T-tubular
membrane

Ryanodine
receptor (RyR)

SR membrane

Caffeine
(agonist)

Ryanodine
Daunorubicin
(antagonists)

Tetracaine
(antagonist)

Perchlorate
Caffeine
(agonists)

A　　　　　*B*　　　　　*C*

Figure 11.11 Experimental strategies investigating effects on q_γ-DHPR1 charge movement: (*A*) RyR antagonists ryanodine and daunorubicin (Huang, 1996), and the RyR agonist caffeine (Huang, 1998*a*). (*B*) The effect on RyR antagonism of the agonists perchlorate and caffeine (Huang, 1998*a*; 1998*b*) and (*C*) the effect of the q_γ-DHPR1 charge movement antagonist tetracaine, and its modification by the agonists perchlorate and caffeine (Huang, 1998*a*, 1998*b*).

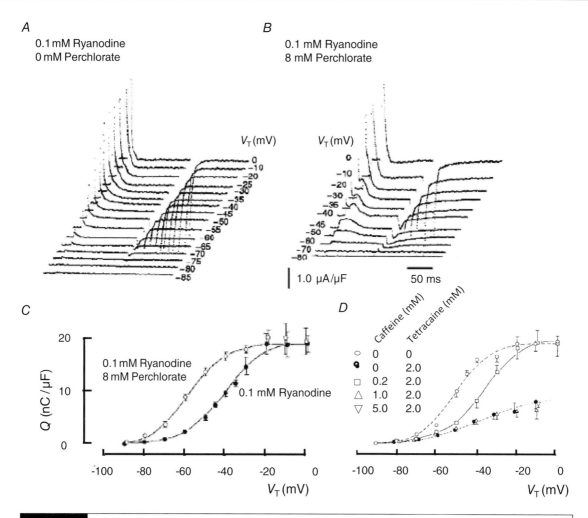

Figure 11.12 Interactions between the effects of perchlorate, caffeine and the RyR inhibitor ryanodine on intramembrane charge movements. (A) Charge movements in 100 μM ryanodine alone, and (B) following further addition of 8.0 mM perchlorate. (C) Charge–voltage curves obtained in the presence of 100 μM ryanodine before and after further introduction of 8.0 mM perchlorate. (D) Paradoxical result in which low, facilitating (0.2 mM), but not high, activating (1.0-5.0 mM), caffeine concentrations rescue the inhibitory action of tetracaine on q_γ charge. ((A) and (B) from Figure 3, and (C) from Figure 4 of Huang, 1998b). (D) from Figure 7 of Huang, 1998a.)

restored (Figure 11.12B). Perchlorate also reversed the positive shift in the charge–voltage curves originally produced by ryanodine (Figure 11.12C).

Thirdly, these RyR agonists also reversed a previously established tetracaine-induced block of DHPR1-mediated q_γ charge movement (Figure 11.11C) that had resulted in a substantially reduced, q_β, charge movement (Section 11.7). Thus, further addition of perchlorate, or caffeine at low (0.2 mM) RyR-potentiating concentrations, restored both delayed q_γ charge movements and the corresponding, steeply voltage-sensitive steady-state charge (Figure 11.12D).

Finally, higher, 1.0 and 5.0 mM, caffeine concentrations known to fully activate the RyR, similarly preserved the steady-state charge. However, they converted the delayed q_γ charge movements to rapid decays. Furthermore, unlike the previous manoeuvres applying RyR agonists, they paradoxically failed to restore q_γ charge movements previously abolished by tetracaine (Figure 11.12D). Full RyR1 activation thus results in a loss of its RyR1–DHPR1 coupling with its accompanying kinetic consequences. This results in RyR1 agonists becoming unable to rescue a previously established DHPR inhibition produced by tetracaine.

11.11 | Cooperative DHPR-RyR Interactions

These findings together suggested a simple model relating delayed DHPR1-mediated q_γ charge movement, cooperative DHPR1-RyR1 interactions and RyR1-mediated SR Ca^{2+} release (Plate 12). Membrane depolarisation normally activates the initially resting T-tubular q_γ-DHPR1 voltage sensor (Plate 12Aa; colour change red to green). This triggers cooperative reciprocal DHPR1-RyR1 interactions leading to DHPR1-RyR1 dissociation (starred) that confer the delayed kinetics on q_γ charge movement (Plate 12Aa', starred), culminating in RyR-mediated SR Ca^{2+} release (Plate 12Aa; colour change, red to green).

Computational analysis demonstrated that modelling these cooperative interactions could reproduce the observed delayed, steeply voltage-sensitive q_γ charge movement kinetics. In the most straightforward situation, charge movements (Figure 11.13Aa) arise from simple transitions between two states separated by an energy barrier with respective energies, G_1, G_2 and G_a, each linearly dependent on membrane potential (Figure 11.13Ab). Charge redistribution between states in response to a voltage step assumes first-order forward and backward rate constants dependent upon the activation energy differences $G_a - G_1$ and $G_a - G_2$ (Section 5.3) producing a first-order, exponential charge movement decay (Figure 11.13Ac).

In contrast, the observed, more complex kinetic properties of q_γ charge movement would result from cooperative processes progressively facilitating both forward and backward reactions by reducing the energy barrier G_a in response to the resulting progressive RyR1 activation (Figure 11.13Ba, c, d, starred). This would result in a delayed increase in the current following a depolarising step (Figure 11.13Ba). In contrast, in both cases (A) and (B) the reverse reaction with termination of the voltage step would reinstate the increased energy barrier whilst giving an 'off' charge with monotonic decaying kinetics (Figure 11.13Ad and Bd).

The DHPR1 blockers tetracaine (at mM concentrations) and nifedipine (Plate 12Ab) both reduce q_γ charge movement (Plate 12Ab', red dot) and consequently its downstream RyR1 dissociation and activation. As indicated above (Section 11.10), the latter actions are rescued by perchlorate or potentiating caffeine concentrations

Figure 11.13 Energetic schemes underlying charge movement kinetics. (A) Charge movements producing (a) simple exponential decays described in terms of transitions between two states separated by an energy barrier whose respective energies, G_1, G_2 and G_a, are altered by membrane potential change from a resting (b) to an activating (c) level, followed by (d) their restoration at the end of the voltage step. This produces first-order, exponential charge movements. (B) More complex kinetic patterns can result if cooperative processes reduce the energy barrier G_a (b) to extents dependent upon prior charge movement (c, starred), resulting in progressive increases in the forward reaction rate constant (Huang, 1984). Note that charge recovery with restoration of the resting potential re-instates the increased energy barrier, giving recovery charge movements with simple decaying kinetics (Ad, Bd).

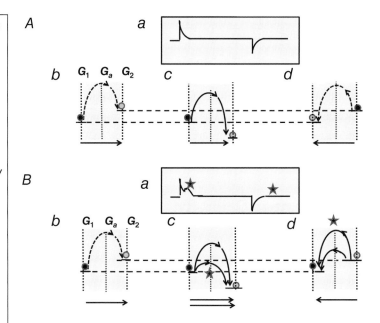

through their agonist action on the RyR1 with which the DHPR1 is coupled. In contrast, challenge by the RyR1 inhibitors ryanodine, daunorubicin or µM tetracaine (Plate 12Ac) leaves total available DHPR1-mediated q_γ charge intact, but alters its kinetics through a DHPR1 interaction with an inhibited RyR1 (Plate 12Ac'). Finally, high concentrations of the RyR agonist caffeine, sufficient to themselves cause RyR1 opening, also cause DHPR1-RyR1 dissociation. They thereby also affect q_γ kinetics (Plate 12Ad) and fail to rescue an established tetracaine-mediated DHPR1 block.

In all these situations, repolarisation (Plate 12Aa''-d'') permits a monotonic q_γ charge movement recovery by its underlying DHPR1 that is now dissociated from, or linked to an inhibited, RyR1 (Plate 12Aa''-Ad'', Plate 12Ba''). In contrast, charge movements do not show delayed q_γ components in cardiac muscle (Plate 12B). Ventricular excitation–contraction coupling involves differing, and uncoupled, DHPR2 and RyR2 isoforms in contrast to their allosteric coupling in skeletal muscle (Section 13.14).

11.12 | Malignant Hyperthermia as an Inherited RyR1 Defect

The clinical condition malignant hyperthermia results from a genetic RyR1 defect. RyR1-mediated SR Ca^{2+} release becomes more easily triggered. There is also increased RyR1 sensitivity to halogenated volatile anesthetics as well as the muscle relaxant, suxamethonium. Both agents are important in anaesthetic practice. The resulting uncontrolled SR Ca^{2+} release induces tetanic muscle contractions and consequent muscle rigidity, potentially leading to muscle tissue breakdown, termed rhabdomyolysis, if sustained. There is a consequent increased ATP consumption resulting from both increased myofilament activity and SR-Ca^{2+} ATPase-mediated SR reuptake of Ca^{2+} released into the cytosol. The result is increased heat generation leading to hyperthermia. Accompanying increases in total O$_2$ consumption and CO$_2$ production can also generate metabolic acidosis. Such episodes are controlled clinically by the RyR1-inhibitor dantrolene (Mickelson and Louis, 1996; MacLennan, 2000).

11.13 | Restoration of Sarcoplasmic Reticular Ca^{2+} Following Repolarisation

Repolarisation restores the DHPR1s to their resting conformation, resulting in a recovery charge movement. This would be expected to close the RyR channels. SR Ca^{2+} release would cease. [Ca^{2+}]$_i$ falls from a ~ 0.1 mM peak within the ~100 ms period of a twitch to levels below those required for significant troponin binding, thereby ending the contraction (Figure 11.14A).

The required resequestration of released Ca^{2+} from the cytosol into the longitudinal SR, which forms a series of membrane-bound sacs between the myofibrils (Section 11.5), is mediated by an active SR Ca^{2+} transport process. This utilises a Ca^{2+}-dependent membrane ATPase (SERCA) localised within the SR membrane from which it has been purified and sequenced by biochemical and ultrastructural techniques (Section 11.2). This intrinsic membrane protein accounts for ~50% of the membrane dry weight and occurs as dimers of a 115 kDa polypeptide. Its hydrophobic sequences permit it to repeatedly cross the SR membrane, with large hydrophilic loops with N- and C-terminals occurring in the cytoplasmic phase. Each subunit has an active phosphorylation site and two Ca^{2+} binding sites. This makes up to a total of four cooperatively interacting Ca^{2+}-binding domains in which initial Ca^{2+} binding enhances subsequent Ca^{2+} binding. This pump maintains or restores the sarcoplasmic [Ca^{2+}]$_i$ to its low resting level. It transports two Ca^{2+} ions for each molecule of ATP hydrolysed from cytoplasm to SR lumen, thereby restoring the resting thousand-fold [Ca^{2+}] gradients across the SR membrane.

Following SR membrane transport, ~90% of the consequently intra-vesicular Ca^{2+} becomes bound, much to the acidic intraluminal Ca^{2+} storage glycoprotein calsequestrin (CASQ1) that accounts for 7% of SR protein content. This can be isolated from heavy microsomal fractions and is localised in terminal cisternal regions of the SR. It exists as an extrinsic membrane protein whose membrane anchorage depends on divalent cation binding. It binds Ca^{2+} with a dissociation constant of 0.5 mM and a 1:45 binding ratio for Ca^{2+}. The latter results in a marked conformational change detectable in circular dichroic or absorption spectra.

Figure 11.14 provides an overview of these SR Ca^{2+} release (B) and re-uptake processes (C), as well as Ca^{2+}-troponin binding initiating

Figure 11.14 Schematic summary of the coupling process in skeletal muscle. (A) Overall scheme summarising underlying exchanges of bound and free cytosolic and SR Ca^{2+}. Following spread of cell surface membrane depolarisation down the transverse (T-) tubules, (B) its detection by dihydropyridine receptors (DHPRs) and coupling of the resulting configurational changes to SR ryanodine receptor (RyR1) Ca^{2+}-release channels permits Ca^{2+} flow down its electrochemical gradient into the cytosol. However, in addition to binding to troponin (forming CaTrop) activating the contractile apparatus, Ca^{2+} binds to cytosolic proteins, including parvalbumin (CaParv). Only a small fraction remains as ionised Ca^{2+}, as detected by experimental dyes (CaD). Repolarisation ends Ca^{2+} release, and (C) Ca^{2+} ions are returned to the SR by Ca^{2+}ATPase; these then associate with skeletal muscle calsequestrin (CASQ1) following their dissociation from the various membrane proteins. (Adapted from Figure 4 of Baylor et al., 1983.)

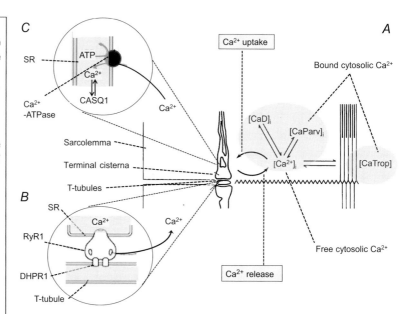

contraction and remaining aspects of cellular Ca^{2+} homeostasis (A). Ultimately skeletal muscle cytosolic [Ca^{2+}]$_i$ strongly depends on the balance between a range of processes. SR Ca^{2+} release follows surface and consequent T-tubular membrane depolarisation (Section 11.5) and its detection by DHPR1-voltage sensors (Section 11.7) in turn coupled to SR RyR1 Ca^{2+}-release channel opening (Section 11.6). The increased cytosolic [Ca^{2+}]$_i$ makes free Ca^{2+} available for rapid binding to troponin (CaTrop), initiating contraction (Section 11.2). However, released Ca^{2+} also distributes into slower binding with parvalbumin ([CaParv]) and reuptake from the cytosol by SR Ca^{2+}-ATPase (Section 11.2), as well as binding to any experimentally introduced Ca^{2+} indicators ([CaD]) when studying these Ca^{2+} homeostatic processes (Section 11.4).

Conversely, restoring the membrane resting potential resets the DHPR1, ending the RyR1-mediated Ca^{2+} release. Cytosolic [Ca^{2+}] now declines to its resting level through continued operation of the [Ca^{2+}]$_{i}$-, rather than voltage-dependent, SERCA activity. This process is facilitated by resequestration of free SR Ca^{2+} by CASQ1. However, free cytosolic [Ca^{2+}] represents only a small fraction of the total Ca compared to the bound Ca^{2+} fractions from which the Ca^{2+} must dissociate for this Ca^{2+}-ATPase-mediated reuptake to occur.

11.14 | Ryanodine Receptor-Na$^+$ Channel Feedback Interactions Related to Excitation–Contraction Coupling

In addition to these feedforward mechanisms (Figure 11.15A) by which membrane depolarisation triggers DHPR-mediated RyR activation, feedback signals from the resulting RyR-mediated

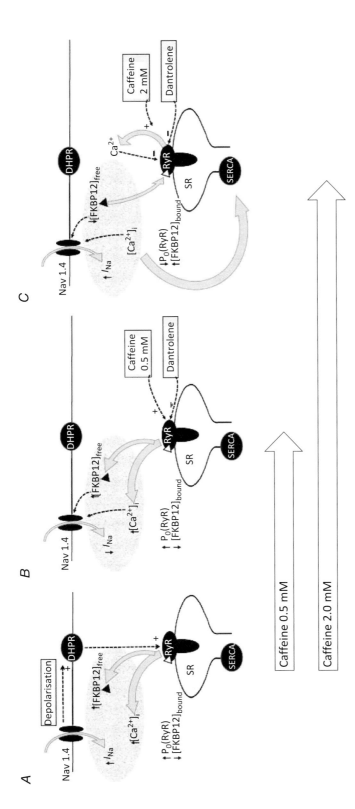

Figure 11.15 Feedforward and feedback interactions between the Na$^+$ channel and the ryanodine receptor (RyR) SR Ca^{2+} release channel. (A) Feedforward Nav1.4 channel mediated inward I_{Na} driving membrane depolarisation, DHPR triggering and opening of SR RyR-Ca^{2+} channels, resulting in increase in cytosolic [Ca^{2+}]$_i$. (B, C) Feedback mechanisms illustrated by the RyR agonists and antagonists caffeine and dantrolene. (B) Low facilitatory (0.5 mM) caffeine concentrations increasing RyR-mediated flux of SR Ca^{2+} into the cytosol and unbinding of FKBP12, downregulating Nav1.4 activity. (C) High (2.0 mM) caffeine concentrations producing an early increased [Ca^{2+}]$_i$ resulting in sustained inactivation of and FKBP12 binding to RyR and subsequent reduction in [Ca^{2+}]$_i$.

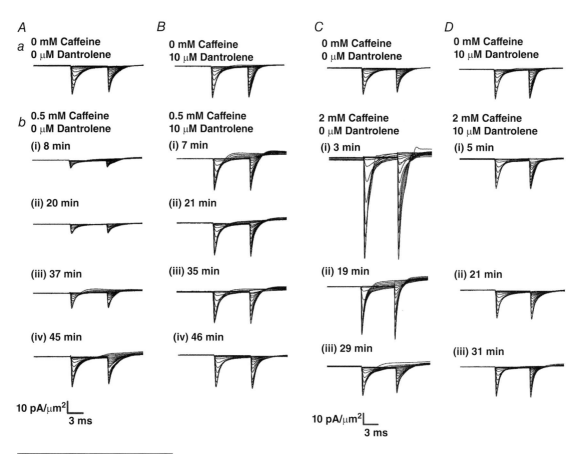

Figure 11.16 RyR-Nav1.4 feedback mechanisms demonstrated by challenge by the ryanodine receptor agonists and antagonists, caffeine and dantrolene. Experiments investigating challenge by 0.5 mM (A, B) and 2.0 mM caffeine (C, D) in the absence (A, C) and presence (B, D) of dantrolene. Families of superimposed I_{Na} records were obtained before (a) or at different times following onset of (b) caffeine challenge. (From Figures 3 and 7 of Sarbjit-Singh et al., 2020.)

SR Ca^{2+} release can reciprocally modify Na^+ channel activation in both skeletal and cardiac muscle (Section 13.19). At the molecular level, both skeletal muscle Nav1.4 and cardiac Nav1.5 possess multiple potential regulatory Ca^{2+} and CaM binding or modulation, as well as phosphorylation sites (Section 5.2). There is also evidence that FKBP12 released by the RyR upon activation of SR Ca^{2+} release (Section 11.5) modifies Nav1.4 activation and inactivation properties.

Rapid photo-release of caged Ca^{2+} and Ca^{2+} overspill from neighbouring Ca^{2+} channels reduced peak I_{Na} in skeletal muscle cell lines and Nav1.4-transfected HEK293 cells. These effects were abolished by mutations in the Ca^{2+}-binding EF hands on CaM, and in the Nav1.4 IQ domain (Ben-Johny et al., 2014). In intact native murine skeletal muscle fibres, loose patch clamp studies (Section 10.2) demonstrated reduced I_{Na} when RyR2-mediated Ca^{2+} release was increased by the Epac activator (Section 13.18), 8-(4-chlorophenylthio)adenosine-3',5'-cyclic monophosphate (8-CPT; Matthews et al., 2019). Further studies utilised the contrasting effects of varied caffeine concentrations in their actions on RyR activation and inactivation. At low concentrations, caffeine (0.5 mM) expectedly increases $[Ca^{2+}]_i$ (Figure 11.15B). At high concentrations (2 mM), it initially induces marked increases in

$[Ca^{2+}]_i$, but this quickly drives the RyR into a sustained inactivation with SERCA-mediated Ca^{2+} reuptake decreasing $[Ca^{2+}]_i$ below even resting levels (Figure 11.15C).

Application of graded depolarising test voltages to obtain current records through a family of test voltage steps revealed I_{Na} alterations that closely paralleled these $[Ca^{2+}]_i$ changes. Both 0.5 and 2 mM caffeine challenge induced early reductions in I_{Na}. When compared to families of I_{Na} obtained in the absence of drug (Figure 11.16Aa), this reduction persisted with 0.5 mM caffeine (Figure 11.16Ab). In contrast, I_{Na} showed increases with 2 mM caffeine (Figure 11.16Cb). Both changes ultimately recovered towards pretreatment levels (Figure 11.16Ab,Cb). The specific RyR antagonist dantrolene (Section 11.10) slightly increased I_{Na} when applied alone (Figure 11.16Ba, Da). But it abrogated all the effects of both 0.5 (Figure 11.16Bb) and 2 mM caffeine (Figure 11.16Db). In combination, all of these actions of caffeine on I_{Na} can be attributed specifically to its effects on RyR-mediated SR Ca^{2+} release. They also suggest that $[Ca^{2+}]_i$ thereby exerts significant inhibitory effects on I_{Na}, even at background RyR activity levels.

These particular feedback mechanisms could complement the hyperpolarising actions of ADP-mediated K_{ATP} and Ca^{2+}-mediated $K_{Ca}1.1$ activation (Section 10.2) that would also reduce membrane excitability under conditions of intense and prolonged activation, including tetanic stimulation and metabolic exhaustion (Sections 12.9–12.14). They could also bear on clinical conditions altering skel-etal muscle excitability. A form of K^+-aggravated myotonia, a Na^+ channel myotonia exacerbated by potassium ingestion, is associated with slowed I_{Na} kinetics, impaired I_{Na} inactivation and Nav1.4 muta-tions in its EF-hand-like domain. Other genetic, K^+ and cold-aggravated myotonias associated with reduced Ca^{2+}-dependent Na^+ channel inhibition are also attributed to Nav1.4 mutations (Ben-Johny et al., 2014; Heatwole and Moxley, 2007). Finally, dystrophic muscle weakness is associated with elevated $[Ca^{2+}]_i$.

12

Contractile Function in Skeletal Muscle

The processes in skeletal muscle described in the preceding chapters culminate in the generation of mechanical activity. The latter has been typically investigated in isolated muscle or nerve–muscle experimental systems such as amphibian gastrocnemius or sartorius nerve–muscle preparations. Experiments on larger mammalian muscles require an intact blood supply and involve study in anaesthetised animals, following isolation of their nerve supply and attachment of their separated tendons to some recording device.

12.1 | Isometric and Isotonic Contractions

Following an electrical stimulus of suprathreshold amplitude applied to the nerve or the muscle itself, the contracting muscle exerts a force upon its attachments equal and opposite to the tension within the muscle. Muscle shortening results if the attachments permit it to do so. The contractile process can be described in terms of two specific variables of length and tension. Experimental studies often maintain one variable constant, permitting the other measured variable to vary. In *isometric* contractions, muscle length is held constant during tension measurements. In *isotonic* contractions, the load on the muscle, equal and opposite to the muscle tension, is held constant while shortening is measured.

An isometric recording device requires a low compliance to minimise shortening during the force measurement. Classical studies used a lever attached to a stiff spring with the output written onto a smoked drum. Modern devices often bond semiconductor strain gauges, whose electrical resistance varies with tension, to a rigid support. Electrical tension readouts can take the form of an oscilloscope, chart recorder or digital display (Figure 12.1). The force exerted by the muscle is usually measured in Newtons. Isotonic recording devices typically utilise a moveable lever whose motion is recorded either directly onto a smoked drum or indirectly via an electrical signal. The lever can be preloaded to different extents, perhaps by hanging weights on it (Figure 12.2).

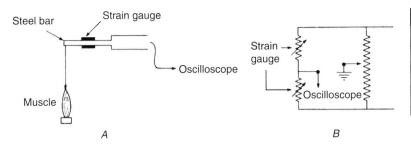

Figure 12.1 Isometric lever system for measuring force exerted by a muscle held at constant length. Semiconductor strain gauges bonded to a steel bar (A) form two arms of a resistant bridge connected to a voltage source (B).

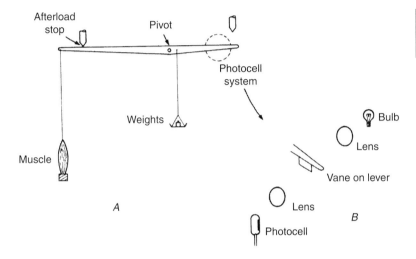

Figure 12.2 (A) Isotonic lever system and (B) photocell system used to record position of the lever.

12.2 | Isometric Twitch and Tetanus

A single high-intensity stimulus applied to a muscle under isometric recording conditions elicits a twitch comprising a rapid tension increase followed by its decay (Figure 12.3A). However, the overall tension transient extends well beyond action potential termination, becoming shorter with increased temperature. Twitch durations also vary between muscles and muscle types. A frog sartorius muscle at 0 °C typically shows a time from onset to peak tension of ~200 ms, which then decays to zero within ~800 ms. Mammalian muscles fall into two distinct types, *fast-twitch* and *slow-twitch*. Fast-twitch muscles contract and relax more rapidly than slow-twitch muscles, their twitches lasting ~50 ms and up to several hundred milliseconds, respectively (Figure 12.4). Fast-twitch muscles make relatively rapid, phasic, movements; slow-twitch muscles mediate long-lasting, tonic, contractions important in maintaining posture. The fast-twitch gastrocnemius extends the ankle joint in walking and running; the slow-twitch soleus muscle stabilises the same joint while the subject is standing still.

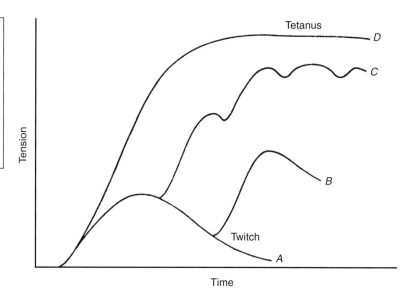

Figure 12.3. Isometric contractions: (A) twitch response to single stimulus, (B) response to two consecutive stimuli, showing mechanical summation,(C) response to a train of stimuli, showing an 'unfused tetanus', (D) response to a train of stimuli at a higher repetition rate, showing a maximal fused tetanus.

A second stimulus applied before the initial twitch fully relaxes elicits a further twitch whose peak tension exceeds that of the first reflecting their mechanical summation (Figure 12.3B). Repetitive stimulation at low frequencies produces a 'bumpy' tension record (Figure 12.3C). Furthermore, whereas the repetitive stimulation produces a train of discrete action potentials, above a critical fusion frequency, it results in fusion of the individual twitches into a sustained and augmented tension generation or *tetanus* (Figure 12.3D). These critical, tetanic, frequencies vary between 300 s^{-1} for fast muscles down to about 40 s^{-1} for slow muscles. The resulting enhanced tension is expressed by the twitch/tetanus ratio between peak isometric twitch tension and maximum tetanic tension. This is typically ~0.2 for mammalian muscles at 37 °C, and rather larger for frog muscles at room temperature or below.

A low-intensity stimulus applied to the motor nerve may produce no muscle contraction because the applied small current flow fails to excite any component nerve fibres. Increasing stimulus intensity excites progressively more nerve fibres and more motor units, increasing total tension. Eventually the stimulus intensity simultaneously excites all available nerves, giving the maximum muscle tension. This does not increase further, despite additional increases in stimulus intensity. These properties form the basis of muscle fibre recruitment underlying normal gradation of muscular force. Gentle movements involve simultaneous activation of a small number of motor units; vigorous movements require activation of many more motor units.

Progressive passive stretch of a resting muscle produces an increasing passive tension reflecting the mechanical properties of its contained connective tissue. This yields a passive length–tension curve (Figure 12.5). The full isometric tetanus tension in a stimulated muscle yields a similarly length-dependent 'total active tension' curve. The 'active increment' curve, derived from the difference between

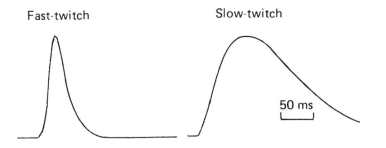

Fast-twitch   Slow-twitch

50 ms

Figure 12.4 Isometric twitches of two types of cat muscle, showing the much longer timecourse of the slow-twitch muscle. (From Buller, 1975.)

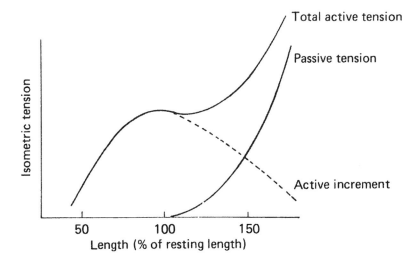

Total active tension

Passive tension

Active increment

Isometric tension

50   100   150

Length (% of resting length)

Figure 12.5 Active and passive isometric length–tension relations of skeletal muscle.

passive and active curves is maximal at near its in situ maximum length, and falls away at greater or shorter lengths, directly reflecting the nature of the contractile mechanism (Section 9.5; Figures 9.4–9.6).

12.3 | Isotonic Contractions

Experimental studies of isotonic contractions (Figure 12.2) typically support the load whilst the muscle is relaxed by including a preload stop. This permits systematic studies of contracting muscle at varied afterloads at predetermined initial muscle lengths. During tetanic stimulation against a moderate load (Figure 12.6A), muscle tension increases soon after onset of stimulation. However, actual shortening requires a further delay, during which the muscle is developing its isometric tension. This delay increases with load P because the muscle takes longer to achieve the tension matching an increased P (Figure 12.6B). Isotonic contraction then takes place with an initial shortening velocity V. This is constant provided that the resting muscle length approximates its maximum in situ length. The V term

Figure 12.6 After-loaded isotonic tetanic contractions: (A) length and tension changes during a single contraction; shortening shown as an upward deflection of the length trace, (B) initial length changes in contractions against different loads.

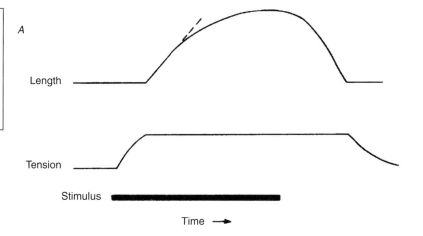

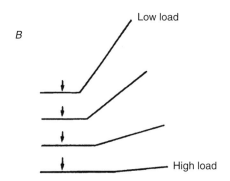

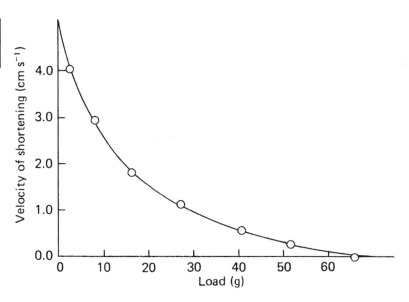

Figure 12.7 Force–velocity curve of frog sartorius muscle at 0 °C. (From Hill, 1938.)

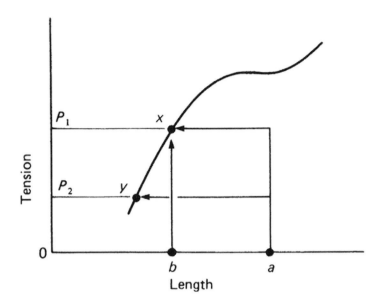

Figure 12.8 Diagram illustrating greater shortening shown by a lightly loaded than by a heavily loaded muscle. Starting from point *a* on the length axis, the muscle contracts isometrically until its tension equals the load. It then shortens until it meets the isometric length–tension curve. With a heavy load (P_1) this occurs at *x*; with a lighter load (P_2) it occurs at *y*. Note that point *x* can also be reached by an isometric contraction from point *b*. When starting from a much extended length, a muscle may in practice stop short of point *x* when lifting load P_1, probably reflecting inhomogeneities amongst the component muscle fibres.

decreases with increasing P, becoming zero when P equals the maximum tension P_0 attainable during isometric contraction at the initial muscle length. The resulting approximately hyperbolic *force–velocity curve* follows the Hill Equation (Figure 12.7). If a and b are constants (Hill, 1938):

$$(P + a)(V + b) = b(P_0 - a) \qquad (12.1)$$

or

$$(P + a)(V + b) = \text{constant} \qquad (12.2)$$

With further shortening, the shortening velocity decreases. The total extent of shortening decreases with increasing load, because isometric tension falls at shorter lengths (Figure 12.5). A more heavily loaded muscle can shorten by a reduced amount before its isometric tension becomes equal to the load (Figure 12.8). When the period of stimulation ends, the muscle is extended by the load until the lever meets the afterload stop. Relaxation becomes isometric and muscle tension falls to its resting level.

12.4 | Energetics of Contraction

Muscle contraction requires conversion of chemical to mechanical energy (work), releasing heat. The law of conservation of energy indicates that the chemical energy equals the mechanical work done plus the heat evolved in that contraction:

$$\text{chemical energy release} = \text{heat} + \text{work} \qquad (12.3)$$

Of the three terms in this equation, the work is the easiest to measure, then the heat, and the chemical change is the most difficult. Let us examine them in this order.

12.5 | Work and Power Output by Muscle

Mechanical energy is measured as work and the rate of performing work is the power. In an isotonic contraction, the work done equals the force exerted by the muscle multiplied by distance shortened. No overall work is generated by an isometric contraction as there is no shortening. Nor is work performed during an unloaded isotonic contraction as this exerts zero force. Power is similarly zero under both these conditions. These limits define the zero points in the resulting dependence upon load of the work performed between the onset and peak of an isotonic contraction (Figure 12.9). Conversely, during relaxation from an isotonic twitch, the load performs work on the muscle. The net work output through the entire isotonic twitch is zero.

Power output at any instant equals force multiplied by shortening velocity. During the initial shortening phase of a tetanic isotonic contraction, it is readily calculated from the force–velocity curve (Figure 12.10). Maximum power is achieved when the load is ~0.3 times isometric tension; the muscle shortens at ~0.3 times its maximum (unloaded) velocity. This has implications for the selection of gears in cycle races. Whatever the speed of the cycle, maximum power is obtained from the leg muscles when they are contracting at about 0.3 times their maximum unloaded velocity. This corresponds to about two revolutions of the pedals per second.

Figure 12.9 Variation of work with load in isotonically contracting muscle.

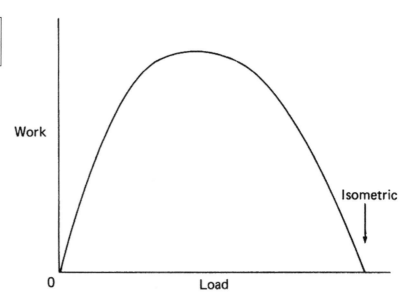

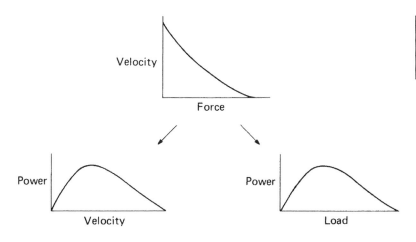

Figure 12.10 Power output during tetanic isotonic contractions plotted against different loads and velocities.

12.6 | Heat Production During Muscle Activity

Our everyday experience confirms that heat production accompanies muscular activity. This must be accounted for when studying muscle energetics. A. V. Hill and his colleagues studied isolated frog muscles laid over a thermopile comprising an array of thermocouples arranged in series to measure small temperature changes (Hill and Hartree, 1920). During isometric tetanus, an *activation heat* is released at a high rate for the first ~50 ms. It then falls rapidly to a lower, more steady, *maintenance heat* level (Figure 12.11). A further, *heat of shortening*, roughly proportional to the distance shortened, is released if the muscle is allowed to shorten. Further heat appears during relaxation, especially if the load does work on the muscle. Finally, after relaxation, a prolonged *recovery heat* reflects muscle metabolism restoring the resting biochemical situation in the muscle.

12.7 | Muscle Efficiency

Muscle efficiency is a measure of the degree to which the energy expended is converted into work, i.e.

$$\text{efficiency} = \text{work}/(\text{total energy release})$$
$$= \text{work}/(\text{heat} + \text{work}) \tag{12.4}$$

If we consider only energy changes during and immediately after contraction, the efficiency works out at ~0.4 for frog muscle and ~0.8 for tortoise muscle. But including the recovery heat gives lower values of ~0.2 and ~0.35. These represent maximum values, obtained by allowing the muscle to shorten at about one-fifth of its maximum velocity. A. V. Hill's calculation of how efficiency varies with velocity

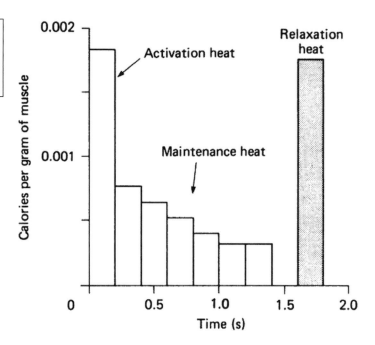

Figure 12.11 Rate of heat production of frog sartorius muscle, stimulated for a period of 1.2 s, during an isometric tetanus. (From Hill and Hartree, 1920.)

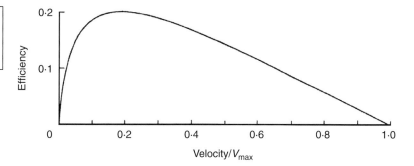

Figure 12.12 Dependence of efficiency upon shortening velocity during isotonic tetanic contractions. (From Hill, 1950a.)

of shortening (Figure 12.12) shows that maximum efficiency occurs at a lower velocity than that at which maximum power output occurs. However, both curves have fairly broad peaks, so the difference between them may not be very important.

12.8 | The Energy Source for Muscle Contraction

Having discussed the output side of the energy balance equation, we now consider the chemical changes supplying energy for muscular contraction. Energy for all bodily activities ultimately derives from food. Food energy is transported to the muscle as glucose or fatty acids and may be stored there as the glucose polymer glycogen. Respiration

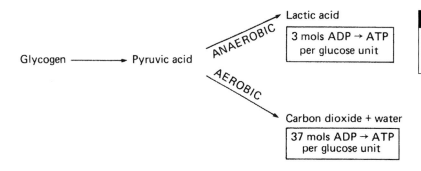

Figure 12.13 Outline of the breakdown of glycogen with the release of energy (in the form of ATP) in respiration.

of these substances within muscle cells generates adenosine triphosphate (ATP) from adenosine diphosphate (ADP; Figure 12.13).

In addition, 'high-energy phosphate' from ATP is transferred to creatine (Cr), catalysed by the enzyme creatine phosphotransferase. This reversible reaction forms creatine phosphate (CrP), providing a short-term store of high-energy phosphate.

$$ATP + Cr \rightarrow ADP + CrP$$

Nevertheless, ATP appears to be the immediate chemical energy source for most cellular activities, including actin–myosin interactions underlying muscle mechanical activation. Experimental evidence demonstrates ATP breakdown in contracting muscle (Hill, 1950b). ATP concentrations were compared in pairs of unstimulated frog muscles and muscles after stimulation following rapid freezing in liquid propane post contraction to prevent any further biochemical change. First, experiments were performed in the presence of the creatine phosphotransferase blocker 1-fluoro-2,4-dinitrobenzene (FDNB), preventing ADP rephosphorylation to ATP from creatine phosphate. The stimulated muscles lost ~0.22 μmoles of ATP per gram of muscle in an isotonic twitch. The heat of hydrolysis of the terminal phosphate bond of ATP is ~34 kJ/mol and so breakdown of 0.22 μmoles ATP should release about 7.5×10^{-3} J of energy. This exceeds the expected work, which amounted to 1.7×10^{-3} J per gram of muscle. Hence the ATP breakdown more than sufficiently accounted for the work done in the twitch, with excess energy expected to appear as heat (Cain et al., 1962).

Secondly, measurements of fuel consumption compared changes in creatine phosphate content of stimulated and unstimulated muscles. These experiments were performed in the absence of FDNB, thereby permitting rapid phosphate transfer from creatine phosphate to regenerate ATP, as in normal muscle, but in the presence of the glycolytic enzyme blocker iodoacetate, in an atmosphere of nitrogen, blocking resynthesis of the creatine phosphate. The stimulated muscle now contained less creatine phosphate than the unstimulated one.

It has proven difficult to draw up precise balance sheets for energy changes in exercising muscle. Careful measurements on energy

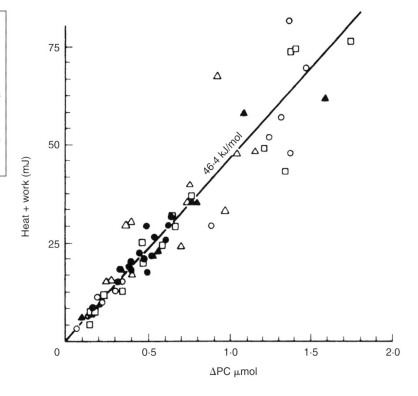

Figure 12.14 Relationship between energy production (heat plus work) and creatine phosphate breakdown in frog sartorius muscles poisoned with iodoacetate and nitrogen. Each point represents a determination on one muscle after the end of a series of contractions, with different symbols for different types of contraction. (From Wilkie, 1968.)

output (heat + work) and creatine phosphate breakdown in frog muscles during a variety of different types of contraction suggested that energy output was linearly proportional to breakdown of creatine phosphate, with 46.4 kJ of energy produced for each mole of creatine phosphate broken down (Figure 12.14; Wilkie, 1968). However, calorific measurements suggested that creatine phosphate hydrolysis should yield only about 32 kJ/mol, leaving an 'unexplained energy' between expectation and observation. Recent results associated this with Ca^{2+}-binding reactions associated with muscle activation processes, but there are still aspects of muscle energy balance that are not understood (Homsher, 1987).

12.9 | Energy Balances During Muscular Exercise

The effectiveness of muscles as generators of mechanical work is determined in part by the speed with which they can convert their stores of chemical energy into mechanical energy. These stores may occur within the muscle cells or outside them. They may or may not require oxygen from outside the cell in order to be utilised. Of the three main energy stores, ATP and creatine phosphate in the muscle first provide a short-term energy store, amounting to about 16 kJ or so in the human body, perhaps enough for a minute of brisk walking. Secondly, glycogen within the muscle and liver provides a medium-term store of variable size, typically ~4000 kJ, providing enough energy for some hours of moderate exercise. Finally, adipose tissue lipid provides a longer term store of ~300 000 kJ.

The high-energy phosphate store can be immediately utilised by the contractile apparatus of the muscle. Some of the muscle glycogen can contribute to the short-term energy store since it can be partially utilised without oxygen supplied from outside. Anaerobic respiration, producing lactic acid, supplies 3 moles of ATP per mole of glucose used. However, this is a much less effective process than aerobic respiration, which supplies 37 moles of ATP for each glucose unit used. There is also some oxygen bound to myoglobin in the muscle, enough for a few seconds of maximal exercise.

However, energy in the fat store and most of that in the glycogen store can only be utilised by aerobic respiration, requiring oxygen transport to the cell. The resulting increased oxygen requirement of the muscles during steady exercise is served by well known physiological changes. Breathing rate and depth, heart rate and stroke volume, and muscle blood supply all increase. Blood concentrations of free fatty acids rise as a result of fat hydrolysis in adipose tissue. There is also mobilisation of liver glycogen reserves. The rate at which these two energy stores can be tapped is limited by the rate at which oxygen can be supplied to the muscle. It is for this reason that athletes' maximum running speeds for short distances cannot be maintained over long distances (Figure 12.15).

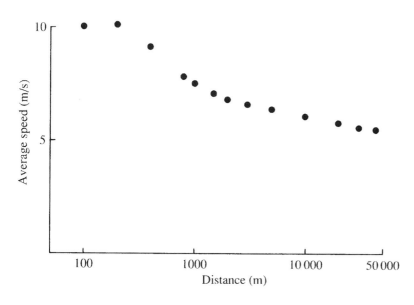

Figure 12.15 Variation of average speed with distance run. Points determined from world records for men as they were in October 1987; distances plotted on a logarithmic scale.

12.10 | Muscle Fatigue

The force exerted during a maximal voluntary isometric contraction in man begins to decline after a few seconds. A similar contraction in which the muscle exerts ~50% maximum tension can be maintained for about a minute, and a 15% maximum tension can be maintained for more than 10 minutes. This inability to maintain tension at a particular level is termed *fatigue*. It is often accompanied by feelings of discomfort and sometimes pain in the muscle.

Fatigue was classically related to imbalances between the energy consumption required by the physiological processes required to generate mechanical activity, and the metabolic capacity to supply such energy. Fatigue occurs more rapidly for isometric than for isotonic contractions. Strong isometric contractions result in increased internal tissue hydrostatic pressure that may reduce or even occlude its blood supply. This depletes muscle oxygen and hence its energy supply. Rhythmic contractions can be maintained for a much longer period since they produce only intermittent interruptions in the blood supply. However, exercise results in further effects potentially contributing to fatigue, discussed in the sections below.

12.11 | Generation of Osmotically Active Metabolites During Muscular Exercise

Both in vivo and in vitro experiments associate exercise with significant transmembrane water movements and cell swelling. This alters the cellular architecture of structures, including the transverse (T-) tubules and sarcoplasmic reticulum (SR), as well as the T–SR junctions between them, that couple tubular voltage-sensing to release of stored SR Ca^{2+}. It can also detach the T-tubules from the surface membrane, as reported following osmotic shock (Section 10.5; Usher-Smith *et al.*, 2006b, 2009). A reduced Ca^{2+} release might also result from Ca^{2+} precipitating as $Ca_3(PO_4)_2$ within the SR, or effects of reduced ATP and raised Mg^{2+} on RyR-mediated SR Ca^{2+} release (Allen *et al.*, 2008).

Sustained muscular activity results in cellular accumulation of lactate, creatine, phosphate and other metabolites. Lactate in particular reaches intracellular concentrations, $[(lactate)^-]_i$, as high as 40–55 mmol/(L cell water) in exercising fast-twitch muscle. Classical work had implicated lactate accumulation in the reduced capacity for tension generation in muscle fatigue. However, more recent reports in skinned fibres indicated that lactate only exerts very small direct inhibitory effects on either the contractile apparatus or SR Ca^{2+} release (Dutka and Lamb, 2000).

12.12 | Osmotic Stabilisation by Cellular H$^+$ Buffering Mechanisms

However, lactate accumulation also increases the quantity of intracellular osmotically active particles. This might cause water influx and consequent cell swelling. Nevertheless, skeletal muscle intracellular osmotic and H$^+$ buffering systems appeared to minimise these effects of lactate and H$^+$ production on the cell volume V_c (Usher-Smith *et al.*, 2006a; 2009). These actions were studied by replicating the effects of exercise-related simultaneous lactate and H$^+$ loading in resting amphibian muscle fibres. Spectroscopic proton nuclear magnetic resonance (^{1}H-NMR) and pH-sensitive microelectrode measurements demonstrated that exposure to ~80 mM extracellular lactate concentrations, [(lactate)$^-$]$_o$, increased intracellular lactate ion concentration, [(lactate)$^-$]$_i$, from ~4.6 to 26.4 mM (Figure 12.16*Aa, c*). It also decreased intracellular pH, pH$_i$, from ~7.24 to ~6.59 (Figure 12.16*B*). These changes replicated the proportional increases of lactate and protons reported in exercising as opposed to quiescent muscle (Figure 12.16*Ab*).

The accompanying cell volume changes, ΔV_c, were measured by laser confocal microscope scanning across the muscle fibre cross-sections. Fibre exposure to extracellular Ringer's solutions to which lactate had been added, resulting in hyperosmotic extracellular conditions, decreased V_c to extents simply reflecting the increased test solution osmolarities (Figure 12.16*C*). However, extracellular solutions to which the same lactate concentrations, [(lactate)$^-$]$_o$, were added isosmotically instead produced transient, rather than sustained, cell volume changes ΔV_c, followed by a return to normal resting V_c (Figure 12.16*Da*). Subsequent withdrawal of the elevated [(lactate)$^-$]$_o$ induced a transient decrease in V_c, similarly followed by a return to normal resting V_c (Figure 12.16*Db*). Thus V_c was stabilised despite both prior loading with, and subsequent withdrawal of, the extracellular osmotically active solute.

Computational reconstruction clarified possible mechanisms for these findings. It employed known values of cell-buffering capacity of the intracellular [H$^+$], ([H$^+$]$_i$), and of the relative membrane permeabilities to un-ionised and ionised lactate, $P_{LacH} : P_{Lac-}$. It successfully replicated the experimentally observed changes in [(lactate)$^-$]$_i$, V_c and pH$_i$ following the [(lactate)$^-$]$_o$ challenge (Figure 12.16*A, B*). The calculations attributed the initial volume increase, ΔV_c produced by the increased [(lactate)$^-$]$_o$ to the direct osmotic effect of an increase in [(lactate)$^-$]$_i$. It attributed the subsequent recovery in V_c to an increase in [H$^+$]$_i$. This would increase the association of H$^+$ with, thereby neutralising charge on, intracellular membrane-impermeant anions Pr^{z-}. This would decrease their effective valency, z_x, reducing their osmotic pressure, permitting a delayed water efflux from the cell (Sections 3.4 and 3.5; Fraser *et al.*, 2005; Figure 12.16*E*). Finally, this model also replicated the relatively small

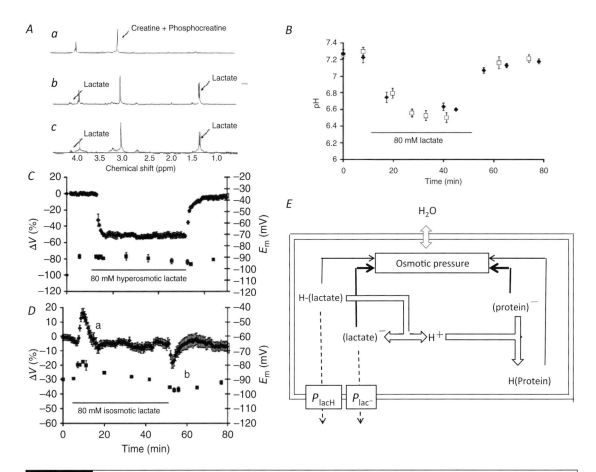

Figure 12.16 Osmotic effects of intracellular lactate. (A) ^{1}H-NMR spectra from control muscle (a), and similar findings in muscle stimulated to fatigue (b) or exposed to isosmotic 80 mM lactate-containing solution for 40 min (c). The only significantly altered peaks correspond to lactate. (B) Intracellular pH (mean ± SEM) in quiescent muscles during and following 40 min applications (horizontal bar) of a Ringer's solution in which extracellular addition followed by withdrawal of 80 mM lactate was either compensated osmotically by removal of 40 mM chloride (open symbols), or left hyperosmotic (filled symbols). (C) Plots of cell volume change, ΔV_c (top trace) and membrane potential E_m against time during (horizontal bar) and following addition of 80 mM lactate to the extracellular solution. Volume changes with lactate addition and withdrawal; but there are no further changes in V_c or E_m. (D) Plot of ΔV_c and E_m against time during and following exposure to an isosmotic 80 mM lactate-containing solution. E_m changes attributable to reductions in $[Cl^-]_o$, and transient volume changes with lactate addition and withdrawal but no further ΔV_c during the remaining 40 min exposure to lactate. (E) Schematic summary of computational modelling of these results. ((A) from Figure 3A, (B) from Figure 4, (C) and (D) from Figure 5 of Usher-Smith et al., 2006a.)

net eventual change in V_c. It demonstrated that the effects on V_c of the increase in $[(lactate)^-]_i$ ultimately balanced the effects produced by the increase in $[H^+]_i$ on the effective valency, z_x, of the impermeant intracellular anions.

These findings together suggest that the intracellular lactic acid, H-(lactate), produced in exercising muscle likely dissociates into H^+ and $(lactate)^-$. H-(lactate) and $(lactate)^-$ both exert positive osmotic effects. However, they also leave the cell at different rates depending on their relative membrane permeabilities, P_{lacH} and P_{lac}. The resulting intracellular concentrations of H-(lactate) and (lactate)$^-$ can influence the amount of H$^+$ liberated by the dissociation reaction H-(lactate) $\rightarrow$ H$^+$ + (lactate)$^-$ and therefore the [H$^+$]$_i$. This influences the extent of H$^+$ binding to charged osmotically active protein, Pr$^-$ within the cell. This binding reduces the effective charge z_x on Pr$^-$, resulting in an opposing reduction in osmotic effect. In summary and importantly, increased [(lactate)$^-$]$_i$ and decreased pH$_i$ exert opposite osmotic actions on V_c. Increased (lactate)$^-$ production causes cell swelling and buffers the intracellular acidification arising from H$^+$ production. Conversely, H$^+$ production opposes the cell swelling promoted by [(lactate)$^-$]$_i$ and promotes acidification that favours lactate efflux (Figure 12.16E).

12.13 | Intrinsic Osmotic Consequences of Altered Ion Balances During Exercise

However, sustained muscle activity also results in major transmembrane electrolyte shifts, and these can affect water balance, thereby osmotically altering cell volume. Repetitive action potential activity causes inward Na$^+$ and outward K$^+$ fluxes that tend to increase both extracellular [K$^+$] ([K$^+$]$_o$) and intracellular Na$^+$ content. Interstitial [K$^+$]$_o$ typically rises from a ~4 mM resting level to >10 mM in intensely exercising human muscle (Sejersted and Sjøgaard, 2000). These alterations in intra- and extracellular ionic concentrations in turn depolarise the cell membrane potential towards values positive to −60 mV. This itself reduces excitability in both Na$^+$ channels initiating the action potential (Sections 4.6 and 10.2) and dihydropyridine receptors (DHPR1s) detecting the resulting T-tubular voltage changes (Sections 11.1 and 11.10) and initiating excitation–contraction coupling.

Recent reports suggested that these cell membrane potential changes, ΔE_m, can also increase cell volume, ΔV_c. This may be attributed to skeletal muscle membranes having a high Cl$^-$ relative to K$^+$ permeability (Section 10.1). Positive membrane potential changes, ΔE_m, likely drive Cl$^-$ entry down the resulting electrical gradients. This may be accompanied by a K$^+$ influx through the inwardly rectifying skeletal muscle background K$^+$ channels. The latter would be facilitated by the shift in the K$^+$ Nernst potential produced by the increased [K$^+$]$_o$. This would correspondingly shift the inward rectifier's current-voltage relationship in a direction allowing K$^+$ to accompany the Cl$^-$ influx (Section 10.1). Together these drive an accompanying osmotic water entry and produce a positive ΔV_c. Cross-sectional, confocal microscopic scanning of repetitively stimulated amphibian cutaneous pectoris muscle fibres demonstrated this predicted ΔV_c (Figure 12.17Aa). Microelectrode measurements demonstrated an accompanying membrane depolarisation ΔE_m (Figure 12.17Ab). Both V_c and ΔE_m recovered with ~25 min half-lives following termination of stimulation (Usher-Smith et al., 2006b).

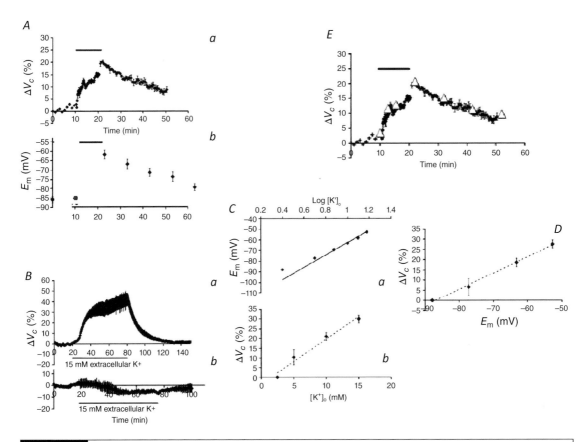

Figure 12.17 Effects of sustained repetitive stimulation on cell volume in amphibian muscle. (A) (a) Mean volume changes ΔV_c expressed as a percentage relative to initial resting volumes (±SEM) during and following 80 sets of intermittent low-frequency stimulation (horizontal bar). (b) Intracellular KCl microelectrode measurements of corresponding membrane potentials, E_m. Fibres bathed in normal Ringer's solution. (B) Experiments in quiescent muscle fibres: (a, b) plots of fibre ΔV_c against time following exposure to 15 mM extracellular K^+ in (a) Cl^--containing and (b) Cl^--deficient extracellular solutions. (C) Plots of (a) E_m, against log $[K^+]_o$ and (b) ΔV_c, against $[K^+]_o$. These were used to derive (D) the relationship between ΔV_c and E_m. The latter furnished (E) predicted ΔV_c based on the experimental E_m (Ab) that could be compared with the experimentally observed ΔV_c changes. ((Aa) from Figure 1, (Ab) from Figure 3, (Ba) from Figure 4b, (Bb) from Figure 7, (Ca, b) from Figure 4a,c, (D) from Figure 4d, and (E) from Figure 5 of (Usher-Smith et al., 2006b).)

The relationships between such altered V_c and ΔE_m in *exercising* muscle were modelled and directly related to electrolyte shifts using *quiescent* muscle fibres whose E_m and V_c were varied by exposures to increased $[K^+]_o$. Increasing $[K^+]_o$ to 15 mM produced rapid increases in V_c to a persistent plateau, ΔV_c, that subsequently recovered to baseline when test solution was replaced with normal Ringer's solution (Figure 12.17Ba). However, these changes were abolished when extracellular Cl^- was replaced with SO_4^{2-}, suggesting their dependence on transmembrane Cl^- fluxes (Figure 12.17Bb).

In such fibres studied in Cl^--containing extracellular solutions, E_m and ΔV_c varied linearly with log $[K^+]_o$ (cf. Adrian, 1956; Hodgkin and Horowicz, 1959) and $[K^+]_o$, respectively (Figure 12.17Ca,b). Combining these two dependences yielded an approximately linear relationship between ΔV_c and E_m (Figure 12.17D). This relationship could be used to predict the ΔV_c expected during stimulation in *exercising* muscle, and test the prediction against experimental results. The E_m values measured in exercising muscle through the stimulation protocol (Figure 12.17Ab) now predicted values of ΔV_c (Figure 12.17E, open triangles) that closely agreed with the experimentally observed volume changes (Figure 12.17E, filled symbols). Thus, the ΔV_c in exercising muscle arises from changes in E_m, in turn producing passive transmembrane Cl^- redistributions, leading to entry of Cl^- and K^+ into the myoplasm.

12.14 | Cellular Ionic Regulatory Mechanisms Ensuring Ion and Osmotic Balances During Exercise

Active skeletal muscle contains a number of cellular mechanisms that counteract such depressant actions of altered intra- and extracellular ion concentrations. First, a high content of membrane Na^+, K^+-ATPase actively extrudes Na^+ from and accumulates K^+ within skeletal muscle. Na^+, K^+-ATPase activity can increase up to 20-fold above the resting level within ~10 s of the onset of exercise. It may be a limiting factor for contractile endurance. Activation of Na^+, K^+-ATPase-driven ionic pumps by β_2-agonists, calcitonin gene-related peptide or dibutyryl cyclic 3,5-AMP restores excitability and contractile force in muscles exposed to elevated (10–12.5 mM) $[K^+]_o$. Conversely, Na^+, K^+-ATPase inhibition by ouabain leads to progressive loss of contractility and endurance (Clausen, 2003).

Secondly, Cl^- membrane permeabilities become reduced by ~60% with the onset of repetitive action potential firing in active muscle through the effects of protein kinase C activation or muscle acidification (Section 10.7; Pedersen *et al.*, 2009). Such a reduced Cl^- permeability increases muscle fibre excitability, even at elevated $[K^+]_o$ and would be expected to reduce the volume changes discussed in Section 12.13 (Nielsen *et al.*, 2001; Pedersen *et al.*, 2004). It could counteract the fatigue caused by electrolyte shifts in active muscle. In contrast, prolonged muscle activity leading to pronounced reductions in the cellular energetic state of the fibres can result in sudden increases in Cl^- and K^+ permeability and reduce fibre excitability. These late elevations of Cl^- and K^+ permeabilities occur particularly in fast- as opposed to slow-twitch muscle. They could constitute a fatiguing mechanism linking cellular energetic state to membrane excitability specific to the fast fibre type (Section 10.7; Figure 10.13; Pedersen *et al.*, 2009).

12.15 | Trophic Changes in Skeletal Muscle

Muscle also shows longer term, clinically important, structural and functional changes measurable through radiological assessments of whole muscle cross-sectional areas, tension generation, and tissue and cellular level needle biopsy. These demonstrated that muscles are markedly affected by the amount of use they receive. Muscles increase in size and strength, particularly with repeated isometric exercise involving generation of high muscle tensions, including weight lifting. The resulting enlargement of individual muscle fibres is accompanied by increased quantities of contractile proteins. In contrast, aerobic endurance exercise, such as distance running, increases muscle blood supply through vascular capillary proliferation. The muscle fibres themselves do not greatly increase in size but increase their quantities of respiratory enzymes. There is also an increase in connective tissue so that the muscles become less susceptible to minor injuries. Conversely, muscle disuse, as occurs when a subject is confined to bed or subject to prolonged weightlessness in space, leads to reduced muscle mass, especially involving the contractile proteins. Hence the need for specific exercise regimes for invalids and astronauts.

These changes involve mechanisms both at the local myocyte level and signalling from other body organs. First, electrical signalling and its consequences for $[Ca^{2+}]_i$, metabolite levels or levels of mechanical stress modify gene expression through excitation–transcription coupling mechanisms in the muscle fibres themselves. In particular, Ca^{2+} signalling processes follow DHPR1 activation by tubular sarcolemmal membrane depolarisation and the consequent ryanodine receptor (RyR1)-mediated release of sarcoplasmic reticular (SR) Ca^{2+} (see Chapter 11). A prolonged entry of extracellular Ca^{2+}, possibly through the DHPR1, independent of sarcoplasmic reticular Ca^{2+} stores, has also been reported following repetitive or prolonged depolarisation.

These resulting $[Ca^{2+}]_i$ alterations likely act through the Ca^{2+}-calmodulin (CaM) complex (Figure 12.18A). Amongst other effects (Sections 5.2, 8.9, 11.6, 11.14), this activates Ca^{2+}/calmodulin-dependent protein kinase II (CaMKII) whose serine-threonine kinase action phosphorylating histone deacetylase 4 (HDAC4) causes its nuclear export, relieving its inhibition of transcriptional pathways driving hypertrophic, mitochondrial and oxidative gene transcription. It also activates calcineurin (CN) whose serine-threonine phosphatase action dephosphorylating nuclear factor of activated T-cells (NFAT) causes its nuclear import, enhancing its activation of such gene transcription.

Secondly, of intercellular signalling molecules (Figure 12.18B), insulin-like growth factor 1 (IGF-1), released by the liver in response to growth hormone, through targetting IGF-1 (IGF1R) and insulin receptors, enhances these anabolic effects. Conversely, the autocrine signalling molecule myostatin, through activin-2B receptors (ACVR2B), inhibits muscle growth. Myostatin gene disruption increases fibre number and size causing muscle enlargement (Chawla et al., 2003; Gundersen, 2011).

12.16 | Age-Related Sarcopaenia

A clinical condition of increasing clinical and public health importance, *sarcopaenia* refers to the skeletal muscle atrophy and weakness often accompanying and associated with loss of mobility and independence,

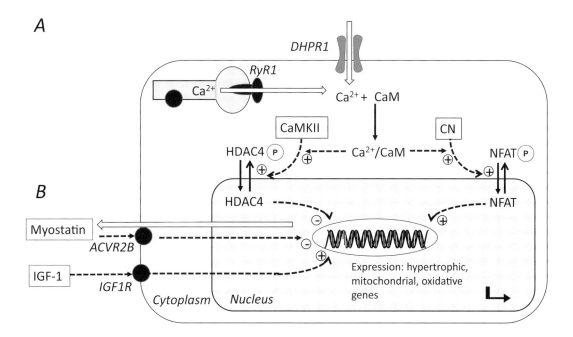

Figure 12.18 Excitation-transcription coupling and intercellular signalling mechanisms influencing trophic changes in skeletal muscle. Signalling from (A) increased $[Ca^{2+}]_i$ arising from RyR1-mediated SR Ca^{2+} release and DHPR1-mediated Ca^{2+} entry. (B) Intercellular signalling molecules. DHPR1: dihydropyridine receptor type 1, RyR1: ryanodine receptor type 1, CaM: calmodulin, CaMKII: Ca^{2+}/calmodulin-dependent protein kinase II, CN: calcineurin, HDAC4: histone deacetylase 4, NFAT, nuclear factor of activated T-cells, IGF-1: insulin-like growth factor 1, ACVR2B: activin-2B receptors, IGF-1R: IGF-1 receptors. P denotes protein phosphorylation.

and frailty, in aging adults. It typically presents as decreased muscle mass. However, the decline in function, measurable as contractile strength normalised to either cross-sectional area or muscle mass for a given muscle, is often worse than expected from the reduced muscle mass.

Its pathophysiology is likely multi-factorial and remains unclear, owing to a lack of robust cellular, histological and basic science animal models. Nevertheless, a number of possible contributing mechanisms have been identified. Electromyographic studies directly recording electrophysiological activity from muscle and techniques estimating motor unit number in proximal and distal limb muscles demonstrated reduced muscle fibre numbers. This likely reflects progressive loss of α-motor neurons, resulting in corresponding losses in entire functioning motor units, typically occurring from the age of 70 years. Anatomical evidence thus suggests age-related reports of losses of anterior horn cells and ventral root fibres in the spinal cord. In addition, changes in the excitation transcription mechanisms outlined above may take place.

Current available management involves modifying lifestyle factors, for which exercise training is the only intervention with established benefit. Resistance training is the most effective in increasing muscle mass and strength. Conversely, physical inactivity accentuates age-related sarcopaenia, but such effects proved reversible with training, even in elderly (>90 years) subjects (Matthews *et al.*, 2011).

Cardiac Muscle

Muscle cells have become adapted to a variety of different functions through their evolution, so that the details of their contractile processes and their control in other muscle types show differences from those in vertebrate skeletal muscle. The final chapters examine the properties of mammalian cardiac and smooth muscle.

13.1	Structure and Organisation of Cardiac Muscle Cells

Cardiomyocytes are considerably smaller than skeletal muscle fibres. Ventricular cardiomyocytes are typically up to 10 μm in diameter and 100 μm in length. Adjacent cardiac cells are mechanically and electrically coupled, both in a branched and an end-to-end manner, by intercalated disks to give a syncytium which transmits both electrical activity and mechanical force (Figure 13.1A). Atrial and ventricular myocytes specialised to generate mechanical activity contain contractile elements whose structure is similar to that found in skeletal muscle. They similarly show transversely aligned thick myosin and thin actin filaments (Figure 13.1B), and are cross-striated in appearance. They also contain mitochondria, sarcoplasmic reticulum (SR) and transverse (T-) tubules.

However, cardiac SR is less developed than that of skeletal muscle. In ventricular cardiomyocytes the SR makes complexes with T-tubular membrane at dyad rather than triad junctions. In atrial myocytes, the T-tubular system is also less developed. The SR makes junctions with caveolae in the membrane surface. Additional cardiac cell types with differing specialisations include cells that primarily generate and conduct electrical impulses rather than contract. These occur in the sino-atrial node (SAN), the atrio-ventricular node (AVN) and the atrio-ventricular (AV) bundle of His. Each cell type assumes a distinct role in cardiac excitation and contraction.

A

Cardiac myocyte

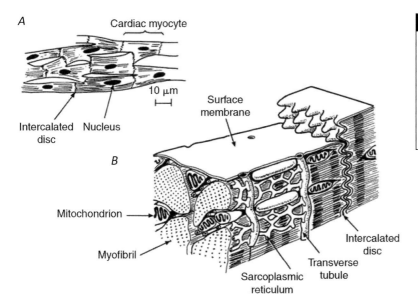

10 μm

Intercalated disc   Nucleus

Surface membrane

B

Mitochondrion

Myofibril

Sarcoplasmic reticulum

Transverse tubule

Intercalated disc

Figure 13.1 (A) Syncytial arrangement of mammalian cardiac muscle cells. (B) Structural components showing relationships between surface and T-tubular membranes, myofilaments and intercalated disks connecting cells. (From Emslie-Smith *et al.*, 1988, after Fawcett and McNutt, 1969.)

13.2 | Electrical Initiation of the Heartbeat

Cardiac excitation occurs through a succession of excitable and conducting structures (Figure 13.2A). Each cardiac cycle includes an atrial followed by a ventricular contraction. Continuous cardiac activity does not require, though is modified by, its autonomic nervous input. Its excitation begins at the SAN pacemaker region, whose component cells are spontaneously active and responsible for rhythmic production of action potentials that determine the frequency of the heartbeat. This pacemaker role of the SAN continues even in the absence of neural control. The sino-atrial (SA) cells excite their neighbours by local current flow, thereby initiating waves of depolarisation across the atria. These trigger the atrial systole that forces blood into the ventricles.

The AV ring electrically isolates the ventricles from the atria, with the AVN providing the only electrical communication between the two pairs of chambers. The AVN comprises small-diameter fibres with slow conduction velocities (~0.2 m/s). This ensures an appreciable delay between atrial and ventricular excitation and contraction. The AVN is connected to large Purkinje fibres in the bundle of His. These cells show more rapid impulse propagation (~2–5 m/s) than the surrounding myocytes (~1 m/s). They carry the excitation on to the main mass of the ventricular muscle, beginning in the septum and spreading from the apex of the ventricle up to its base, running along the ventricular endocardial surface. Purkinje cells can also function as

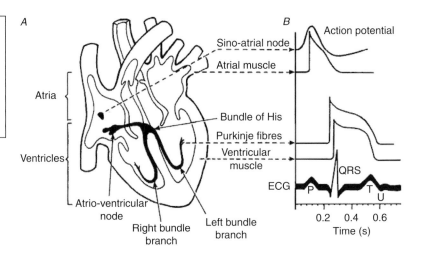

pacemakers in the presence of pathological conditions leading to conduction block, but have substantially lower intrinsic firing rates. It is the SAN cells with the highest automaticity that normally determine overall heart rate. Finally, the main mass of the cardiac ventricular muscle acts as an electrical syncytium in which the cells are separated by low electrical impedances and structures potentially mediating ephaptic couplings between adjacent cell membranes (Sections 8.8 and 13.8). This permits propagation of electrical changes from cell to cell by local current spread and possible cell-cell ephaptic excitation. This resulting activation pattern leads to a ventricular contraction that optimises ejection of blood from the chambers.

13.3 | The Cardiac Action Potential

Cardiac action potentials were first studied by intracellular recordings from cardiac muscle fibre preparations made using isolated bundles of canine Purkinje fibres (Weidmann, 1952; 1956). Subsequent availability of isolated cardiac myocytes representing different cardiac cell types permitted closer study of both their ion currents and the subsequent signalling events (Powell *et al.*, 1980). After being isolated for a short time, Purkinje fibres begin to produce rhythmic spontaneous action potentials, of the sort shown diagrammatically in Figure 13.3. In common with nerve membranes, typical atrial or ventricular action potential waveforms begin with a rapid (~400 V s^{-1}) initial depolarisation from the normal negative resting potential (around -90 mV) to a positive voltage ($+40$ to $+60$ mV) close to the Na$^+$ Nernst potential. However, the subsequent forms of these action potentials differ from those of nerve axons and skeletal muscle fibres. Furthermore, each cardiomyocyte type possesses specific electrophysiological properties and detailed action potential waveform features related to its specific function (Figure 13.2*B*).

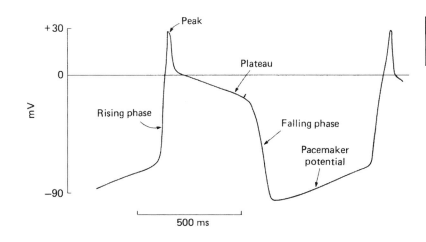

SAN pacemaker cells show less polarised resting potentials, marked pacemaker potentials, but an absence of an action potential plateau phase. In contrast, neither atrial nor ventricular cells show pacemaker potentials. Whereas atrial action potentials have a triangulated waveform, ventricular action potentials show a prolonged 'plateau' between the action potential peak and the repolarisation phase. Ventricular action potentials are divided into five phases. In Purkinje fibres, a preceding slowly rising *pacemaker potential* acts as a trigger for the action potential when it crosses a threshold level. There follows a rapid depolarisation (phase 0), an initial brief rapid repolarisation (phase 1), a plateau (phase 2) and a terminal repolarisation (phase 3) that restores the membrane potential to the resting level, which persists for a period termed the electrical diastole (phase 4).

The long duration of the cardiac action potential compared to that in twitch skeletal muscle fibres relates to important differences in the roles of their respective action potentials in excitation–contraction coupling. In skeletal muscle the action potential acts simply as a trigger initiating the resulting contraction, but has no further control over it (Sections 11.1 and 11.13). In cardiac muscle the action potential timecourse is coincident with much of the contraction phase, with relaxation beginning during its repolarisation phase (Sections 13.10 and 13.14). If the action potential is shortened in some way, relaxation also begins earlier and so the tension reaches a lower peak level. The reverse happens if the action potential is lengthened. Hence the cardiac action potential both triggers and controls the process of contractile activation.

13.4 | Extracellular Measurement of Cardiac Electrophysiological Activity

In clinical practice, these cardiac electrophysiological events are measured by non-invasive extracellular, electrocardiographic (ECG), recording using body-surface electrodes. Einthoven introduced the ECG to detect and localise clinical abnormalities arising from altered function in particular cardiac regions (Einthoven, 1924). The ECG represents voltage gradients produced by extracellular current flow between different parts of the heart as they successively undergo electrical excitation and recovery. Different components of the ECG waveform correlate with, thereby throwing light upon, atrial and ventricular activation and recovery, and their propagation (Noble, 1979). The density of axial current flow, i_a, along the length, x, of any given excitable cell or fibre structure is related to its intracellular resistance of unit length r_a, and the voltage gradient, $\partial V/\partial x$, along its length by Ohm's law giving (Section 6.1, Figure 6.1, Equation (6.2)):

$$\frac{\partial V}{\partial x} = -r_a i_a \tag{13.1}$$

If the action potential conducts at conduction velocity, $\theta = \partial x/\partial t$, the chain rule gives:

$$\frac{\partial V}{\partial x} = \frac{\partial V}{\partial t} \div \frac{\partial x}{\partial t} = \frac{1}{\theta}\frac{\partial V}{\partial t} \tag{13.2}$$

Combining Equations (13.1) and (13.2) demonstrates that i_a is proportional to the rate of change ($\partial V/\partial t$) of the transmembrane voltage, V:

$$i_a = \frac{1}{r_a \theta}\frac{\partial V}{\partial t} \tag{13.3}$$

The density of the extracellular current producing the measured surface potentials detected by the ECG is proportional to the sum of these intracellular current flows, i_a, since the extracellular and intracellular fluids form a closed loop for the flow of a conserved current. ECG deflections from the baseline, or isoelectric line, at any time, each reflect the rate of transmembrane voltage change from active tissue. A positive deflection indicates an effective cardiac dipole with the positive pole facing the electrode, which typically occurs when a depolarising impulse is conducted towards the electrode or if a repolarising wave propagates away from the electrode. The opposite interpretation applies for a negative deflection.

The magnitude of the ECG deflection also depends upon the direction of the current i_a relative to an axis joining the recording electrodes. This is the basis for current clinical practice measuring and comparing electrocardiographic potentials between differently oriented pairs of electrodes. Each optimises visualisation of electrophysiological function in different cardiac regions, depending on the angle that propagation of its electrical activity subtends relative to the axis of their electrical changes. The respective recordings from right and left arm (lead I) or left leg (lead II), or left arm and left leg (lead III) or the average of two leads relative to the remaining lead (augmented limb leads aVF, aVR, aVL) view cardiac activity taking place along different directions within the coronal plane. In contrast, the precordial leads (leads V_1–V_6) compare potentials between differently located chest leads and the averaged voltage of the three limb leads forming a common reference terminal (of Wilson) posterior to the heart, effectively detecting voltage differences across the heart cross-section.

Finally, the contribution to the i_a term of each component of the heart to the ECG varies with its tissue mass and consequent current generating capacity. This would lead to the expectation that coincident activation in more extensive regions of an area would generate larger deflections than regions accounting for a small fraction of the overall cardiac tissue mass.

13.5 | Physiological Interpretation of the Electrocardiogram

The ECG can readily be interpreted by comparing its successive waveform components with the expected intracellular action potential waves in successively excited atrial and ventricular components of the heart (Figure 13.2*B*). The initial ECG deflection from the isoelectric baseline is the P wave. This coincides with and likely represents electrical events associated with atrial depolarisation. Much of the atrial repolarisation phase occurs after the onset of ventricular depolarisation. This is followed by conduction through the specialised conducting AVN and the remaining Purkinje conducting system. The latter form only a small fraction of the total cardiac mass, and generate negligible axial current. The period of AV conduction, the PR interval, is represented by the duration the ECG trace has spent at the isoelectric line.

The QRS complex that follows is triphasic in waveform. This may reflect not only the timecourse of the single ventricular action potential followed by its initial early, phase 1 repolarisation, but also its propagation pattern through the right (RV) and left (LV) ventricles. The plateau phase that follows, resulting in zero rate of change of membrane potential (dV/dt), explains the return of the ECG trace to baseline during the ST segment. During this time, the ventricular muscle is contracting to pump blood through the aorta and pulmonary artery.

The T-wave coincides with the downward deflection in the action-potential waveform in its recovery to baseline. Conditions in which the propagated depolarisation and its recovery are uniform through the ventricle would result in depolarisation and its recovery propagating with the same direction and magnitude through the heart. This would predict a downward deflecting T wave. In contrast, the normally upright T wave may reflect a heterogeneity in recovery times resulting in the recovery wave taking different, possible opposite, directions to the depolarisation wave. For example, if the apex repolarises before rather than after the base of the heart, this will give the propagation of the recovery wave a direction opposite to the direction of propagation of the preceding depolarisation. The velocity of repolarisation becomes negative relative to the velocity of conduction, θ. The current during repolarisation i_a, and therefore the T-wave, consequently typically assumes a positive rather than a negative sign.

The normal ECG thus conforms to a standard recognisable pattern. Deviations from this permit diagnosis of different cardiac electrophysiological disorders (Section 14.1, Figure 14.1). Major diagnostic categories for which the ECG is useful include: (a) conduction disorders; these may take place in different structures ranging from the pacemaker cells in the SA node to the ventricular myocardium, (b) rhythm disorders, whether in the atria or ventricles and (c) metabolic disorders, involving electrolyte balance, or ischaemia or infarction. Any of these could result in departures from criteria for normality in an adult. The latter include: (a) For the P wave: one before each QRS complex and ≤ 0.12 s wide and ≤ 0.3 mV high in lead II. (b) For the PR interval: a consistent duration between 0.12 and ≤ 0.24 s. Shorter PR intervals suggest an existence of abnormal, accessory, conduction pathways; longer intervals suggest first-degree heart block. (c) The duration of the QRS complex should be ≤ 0.12 s. Longer complexes can result from intraventricular conduction defects. (d) The ST segment is normally isoelectric. (d) The QT interval is frequency dependent, but should generally be ≤ 0.45 s. Longer QT intervals may suggest long QT syndrome (Sections 14.8–14.10).

13.6 | Ionic Currents in Cardiac Muscle

Clarifying the ionic basis of the cardiac muscle action potential has proven a complicated task. The small size and complex geometry of cardiac muscle fibres, in comparison to squid axons, makes them more difficult to voltage clamp, a problem solved through the introduction of isolated cardiomyocyte preparations (Powell et al., 1980). Furthermore, a considerably larger number of different ion channels are involved than in either nerve or skeletal muscle. Plate 13 summarises the ionic channel contributions to cardiac muscle action potentials. It identifies the membrane currents that they generate with their underlying protein molecules and encoding genes and distinguishes inward depolarising (A) and outward repolarising currents (B) acting at different stages of the atrial (C) and ventricular (D) action potential.

This wide range of ion channel types shows a wide diversity of protein quaternary structures extending beyond those of the voltage-dependent Na^+ and Ca^{2+} channels discussed in Chapter 5. They vary in the number of their component domains and of transmembrane segments making up each domain (Plate 6). Clarification of each of their associated ionic currents and analysis of their roles in cardiac electrophysiological activity required their own separate voltage- or patch clamp characterisations.

The interactions between the currents that produce the atrial (C) and ventricular action potential (D) can be clarified by computational reconstruction. Figure 13.4 illustrates this for the Purkinje fibre action potential (top panel) (DiFrancesco and Noble, 1985), breaking the groups of currents into four major ionic conductances (Figure 13.4, bottom panel). The Na^+ conductance g_{Na} is rapidly activated and then inactivated by depolarisation, and is carried by a specific cardiac Nav1.5 isoform. Nevertheless, this is similar in structure and function to the Na^+ conductances of nerve and skeletal muscle cells, carried by Nav1.1 and Nav1.4, respectively, apart from slower kinetics and reduced tetrodotoxin (TTX) sensitivity. The K^+ conductance g_K arises from a large number of channel types with specific properties and functions, some activated by hyperpolarisation and others by depolarisation. An appreciable Ca^{2+} conductance g_{Ca}, carried by the cardiac Cav1.2, is activated by depolarisation and produces an inward current during the plateau. Finally, repetitive atrial and ventricular excitation cycles normally depend upon SAN automaticity driven by pacemaker cells. These show a specific conductance g_f, carried by a hyperpolarisation-induced cyclic nucleotide activated channel isoform, HCN4, that conducts a slow inward movement of Na^+ and other ions. It is activated by hyperpolarisation and is important during the pacemaker potential.

13.7 | Pacemaker Activity in Specialised Cardiac Regions

The sequence of events in the DiFrancesco–Noble model (Figure 13.4) begins at the point in the cycle where the membrane potential is at its most negative, at about 0.4 ms on the time trace. It has reached this negative value because the K^+ conductance g_K is high. However, the pacemaker conductance g_f carried by HCN4 has been switched on by the hyperpolarisation, and it increases steadily for the next second or so (Mangoni and Nargeot, 2008). The slow Na^+ inflow which this permits results in a steady progressive depolarisation, the pacemaker potential. Subsequent work has implicated a range of additional ion channels as well as mechanisms involving Na^+/Ca^{2+} exchange currents and intracellular Ca^{2+} homeostasis in this pacing process (Section 13.16; Lei *et al.*, 2005; 2007).

Pacemaker activity is important in regulating the frequency of contractile activity. Pacemaker conductances occur in the SAN and AVN. Of these, under normal conditions, the high background leak conductance of SAN cells results in a less hyperpolarised background resting potential and confers the highest intrinsic firing frequency. This primary pacemaker determines the normal frequency of the human heartbeat of around 70 min^{-1}. Sinus node disorder compromising SAN function permits other sites with slower intrinsic rates to act as substitute pacemakers, thereby establishing an escape rhythm. This is most frequently the AVN. This paces at slower (~40–60 min^{-1}) rates that nevertheless permit a normal spread of

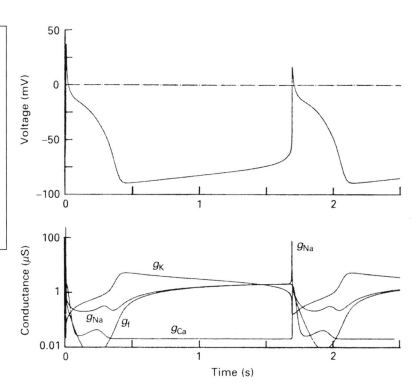

Figure 13.4 Computer simulation of the cardiac action potential. The associated conductance changes are shown in the lower graphs. g_f is the inward current which becomes apparent during the pacemaker potential. The Na$^+$ conductance g_{Na} includes both the conductance due to fast Na$^+$ channels and the Na$^+$ component of g_f. Activation of the Ca^{2+} conductance g_{Ca} coincides with the action potential plateau phase. (From DiFrancesco and Noble, 1985.)

electrical activity through the Purkinje fibres into the ventricles. Genetic alterations involving HCN channels are associated with clinical arrhythmic conditions (Baruscotti *et al.*, 2010). Myocardial cells whose main function is to contract usually cannot produce pacemaker activity.

13.8 | Phase 0 Depolarisation and Early Phase 1 Repolarisation: Na$^+$ and Transient Outward K$^+$ Currents

The pacemaker potential eventually drives a membrane depolarisation sufficient to open the fast-activating cardiac Nav1.5 Na$^+$ channels. In other cardiac regions propagation of excitation from previously activated regions of the heart results in a local cable current spread to such initially quiescent regions, similarly depolarising their resting potentials (Sections 6.1 and 6.2). Reaching the excitation threshold initiates a regenerative cycle of Na$^+$ channel opening, with further membrane depolarisation causing further channel opening and a rapid increase in membrane Na$^+$ conductance, g_{Na}. This produces the initial, rapid phase 0 depolarisation phase in common with the action-potential excitation phase in nerve axons and skeletal muscle (Section 4.3). The rapidly increasing Na$^+$ current (I_{Na}) drives the myocardial membrane potential from its normal negative resting potential to a positive (+40 to +60 mV) voltage close to the Na^+ Nernst potential. The peak membrane potential deflection is reduced when the extracellular Na^+ concentration $[Na^+]_o$ is lowered.

The resulting depolarisation causes subsequent Na^+ channel inactivation into a refractory state. It also activates early transient outward K^+ currents, I_{to}, comprising fast, $I_{to,f}$, carried by Kv4.2 and Kv4.3, and slow $I_{to,s}$, carried by Kv1.4. These drive early phase 1 repolarisation of the membrane potential from its positive peak. Their greater prominence in atrial compared to ventricular cardiomyocytes results in the shorter, triangulated form of the atrial compared to the ventricular action potential (see Plate 13).

The action potential drives local electrical excitation between active and quiescent cells. Electrotonic excitation takes place through intercellular gap junction, atrial (Cx40, Cx43) and ventricular (Cx43), connexin channels (Section 4.2). These in turn depolarise neighbouring initially quiescent cells. In addition, ephaptic, coupling across narrow extracellular clefts separating adjacent cells (Section 8.8; Salvage *et al.*, 2020a; Veeraraghavan *et al.*, 2014) may also contribute to propagating the wave of excitation onwards. Effective cardiac function depends upon repetitive cycles of these excitation events within individual cardiomyocytes propagating as coherent electrical waves successively through the SANs, atria, AV bundles, Purkinje conducting tissue, and ventricular endocardial and epicardial myocardium.

13.9 | The Phase 2 Plateau: Inward Ca^{2+} Current

In contrast to the situation in nerve membrane, rather than an immediate return to resting potential, early Phase 1 repolarisation in ventricular myocytes and Purkinje cells is followed by a plateau phase. This maintains a depolarised membrane potential near 0 mV for as long as 500 ms after the rapid early upstroke. It arises from prolonged inward Ca^{2+} current (I_{CaL}) through voltage-gated L-type Ca^{2+} channels, often termed dihydropyridine receptors (DHPRs; Section 11.7). The currents are initially activated by the early depolarisation phase, and subsequently maintained by the sustained depolarisation that they themselves produce (Bers, 2001). They are carried by a specific cardiac, Cav1.2, DHPR2, isoform whose activation timecourse is substantially more rapid than the corresponding I_{Ca} (Cav1.1) activation in skeletal muscle (Section 10.2). Its amplitude varies with extracellular Ca^{2+} concentration, $[Ca^{2+}]_o$, and is diminished by Ca^{2+}-channel blockers such as verapamil and nifedipine.

In addition, inward ('anomalous'; Section 10.1) rectification properties of I_{K1} mediated by Kir2.1, Kir2.2 and Kir2.3 K^+ channels (Section 10.3; Plate 6) reduce their K^+ conductances, g_K, during Phases 0–2, when membrane potential is depolarised to ~ -20 mV (Corrias *et al.*, 2011). This increases the background membrane resistance, in turn minimising the inward I_{CaL} required to maintain the plateau potential. This similarly minimises its associated dissipation of trans-membrane $[Ca^{2+}]$ gradients. These I_{K1} channels contrastingly enhance the subsequent late Phase 3 outward currents with repolarisation to membrane potentials more negative than -40 mV. Finally, they stabilise the diastolic, Phase 4, resting potential. Furthermore, as in skeletal muscle (Section 10.6), K^+ accumulation following ventricular depolarisation within the restricted T-tubular luminal space shifts the local tubular K^+ Nernst potential, activating I_{K1}, and further modulating action potential recovery.

13.10 | Phase 3 Repolarisation: Voltage-Dependent Outward K^+ Currents

Phase 3 repolarisation results from gradual activation of voltage-dependent outward K^+ currents. The rapid I_{Kr} (Kv11.1) current is important both for completing Phase 3 repolarisation and preventing pro-arrhythmic early afterdepolarisation phenomena (Sections 14.5 and 14.10). The underlying Kv11.1 channel becomes rapidly activated by Phase 0 depolarisation early in the action potential. However, it quickly transitions from this open to an inactivated state, minimising outward I_{Kr} during the action potential plateau phase. Repolarisation produces a prompt recovery from this inactivated state, via the open state before final return to the resting state. This results in an outward repolarising tail current (Lu *et al.*, 2001; 2003; Vandenberg *et al.*, 2006).

In addition, a later contribution from the more slowly activating Kv7.1-mediated I_{Ks} drives the membrane potential back towards the resting level. This current is also sensitive to adrenergic stimulation through which it adjusts the action potential duration inversely with the heart rate. This ensures that the relative durations of systole and diastole are appropriate to changes in the diastolic interval between successive action potentials.

13.11 | Phase 4 Electrical Diastole: Inward Rectifying K^+ Currents

By Phase 4 electrical diastole, the cell has regained the resting potential. The inward rectifying current I_{K1} is essential for maintaining a fully polarised (~-90 mV) resting potential. Additional contributions to such repolarisation reserve come from two-pore domain K^+ leak currents (I_{K2p}). Any Ca^{2+} and fast Na^+ channels remaining open are closed during this repolarisation phase. Na^+ channels recover their capacity for re-excitation. This recovery takes place during both absolute and relative (effective) refractory periods, ERPs. Only after such recovery do Na^+ channels become capable of subsequent re-excitation. Finally, by the end of the action potential the pacemaker conductance g_f has already begun to rise, initiating the next cardiac cycle.

13.12 | The Prolonged Refractory Period in Cardiac Muscle

The presence of a prolonged action potential plateau depolarisation phase results in a delayed recovery of g_{Na} from inactivation. Cardiac muscle consequently shows a refractory period (Section 2.4) following the initial excitation that is substantially longer (up to 100 ms) than in nerve. The membrane is absolutely refractory between the early rapid depolarisation to the point when the membrane potential is repolarised to ~−40 to −50 mV, largely owing to Na$^+$ channel inactivation. It is relatively refractory beyond that, during which an evoked action potential has a smaller amplitude and rate of rise, and consequently is conducted more slowly (Figure 13.5).

The relatively long effective refractory period (ERP) prevents tetany in cardiac muscle following repetitive stimulation, since the refractory period is long enough to allow the muscle to relax after each action potential. This contrasts with the situation in skeletal muscle (Section 12.2). It is important to the function of the heart as a pump: the relaxation (diastolic) phase permits refilling of the heart with venous blood, permitting its expulsion into the arteries during the contraction (systolic) phase. It also prevents premature or re-entrant excitation occurring in cardiac cells elsewhere from initiating inappropriate re-excitation and consequent arrhythmias (Section 14.3).

13.13 | Varying Ionic Current Contributions in Different Cardiomyocyte Types

The analysis above can be generalised to guide expectations for ion-channel mechanisms underlying excitable activity in other

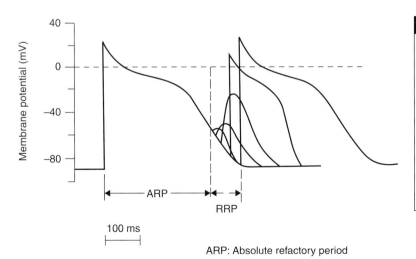

Figure 13.5 The absolute and relatively refractory periods of ventricular muscle in relationship to the action potential waveforms of ventricular muscle. Extra-stimulation within the absolute refractory period results in a failure or re-excitation. Within the relatively refractory period it results in an action potential with diminished initial slope and reduced amplitude. (From Emslie-Smith et al., 1988.)

ARP: Absolute refactory period

RRP: Relative refactory period

cardiomyocyte types. The different phases of the action potential can be identified with particular groups of ionic currents. The inward I_{Na} or I_{Ca} drive the rapid depolarising Phase 0 and I_{Ca} maintains the Phase 2 plateau. The wide range of outward K^+ voltage-dependent currents drives the early repolarising Phase 1 and the Phase 3 repolarisation that terminates the plateau. Inwardly rectifying currents and the refractory period, which reflects the time required for Na^+ channel recovery from refractoriness, are associated with Phase 4 electrical diastole. Finally, the gap junction, connexin, proteins conduct the resulting wave of electrophysiological changes through successive cardiomyocytes. Ventricular action potentials resemble those in Purkinje cells, as outlined above, apart from their physiological activity, not including pacemaker currents.

Atrial action potentials show similar early repolarising Phase 0 mechanisms, but begin from more depolarised resting potentials, reflecting their smaller I_{K1}. They show a more prominent Phase 1 repolarisation owing to their relatively larger I_{to}. Atrial myocytes also express particular ultra-rapid delayed rectifier, I_{Kur}, acetylcholine-activated, I_{KACh}, and Ca^{2+}-activated K^+ currents, I_{KCa}, but smaller I_{Kr}, I_{Ks} and I_{K1}, resulting in a more prolonged Phase 3 repolarisation. Together these result in triangular repolarisation waveforms and less prominent Phase 2 plateau phases than shown by ventricular action potentials (Plate 13). Finally, an adenosine triphosphate (ATP)-sensitive K^+ current, I_{KATP} (Kir6.2) occurs throughout the heart. It generally contributes little current as it is inhibited by the existing levels of intracellular ATP, but this current may increase under conditions of energetic stress.

13.14 | Cardiac Excitation–Contraction Coupling

In common with skeletal muscle, cardiac muscle mechanical activity is initiated by increases in $[Ca^{2+}]_i$ following T-tubular membrane depolarisation. Transient increases in $[Ca^{2+}]_i$ have been observed in cardiac muscle cells just as in skeletal muscle. The SR remains the most important source of this activator Ca^{2+}. However, cardiac muscle expresses a specific dihydropyridine receptor DHPR2 isoform and a specific ryanodine receptor RyR2 isoform, in contrast to the corresponding skeletal muscle DHPR1 and RyR1. These DHPR2 and RyR2 isoforms similarly act as voltage sensors and SR Ca^{2+} release channels. However, they show important differences in properties leading to distinct mechanisms of excitation–contraction coupling that contrast with those operating in skeletal muscle (Section 11.6).

The direct skeletal muscle DHPR1–RyR1 allosteric coupling (Section 11.9, Plate 12) does not exist in cardiac muscle. Instead, cardiac excitation–contraction coupling is initiated by opening of T-tubular voltage-gated Ca^{2+} channels mediating the inward I_{Ca} responsible for the Phase 2 action potential plateau (Section 13.9). The resulting extracellular Ca^{2+} influx elevates local $[Ca^{2+}]_i$ in regions of the sarcolemmal–SR junctions. This elevated Ca^{2+} triggers opening of cardiac SR RyR2 Ca^{2+} channels through Ca^{2+}-induced Ca^{2+} release (Section 11.8). The consequent release of intracellularly stored SR Ca^{2+} elevates $[Ca^{2+}]_i$, in turn triggering mechanical activation (Bers, 2001).

RyR2-mediated Ca^{2+}-induced Ca^{2+} release is likely to reflect a summation of discrete elementary Ca^{2+} release events measurable as spontaneously occurring Ca^{2+} sparks, observable even in resting cardiac myocytes. These were first demonstrated by field- or line-scan laser scanning confocal microscopy in mammalian cardiac myocytes loaded with the Ca^{2+} fluorescence indicators fluo-3 or fluo-4 (Section 11.3). They appeared as brief (~10–20 ms), localised (~2–3 μm) myoplasmic free $[Ca^{2+}]_i$ elevations with rapidly rising followed by exponentially falling timecourses. They also occurred in resting amphibian skeletal muscle, which expresses both RyR3 and RyR1, but not adult mammalian skeletal muscle, which only expresses RyR1 (Cheng et al., 1993; Hollingworth et al., 2002; Zhou et al., 2003).

Studies in cellular expression systems demonstrated and characterised this dependence of the skeletal or cardiac muscle mechanism of excitation–contraction coupling upon expression of their particular DHPR or RyR isoforms. Myotubes from genetically abnormal muscular dysgenic (mdg) mice showed deficient excitation–contraction coupling. This was associated with reduced DHPR expression and deficiencies in the slow I_{Ca} characteristic of in vivo skeletal muscle. However, injecting an expression plasmid carrying DNA complementary to the DHPR1 in dysgenic myotubes restored both an excitation–contraction coupling mechanism resembling the form occurring in skeletal muscle and such I_{Ca}. In contrast, DNA complementary to the cardiac DHPR2 restored an excitation–contraction coupling mechanism resembling that of cardiac muscle. This was dependent on Ca^{2+} entry and was associated with the more rapidly activating I_{Ca} resembling that found in cardiomyocytes (Adams and Beam, 1990).

These functional differences between cardiac and skeletal muscle correlated with structural differences in specific regions of the two DHPR isoforms. This emerged from experiments on constructs of different chimaeric Ca^{2+} channels (Section 5.2) successively replacing each of the four skeletal muscle DHPR domains by the corresponding cardiac domain. They implicated domain I in determining whether channels showed slowly activating I_{Ca} characteristic of skeletal muscle or rapidly activating I_{Ca} typical of cardiac muscle. The cytoplasmic loop between domains II and III was important in conferring either skeletal muscle-type excitation–contraction coupling involving allosteric DHPR–RyR interactions or cardiac muscle-type excitation–contraction coupling involving Ca^{2+}-induced Ca^{2+} release. The region linking domains III and IV were highly conserved between skeletal DHPR1 and cardiac DHPR2, as well as with corresponding regions in the Na^+ channel, which are thought to be involved in channel inactivation. The DHPR distal N-terminus is weakly conserved and does not appear critical for either excitation–contraction coupling or Ca^{2+} channel function. However, the proximal N-terminus may be important in sarcolemmal expression of the DHPR (Tanabe *et al.*, 1990; Grabner *et al.*, 1999).

13.15 | Forms of Ca^{2+}-Induced Ca^{2+} Release in Cardiac Myocyte Subtypes

In ventricular cells, DHPR2 activation involves propagation of the cardiac action potential into the T-tubules in which these receptors reside, in parallel with tubular activation processes previously outlined for skeletal muscle (Section 10.3; Figure 13.6). In contrast, the smaller atrial myocytes have less developed T-tubular networks, particularly in small mammals such as the mouse. However, they then show an SR membrane differentiation. Corbular SR close to the cell periphery forms junctional regions containing flanked surface-membrane Ca^{2+} channels and SR membrane RyR2-Ca^{2+} release channels. Non-corbular SR occurs within the cell interior. This also contains RyR2-expressing membrane regions, but these do not show the same proximity to the cell surface. Cell surface depolarisation triggers a Ca^{2+} channel mediated membrane entry of extracellular Ca^{2+}, locally increasing $[Ca^{2+}]_i$. The latter induces a Ca^{2+}-induced RyR2-mediated Ca^{2+} release by corbular SR. The resulting further local elevation of $[Ca^{2+}]_i$ initiates a wave of Ca^{2+}-induced Ca^{2+} release that is propagated through adjacent, successively deeper regions of non-corbular SR. The overall result is an inward, centripetal, propagation of a wave of Ca^{2+}-induced Ca^{2+} release by successive non-corbular cytoplasmic SR elements into the cell interior (Figure 13.7).

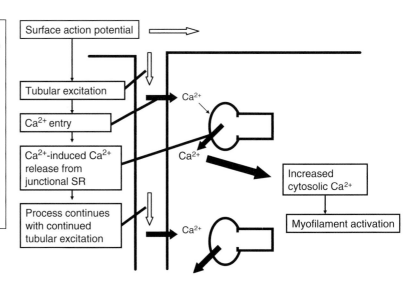

Figure 13.6 Scheme summarising excitation–contraction in ventricular muscle. An inward wave of T-tubular excitation triggers Ca^{2+} currents carried by dihydropyridine receptor type 2 (DHPR2) Cav1.2 Ca^{2+} channels. This both maintains the action potential plateau and triggers Ca^{2+}-induced Ca^{2+} release by ryanodine receptor type 2 (RyR2) sarcoplasmic reticular Ca^{2+} release channels, thereby elevating cystolic Ca^{2+}.

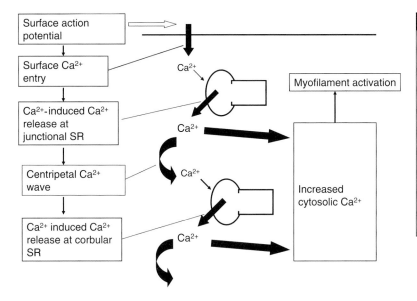

Figure 13.7 Atrial cells, particularly those in small mammals, show less developed transverse (T-) tubules, but express cell-surface dihydropyridine receptors (DHPR2s), permitting Ca^{2+} entry that activates Ca^{2+} release by junctional sarcoplasmic reticular (SR) elements. This initiates an inward centripetal propagation of Ca^{2+}-induced Ca^{2+} release by cytoplasmic, corbular SR that also contains ryanodine receptors (RyR2s), thereby elevating cystolic Ca^{2+}.

13.16 | Cardiomyocyte Recovery from Excitation

Cardiomyocyte relaxation involves dissociation of Ca^{2+} from troponin, requiring return of the elevated $[Ca^{2+}]_i$ to baseline in a sequence of events that shows particular similarities and differences from skeletal muscle (Section 11.13; Figure 11.14). First, the released cytosolic Ca^{2+} is returned to its SR store. This is mediated by a specific cardiac (SERCA2), as opposed to skeletal muscle, SR Ca^{2+}-ATPase (SERCA1) isoform, distinguished by its association with the inhibitory regulatory protein phospholamban (PLN). The latter inhibitory effect is relieved by PLN phosphorylation (Section 13.18). SERCA2 activity accounts for a greater proportion of the transport of Ca^{2+} out of the cytosolic compartment.

Secondly, in contrast to skeletal muscle, cardiomyocytes show significant sarcolemmal Na^+/Ca^{2+} exchange with the extracellular space. Between action potentials, the underlying Na^+-Ca^{2+} exchanger (NCX) normally drives Ca^{2+} efflux. This utilises energy from the exchanging Na^+ influx down its electrochemical gradient previously established by Na^+, K^+-ATPase-mediated outward Na^+ and inward K^+ transport. Thirdly, slower transport systems are provided by sarcolemmal Ca^{2+}-ATPase (PMCA) and mitochondrial Ca^{2+} uniport.

The NCX has additional important electrophysiological effects, some of clinical significance (Sections 14.5 and 14.15). Each NCX activity cycle translocates $1Ca^{2+}$ from the intracellular to the extracellular space, in return for an opposite transfer of $3Na^+$. This generates an exchange current, I_{NCX}, whose reversal potential falls between the Na^+ and Ca^{2+} Nernst potentials, E_{Na} and E_{Ca}, giving $E_{NCX} = (3E_{Na} - 2E_{Ca})$. Accordingly I_{NCX} is inward in direction with a depolarising effect under conditions of elevated $[Ca^{2+}]_i$, which thereby drives Ca^{2+} efflux. In contrast it takes an outward hyperpolarising direction under conditions of low intracellular $[Ca^{2+}]_i$ to drive Ca^{2+} influx. The NCX thus has electrogenic effects with potentially important impacts on membrane potential and arrhythmic tendency.

Patterns of I_{NCX} activity also vary through the course of the action potential (Plate 13). The plateau phase is accompanied by a sustained I_{NCX}-mediated Ca^{2+} influx. Following repolarisation, a large outward electrochemical gradient drives NCX-mediated Ca^{2+} efflux. Either of these NCX effects in turn can modify action potential duration. In SA cells, I_{NCX} may also contribute depolarising pacemaker currents activated by RyR2 and/or RyR3-mediated Ca^{2+} release after SR Ca^{2+} stores are refilled following each SAN action potential. This gives rise to a SAN 'Ca^{2+} clock' complementing the 'membrane clock' driven by I_f in SAN-mediated pacemaker activity (Vinogradova et al., 2006). Finally, increases in $[Na^+]_i$ decrease the inward electrochemical gradient driving Na^+ entry. This can result in an increased $[Ca^{2+}]_i$ and consequently increased contractile force. Such an effect can also follow Na^+ pump block by digitalis and other cardiac glycosides and is the basis for their historic use in management of cardiac failure.

Each cardiomyocyte activity cycle carries a significant energetic cost reflected in ATP consumption. Around ~60–70% of this is accounted for by contractile activity. The remainder is taken by excitable activity, alterations in Ca^{2+} homeostasis and restoration of transmembrane ion gradients. Aerobic mitochondrial oxidation, deficiencies in which have important clinical implications (Section 14.16), constitutes the main cardiac cellular source of the required ATP.

13.17 | Cardiac and Skeletal Myocyte Activation Characteristics Compared

The differing cardiac and skeletal muscle activation mechanisms result in functionally important contrasts in electrophysiological and excitation-contraction coupling characteristics. Skeletal muscle surface action potential activation extends over only tens of ms (Figure. 13.8*Aa*) producing Ca^{2+} transients and tension traces (Figure 13.8*Ab*) continuing well after repolarisation, all relatively unaffected by $[Ca^{2+}]_o$. The more prolonged cardiac excitation (Figure 13.8*Ba*) causes a more gradual development of Ca^{2+} release and tension (Figure 13.8*Ba*), both affected by action potential timecourse. They decline with premature action potential plateau termination produced by timed voltage clamp repolarising steps (Fig. 13.8*Bc*, arrowed) reducing the maximum developed tension (Fig. 13.8*C*). They are also profoundly influenced by both $[Ca^{2+}]_o$ and factors affecting inward I_{Ca} through actions on Ca^{2+}-induced Ca^{2+} release. This importance of $[Ca^{2+}]$ in cardiac muscle function was first discovered by Sydney Ringer in 1883 (Miller, 2004). His name has since been used to describe the saline solutions that maintain

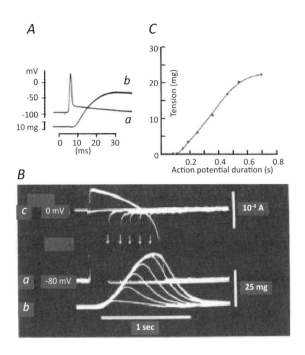

Figure 13.8 Comparisons of action potential excitation and tension transients in skeletal and cardiac muscle. (A, B) (a) action potential and (b) tension traces in amphibian (A) skeletal and (B) ventricular muscle. Timecourses of the latter tension traces are markedly affected by (c) premature action potential termination produced by timed repolarising voltage clamp steps, thereby altering the consequent (C) maximum developed tension, shown plotted against action potential duration. ((A) from Fig. 1 of Hodgkin and Horowicz, 1957; (B), (C) from Figure 3 of Morad and Orkand, 1971).

frog tissues in isolation from the body. Drugs which reduce Ca^{2+} influx reduce myocardial mechanical activity. Finally, restoration of regular stimulation following periods of cardiac quiescence that would reduce in vitro Ca^{2+} entry initially results in twitches with a reduced amplitude. However, subsequent stimulation resulting in resumption of Ca^{2+} entry, causes a successive restoration of twitch tension, as SR Ca^{2+} is restored to its equilibrium levels.

These cellular differences in excitation–contraction coupling mechanisms form the basis of a number of major physiological differences between activation of cardiac and skeletal muscle at the organ level. Skeletal myocyte activation results in a release of a relatively constant quantity of intracellularly stored Ca^{2+} from a relatively constant, normally fully loaded, SR Ca^{2+} store (Section 11.2). This results in a relatively constant, even if fatiguable (Section 12.10), tension transient. Skeletal muscle modulates overall contraction strength by varying the recruitment of individual motor units and their component muscle fibres by the central nervous system (Section 7.1). In contrast, cardiac myocytes are linked by intercalated discs into a syncytium. Each excitation stimulates all its component muscle cells. However, SR Ca^{2+} release in cardiac myocytes varies with the extent of filling of this intracellular store and a range of other intrinsic and extrinsic factors related to cardiac innervation (Section 13.18). The overall strength of cardiac contraction is accordingly regulated by the amount of Ca^{2+} made available to the myofilaments during excitation–contraction coupling.

These properties complement further differences between cardiac and skeletal muscle in their intrinsic sensitivity of tension generation to preload. As in skeletal muscle, mechanical activity in isolated cardiac papillary muscles can be studied through either their tension generation during isometric contraction or their velocity of isotonic shortening under conditions of constant load (cf. Sections 12.1–12.3). This reveals length–tension relationships in resting cardiac muscle considerably steeper than that of skeletal muscle (Section) likely reflecting differing properties in their intracellular titin (Section 9.11). This is the basis for the Frank–Starling Law of the whole heart, which states that the heart adjusts its energy of contraction in response to variations in diastolic stretch of its component muscle fibres. The heart consequently acts as a self-regulating pump that intrinsically responds to the presystolic filling of its cardiac chambers by returning venous blood from the peripheral circulation and balancing pumping by the right and left sides of the heart.

13.18 | Nervous Control of the Heart: Feedforward Modulation

The physiological processes of cardiac pacing, ion current activation in action potential generation and their coupling to the release of Ca^{2+} that drives mechanical activity are all modulated by autonomic, sympathetic and parasympathetic, innervation to the heart. These nerves release transmitters and cotransmitters, activating regulatory biochemical cascades with complex, often G-protein-mediated, effects on cell function (Section 8.7; Figure 8.9).

The sympathetic nervous system shows a widespread distribution and innervation of the different cardiac regions. Cardiac sympathetic terminals release noradrenaline (norepinephrine). Sympathetic activation also triggers adrenomedullary adrenaline (epinephrine) release into the circulation. Both transmitters act as β_1- and β_2-adrenergic receptor ligands (Figure 13.9). The heart expresses β_1-adrenergic receptors, whose activation triggers widespread cardiomyocyte actions (Figure 13.9A–C). Noradrenaline binding activates a stimulatory G-protein (G_s) so that its $G\alpha$-subunit binds GTP and is released from the receptor and the $\beta\gamma$-subunit (see Section 8.7; Figure 13.9D). The $G\alpha$-subunit activates the enzyme adenylyl cyclase, enhancing its cyclic 3′,5′- adenosine monophosphate (cAMP) production. Increased cellular cAMP has three regulatory effects.

First, cAMP opens g_f channels, particularly in SA cells increasing pacemaker current and therefore heart rate.

Secondly, cAMP activates protein kinase A (PKA). This has a wide range of strategic actions. PKA catalyses phosphorylation of C-terminal tail regions of Cav1.2 L-type Ca^{2+} channels, increasing their open probability, accentuating the amplitude and duration of the ventricular action potential plateau and accelerating SAN pacemaker potentials. The increased net entry of Ca^{2+} into the cell also increases

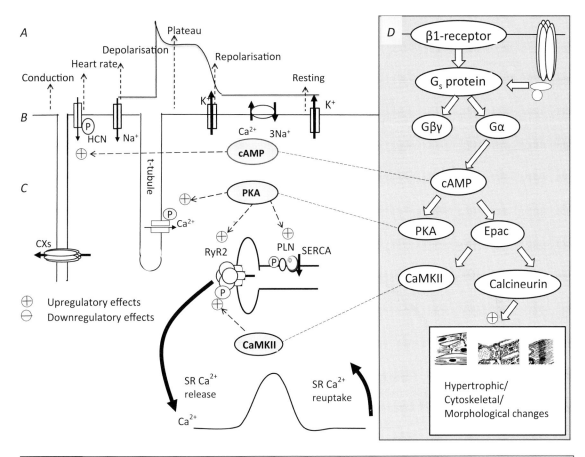

Figure 13.9 Feedforward regulation by the sympathetic nervous system matching (A) electrophysiological events, (B) principal underlying surface membrane ion channels, and (C) consequences for intracellular ion homeostasis to their regulatory (D) cellular signalling pathways. Signalling entities illustrated: Cx: connexin; HCN, hyperpolarisation-induced cyclic nucleotide-activated channel; cAMP: cyclic 3',5'-adenosine monophosphate; PKA: protein kinase A; CaMKII: calmodulin kinase II; RyR2: ryanodine receptor, type 2; PLN: phospholamban; SERCA: sarcoplasmic reticular Ca^{2+} ATPase; G_s: stimulatory G-protein; Epac: exchange protein directly activated by cAMP. P denotes phosphorylatable proteins. Upregulatory and downregulatory effects are annotated by '+' and '-' symbols.

the rate and force of contraction in subsequent beats. PKA-mediated RyR2 phosphorylation reduces binding of its regulatory ligand FK506-binding protein type 12 (FKBP12) that normally stabilises its closed state. Its dissociation increases the Ca^{2+} sensitivity of the RyR2, thereby enhancing Ca^{2+}-induced Ca^{2+} release. PKA-mediated phosphorylation of phospholamban (PLN) removes its inhibition of SERCA2-mediated reuptake of previously released cytosolic Ca^{2+}. This enhances diastolic reloading of SR Ca^{2+} stores.

Thirdly, of isoforms of cAMP-dependent exchange proteins directly activated by cAMP (Epac), Epac2 activates Ca^{2+}/calmodulin-dependent protein kinase II (CaMKII) activity, in turn enhancing RyR2-mediated SR Ca^{2+} release. The Epac activator 8-(4-chlorophenylthio)adenosine-3',5'-cyclic monophosphate (8-CPT) induces both ectopic Ca^{2+}

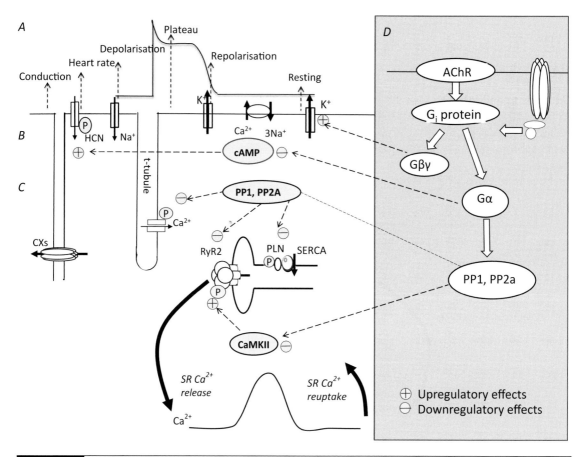

Figure 13.10 Feedforward regulation by the parasympathetic nervous system matching (A) electrophysiological events, (B) principal underlying surface membrane ion channels and (C) consequences for intracellular ion homeostasis to their regulatory (D) cellular signalling pathways. Signalling entities illustrated: Cx: connexin; HCN, hyperpolarisation-induced cyclic nucleotide-activated channel; cAMP: cyclic 3′, 5′-adenosine monophosphate; PP1 and PP2A: protein phosphatase isoforms; CaMKII: calmodulin kinase II; RyR2: ryanodine receptor, type 2; PLN: phospholamban; SERCA: sarcoplasmic reticular Ca^{2+} ATPase; AChR: acetylcholine receptor; G_i: inhibitory G-protein. P denotes phosphorylatable proteins. Upregulatory and downregulatory effects are annotated by '+' and '−' symbols.

transients in stimulated cells and spontaneous Ca^{2+} transients in resting cells. The latter forms Ca^{2+} waves propagating at a constant ~100 μm/s velocity along the lengths of the cells (Plate 14A). These result in rising and falling transients of similar waveforms and amplitudes but successively larger latencies in successive regions of interest (Plate 14B, C). Epac1 activation induces programmes of hypertrophic, morphological and cytoskeletal changes accompanied by increased protein synthesis and induction of cardiac hypertrophic markers mediated by Ca^{2+}-dependent calcineurin activation.

Parasympathetic inhibitor nerve fibre activity contrastingly slows heart rate and decreases contractile force. The underlying neurotransmitter, acetylcholine (ACh), acts through muscarinic (M_2) receptors

(Figure 13.10). ACh-receptor binding activates the G-protein G_{i2}. Its $G\alpha$ subunit binds GTP and splits off from the receptor and the $G\beta\gamma$ subunit. The $G\beta\gamma$ subunit binds to and opens a particular K^+ (GIRK1) channel, driving the membrane potential towards E_K, particularly in the SAN, but also the atria and ventricles. In addition, the dissociated $G\alpha$ binds to and inhibits adenylyl cyclase, reducing cAMP production in pacemaker cells (DiFrancesco, 1993). G_i activation may also upregulate activity in the protein phosphatases PP1 and PP2A. These dephosphorylate PKA-phosphorylated proteins at the same serine/threonine phosphorylation sites. The two, PP2A and PP1, isoforms reverse the effects of PKA on L-type Ca^{2+} channels and RyR2s, and the SERCA2a inhibitor PLN, respectively.

Finally, adenine nucleotides are excitatory postganglionic sympathetic cotransmitters that activate metabotropic P2Y receptors. The latter in turn activate phosphokinase C (PKC) through phospholipase C-mediated production of diacylglycerol (Section 8.7; Figure 8.9). PKC targets L-type Ca^{2+} channels, RyR2s, and voltage-gated Na^+ and K^+ channels.

All these systems show considerable amplification. Activation of one β-adrenergic receptor activates many G-proteins, each activating an enzyme molecule, in turn producing many cAMP molecules, and each activated protein kinase A molecule will phosphorylate several Ca^{2+} channels. Similarly, activation of one muscarinic receptor produces many $G\beta\gamma$ subunits and so opens many GIRK1 channels.

13.19 | Feedback Actions on Excitation–Contraction Coupling Related to Ca^{2+} Homeostasis

Both skeletal and cardiac muscle excitation–contraction coupling involves feedforward mechanisms in which Na^+ channel activation produces membrane depolarisation detected by L-type Ca^{2+} channels whose activation initiates RyR-mediated SR Ca^{2+} release, triggering contractile activation. However, feedback mechanisms by which downstream effects of such activation modulate this triggering have attracted recent attention (See also Section 11.14). In particular, both Nav1.4 and Nav1.5 possess potentially multiple regulatory Ca^{2+} and CaM binding or modulation, as well as phosphorylation sites (Section 5.2).

Patch clamped cultured neonatal rat cardiomyocytes respectively showed increased and decreased I_{Na} with $[Ca^{2+}]_i$ elevations and reductions produced by increased pipette extracellular $[Ca^{2+}]$ and applications of the Ca^{2+} chelator BAPTA (1,2-bis(o-aminophenoxy)ethane-N,N,N′,N′-tetraacetic acid tetrakis-acetoxymethyl ester), respectively (Chiamvimonvat *et al.*, 1995).

Loose patch clamp studies (Section 10.2) examined myocytes in both intact murine skeletal muscle (Section 11.14; Matthews *et al.*, 2019) and atrial and ventricular preparations (Valli *et al.*, 2018b). They used the

Epac activator 8-(4-chlorophenylthio)adenosine-3′,5′-cyclic monophosphate (8-CPT) to induce RyR2-mediated Ca^{2+} release (Hothi *et al.*, 2008; Section 13.18; Plate 14). Plate 14 (*D–H*) illustrates the consequent effects on I_{Na} in ventricular myocytes. A series of incrementally depolarised test voltages (Plate 14*Da*) examined voltage-dependent I_{Na} activation, giving families of I_{Na} records (Plate 14*Db*, *Ea*). 8-CPT reduced their amplitude (Plate 14*Eb*), but the RyR blocker dantrolene (Section 11.12), whether applied following (Plate 14*Ec*) or in parallel with the 8-CPT challenge (Plate 14*Ee*), abrogated this effect. Dantrolene alone was without effect (Plate 14*Ed*). Concordant results emerged from assessments of Na^+ channel inactivation by the current persisting in subsequent steps to a fixed voltage 100 mV positive to the resting potential (Plate 14*F*). These findings were confirmed in quantifications of peak current activation (Plate 14*G*) or inactivation plotted against voltage (Plate 14*H*).

This acute, reversible, Na^+ channel inhibition was complemented by observations of reduced Nav1.5 function and expression and pro-arrhythmic slowing of action potential propagation with chronic alterations in Ca^{2+} homeostasis in genetically modified *RyR2*-P2328S murine hearts (King et al., 2013b). The latter carry genetically altered RyR2-Ca^{2+} release channels known to model the clinical pro-arrhythmic condition of catecholaminergic polymorphic ventricular tachycardia (CPVT; Sections 14.14 and 14.15).

The above feedback effects related to $[Ca^{2+}]_i$ would modulate action potential excitation and propagation following RyR-mediated Ca^{2+} release. RyR opening is also accompanied by dissociation of its bound FKBP12 (Section 11.6), reported to shift the voltage dependence of Na^+ channel properties in an inhibitory direction. Both these effects add to the electrogenic consequences of increased cytosolic $[Ca^{2+}]$ through its actions on NCX activity (Section 14.13; Figure 13.11).

13.20 | Feedback Actions on Excitation–Contraction Coupling Related to Cellular Energetics

Feedback effects from more downstream, metabolic and energetic, outcomes of cellular excitation also modify cell excitability and action potential propagation. Cellular activity involves ATP consumption that increases cellular ADP levels; the ATP is normally restored by mitochondrial activity. Excessive energetic demand, compromised vascular oxygen supply or pathological energetic disorders associated with mitochondrial dysfunction increase production of reactive oxygen species (ROS). These inhibit Na^+ channel function. They also oxidise RyR2, increasing SR Ca^{2+} leak, thereby increasing $[Ca^{2+}]_i$. The latter itself downregulates Na^+ channel function (Section 13.19). Both ROS and the ATP depletion associated with energetic deficiency also open sarcolemmal ATP-sensitive K^+ channels (sarcK$_{ATP}$), which occur

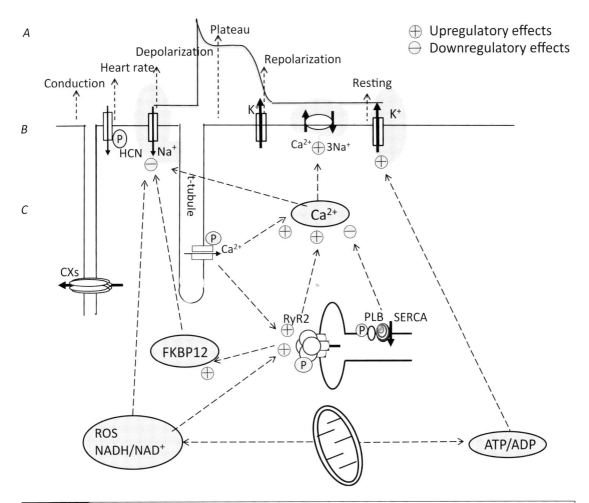

Figure 13.11 Feedback regulation at the cellular level matching (A) physiological events, (B) their principal underlying surface membrane ion channels and (C) intracellular feedback signalling mechanisms. Signalling entities illustrated: Ca^{2+}: intracellular Ca^{2+}; FKBP12: FK506 binding protein; ROS: reactive oxygen species; NAD: nicotinamide adenine dinucleotide; ATP/ADP: ATP/ADP ratio; RyR2: cardiac ryanodine receptor, type 2; PLN: phospholamban; SERCA: sarcoplasmic reticular Ca^{2+}-ATPase; Cx: connexin; HCN, hyperpolarisation-induced cyclic nucleotide-activated channel; cAMP, cyclic 3', 5'-adenosine monophosphate. Upregulatory and downregulatory effects are annotated by '+' and '−' symbols.

at relatively high densities in myocyte surface membranes. Finally, relative cellular levels of reduced or oxidized nicotinamide adenine dinucleotides NADH and NAD^+, reflecting cell oxidative state, respectively inhibit and enhance Nav1.5 activity, in the presence of normal overall Nav1.5 expression (Liu *et al.*, 2009; Figure 13.11).

These effects similarly affect intact myocytes studied in situ, with potential clinical implications. In intact skeletal myocytes under loose patch clamp, elevations of $[Ca^{2+}]_i$ by carbonyl cyanide-3-chlorophenyl-hydrazone-mediated mitochondrial Ca^{2+} release compromised Na^+ channel function. This was reversed by the Ca^{2+} buffer, BAPTA (Filatov *et al.*, 2009).

13.21 | Recapitulation

Figures 13.9–13.11 usefully recapitulate the sequence of physiological processes underlying cardiac activation. Surface membrane excitable activity is normally initiated by automaticity from SAN pacemaker, HCN4-mediated, I_f, currents. Subsequent cardiomyocyte activation involves I_{Na}- or I_{Ca}-mediated membrane depolarisation, in turn driving local circuit current, resulting in conduction of the action potential to hitherto quiescent myocardial cells at the tissue level. It also activates inward L-type Ca^{2+} currents producing the action potential plateau. Excitation is terminated by a range of outward I_K driving action potential recovery (Figures 13.9A,B and 13.10A,B). Excitation–contraction coupling follows Ca^{2+} channel mediated Ca^{2+} influx into the intracellular space. This triggers a Ca^{2+}-induced RyR-mediated release of intracellularly stored SR Ca^{2+} (Figures 13.9C and 13.10C). Following repolarisation, the released Ca^{2+} is restored to the SR or expelled into the extracellular space by Ca^{2+}-ATPase-mediated transport, and into the extracellular space by Na^+–Ca^{2+} exchange, thereby returning cytosolic Ca^{2+} to its resting level.

Of feedforward, autonomic, regulatory inputs, sympathetic stimulation triggers cAMP-activated accelerations of heart rate and enhancement of excitation–contraction coupling by phosphorylation of proteins regulating intracellular Ca^{2+} balance and SR Ca^{2+} cycling. These proteins include L-type Ca^{2+} channels, RyR2 and the SR regulatory protein phospholamban (PLN; Figure 13.9D). Parasympathetic stimulation triggers a complementary messenger cascade acting on background atrial K^+ conductance and reversing some of the phosphorylation events (Figure 13.10D). Feedback, cellular regulatory mechanisms may act through alterations in cellular Ca^{2+} homeostasis, or further downstream, energetic changes related to alterations in mitochondrial function. These modify SA automaticity, Na^+ channel function and electrogenic Na^+/Ca^{2+} exchange (Figure 13.11).

14

Ion Channel Function and Cardiac Arrhythmogenesis

Disruption of the cardiomyocyte electrophysiological activation and recovery processes underlying normal electrical and mechanical atrial and ventricular activation (Chapter 13) leads to cardiac arrhythmias. These compromise cardiac function and constitute a major public health problem. The abnormalities may involve initiation of excitation, or the subsequent action potential propagation or recovery processes. These are often diagnosed through deviations from the normal electrocardiographic (ECG) P, QRS and T components, signalling atrial and ventricular activation, and ventricular repolarisation, respectively (Sections 13.4–13.5; Figure 14.1A).

Of arrhythmias, sino-atrial (SA) disorders (Section 13.2) increase (Figure 14.1Ba), compromise (Figure 14.1Bb) or pause the heart rate (Figure 14.1Bc) or its autonomic response. They constitute the indications for ~50% of the million permanent pacemaker implants per year

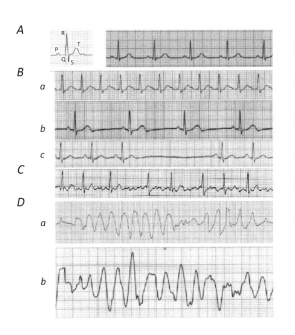

A   Sinus rhythm

B   a   Sinus tachycardia

    b   Sinus bradycardia

    c   Sinus arrest

C   Atrial fibrillation

D   a   Torsade de pointes

    b   Ventricular fibrillation

Figure 14.1 Electrocardiographic (ECG) manifestations of cardiac arrhythmogenesis. (A) Typical P, QRS and T components making up the normal ECG complex, and typical lead II trace from heart in sinus rhythm. (B–D) Rhythm strips illustrating (B) abnormal sino-atrial function resulting in (a) increased, (b) compromised or (c) pauses in heart rate, (C) atrial arrhythmia exemplified by atrial fibrillation, (D) ventricular arrhythmias, exemplified by (a) torsade de pointes (TdP) and (b) fast ventricular fibrillation.

worldwide. Of atrial arrhythmias, the commonest sustained cardiac arrhythmia, atrial fibrillation (AF) (Figure 14.1*C*), a chaotic breakdown of atrial excitation, affects 1– 2% of the general population. It predisposes to further, major, cardiovascular morbidities and mortality, including stroke. Ventricular arrhythmias, including torsades de pointes (TdP) (Figure 14.1*Da*) and fast ventricular fibrillation (VF) (Figure 14.1*Db*) can follow ventricular tachycardias (VT). They cause >300 000 and ~70 000 cardiac deaths/year in the United States and United Kingdom, respectively. This chapter discusses electrophysiological events and properties associated with arrhythmic phenomena and relates these to particular examples of alterations in ion channel function.

14.1 | Experimental Studies of Cardiac Arrhythmogenesis

Arrhythmic phenotypes likely involve alterations in ion channel function amenable to experimental analysis of their underlying physiological mechanisms that could identify potential therapeutic targets. This requires experimental models that replicate the specific clinical arrhythmic situations in which the underlying channels and their abnormal function can be identified and characterised. This could involve investigating the effects of reversible, often pharmacological, manipulations replicating acute clinical situations. Alternatively, models for chronic monogenic disease conditions replicating both the genetic and clinical abnormalities could relate the pro-arrhythmic properties with the underlying specific ion channel abnormalities (Huang, 2017).

A range of animal species show cardiac electrophysiological properties resembling those in human heart, and provide potential specific arrhythmic disease exemplars. Rabbit and canine ventricular cardiac action potentials share similar ionic currents and positive plateau phases as human hearts. Transgenic rabbits lacking I_{Ks} and I_{Kr} due to *KCNQ1* and *KCNH2* mutations recapitulate the human pro-arrhythmic long QT syndromes LQTS1 and LQTS2 that show high incidences of sudden cardiac death preceded by polymorphic VT. Canine hearts with chronic atrioventricular (AV) block showed complex structural hypertrophic and electrophysiological remodelling, downregulated I_{Ks} and I_{Kr}, upregulated Na^+-Ca^{2+} exchanger (NCX) activity and predispositions to TdP and sudden cardiac death. A canine arrhythmogenic right ventricular cardiomyopathy (ARVC) model shows VT, an enlarged RV, myocyte loss and fibrofatty replacement, and predisposition to sudden death.

More recently, transgenic mouse models amenable to insertion of specific gene mutations or deletions have furnished targetable human genetic disease paradigms. Knock-in models contain alterations in the genetic code giving a modified mRNA with reduced, increased or modified function in its translated protein. Knock-out models produce no mRNA from the altered gene. The resulting animals often recapitulated their corresponding human arrhythmic conditions, particularly when these involved clinically established and well-defined monogenic mutations. They furnished well-defined platforms for assessing the role of particular ion channels in human cardiac arrhythmias.

Murine hearts have similar overall anatomy to human hearts. They show rapid, Na^+ channel mediated ventricular action potential depolarisation propagating with similar transmural conduction velocities. Despite their small size, they can sustain the monomorphic and polymorphic arrhythmias clinically observed in arrhythmic human hearts (Sabir et al., 2008a). Admittedly, mouse ventricular action potentials show short, triangulated, as opposed to prolonged plateau, recoveries (~30–80 ms) dominated by fast, Kv4.3 and Kv4.2-mediated $I_{to,f}$ and slower Kv1.4-mediated $I_{to,s}$ currents (Section 13.8). Nevertheless, the experimental flexibility of murine hearts has made them effective models for assessing ion channel roles in a wide range of human arrhythmic conditions, particularly in situations involving action potential activation and conduction, altered Ca^{2+} homeostasis and energy metabolism, at all levels of biological organisation.

Available experimental methods for murine studies and their experimental readouts range from the whole organ through tissue to the cellular levels (Plate 15): (A) Electrocardiographic (ECG) studies (Sections 13.4–13.5) in ambulatory or anaesthetised intact animals detect spontaneous rhythm disruptions and acute or chronic ECG abnormalities. (B) Isolated Langendorff-perfused preparations permit closer investigations for spontaneous pro-arrhythmic triggering events during intrinsic activity or regular pacing. They can also be used to assess arrhythmic substrate sustaining arrhythmias elicited by extrasystolic stimulation interposed during regular pacing or high-frequency burst pacing. Such stimuli have classically been imposed by direct electrical stimulation. However, optical excitation of genetically expressed photosensitive channelrhodopsin molecules has recently proven promising for non-invasive, anatomically targeted, stimulation (Ferenczi et al., 2019).

Readouts in the form of extracellular, unipolar or bipolar recordings detect both latencies and durations of excitation, and arrhythmic phenomena. Action potential waveforms themselves can be determined by (a) surface monophasic action potential (MAP) recordings (Section 14.9), particularly useful for quantifying action potential durations (APDs), typically their times to 90% full repolarisation, APD_{90}, and effective refractory periods (ERPs) (Section 13.12). These complement (b) sharp single-cell intracellular microelectrode impalements that additionally can quantify maximum absolute action potential upstroke velocities $(dV/dt)_{max}$ and durations. (c) Multi-electrode array mapping methods provide patterns and timings of excitation wavefront propagation in the form of isochronal activation and recovery maps providing conduction velocity magnitudes and directions, and activation–recovery intervals. Fluorescence mapping involving voltage- (e.g. di-4-ANEPPS, RH237) or Ca^{2+}-sensitive indicators (such as Rhod-2-AM; see also Section 11.3) similarly signal spatio-temporal characteristics of voltage change and Ca^{2+} cycling.

At the cellular level, (C) loose patch clamp techniques (Section 10.2) bridged tissue- and cellular-level studies of Na^+ and K^+ current, I_{Na} and I_K, in superfused atrial or ventricular tissue. (D) Single-cell techniques in (a) intact or perforated whole-cell patch clamped (Section 7.6) enzymically isolated myocardial cells have studied ionic current activation and inactivation. Isolated fluorophore-loaded (Section 11.3) atrial and ventricular myocardial cells were also used to examine (b) Ca^{2+} signalling properties (Section 11.4), deriving measures of SR Ca^{2+} release or reuptake, or cellular Ca^{2+} entry or expulsion, and related pro-arrhythmic effects.

Longer-term molecular-level assessments involve quantitative reverse-transcriptase polymerase chain reaction (qPCR) and western blotting, of ion channels and their regulatory genes. Use of such cell physiological platforms has been supplemented by induced pluripotent stem cells derived from carriers of clinical conditions. Additionally, morphological changes at all biological levels from organ architecture, through tissue hypertrophy and fibrosis, to cellular changes may contribute pro-arrhythmic effects. Finally, as in other areas bearing on nerve and muscle, quantitative computational reconstructions have explored consequences of the individual changes in cellular properties at the systems level, for both normal and abnormal function.

14.2 | Arrhythmic Substrate Arising from Spatial Heterogeneities in Action Potential Generation and Propagation at the Tissue Level

Sino-atrial node (SAN) pacing is normally followed by a spatially and temporally ordered sequence of atrial, atrio-ventricular node (AVN), Purkinje cell and ventricular action potential repolarisation and recovery (Section 13.2). Ventricular action potentials finally propagate transmurally from endo- to epicardium and from base to apex, with a particular velocity θ through any given tissue region, and specific temporo-spatial APD and ERP gradients. Re-excitation is only possible with triggering timed at diastolic intervals, DI, that exceed the ERP (Figure 14.2A).

These variables map spatially onto an active action potential wavelength ($\Lambda' = \theta \times$ APD or $\theta \times$ ERP) (Figure 14.2B) within which there is a reduced likelihood of abnormal, pro-arrhythmic, re-excitation. Such a region is followed by a resting wavelength Λ_0' representing the recovered tissue behind the advancing excitation wave (Section 6.4). Triggering events initiate sustained arrhythmia where tissue-level instabilities produce arrhythmic substrate. These can involve contributions from altered propagation, initiation of, or recovery from, excitation (Huang, 2017).

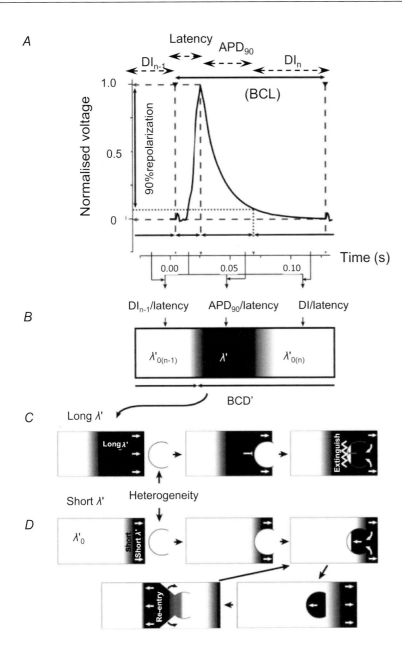

Figure 14.2 Action potential timecourse and its relation to wavelength with conditions for wave-break and re-entry. (A) Typical murine right ventricular action potential waveform characterised by basic cycle length (BCL), action potential duration at 90% recovery (APD$_{90}$), latency and diastolic interval (DI) of the current (nth) and preceding (($n-1$)th) action potential. (B) Spatial mapping of propagating action potential waveform onto active and resting wavelengths Λ' and Λ_0' and basic cycle distance, BCD' = Λ' + Λ_0'. (C) With a sufficiently long Λ', orthograde action potential propagation over a heterogeneity results in the back of the propagating wave blocking activation of a retrogradely propagated action potential. (D) With a short Λ', the back of the advancing action potential wave passes the heterogeneity before any retrograde excitation has crossed the unidirectional block. This initiates a new propagating retrograde wave producing re-entrant excitation.(From Figure 1 of (Matthews et al., 2013.)

Normally, a sufficiently long Λ' ensures that the action potential wave is completely extinguished following its propagation through any given tissue region without re-excitation of any previously recovered tissue behind the advancing action potential wavefront (Section 6.4). It also ensures that should the action potential propagate over a heterogeneity that can potentially initiate ectopic conduction, the back of the advancing wave blocks such retrograde excitation. The orthograde excitation wave survives to continue the process of normal unidirectional action potential conduction (Figure 14.2C). An inappropriately short $\varLambda'$ permits the back of the wave to traverse the heterogeneity before any retrograde excitation has passed through the unidirectional block. This initiates a new, inappropriate, pro-arrhythmic propagating retrograde excitation wave (Figure 14.2D).

Pro-arrhythmic *spatial heterogeneities* in action potential propagation can occur in the presence of (a) an obstacle around which the action potential has to circulate with (b) conduction velocities sufficiently slowed to permit recovery of excitation before the wave returns to the region involved, in the presence of (c) a unidirectional conduction block. Plate 16 illustrates the effects of a slow conducting pathway traversing a region of non-conducting myocardium (path 1: coloured gray) in parallel with a second pathway with normal conduction (path 2: white).

A normal propagated action potential (blue arrow) along path 2 (Plate 16Aa) has an excitation wavelength $\varLambda$ (yellow region) sufficiently long to prevent re-entry from impulses travelling along path 1 as they would collide with refractory tissue along path 2 (Plate 16Ab). Conversely, an abnormal impulse arising from triggered activity immediately following the normal action potential arising from an ectopic focus within path 1 cannot enter path 2 as this remains refractory (Plate 16Ba). The ectopic action potential splits at the end of path 1 to conduct retrogradely along path 2 and orthogradely along path 1 (Plate 16Bb). However, self-perpetuating re-entrant excitation results when a retrogradely conducting action potential along path 1 enters the beginning of path 2 (Plate 16Ca). This transmits an action potential with an excitation wavelength $\varLambda'$ reduced to values smaller than the dimensions of the available conducting circuits. This could arise from abnormally reduced θ and/or ERP: $\varLambda' = \theta \times$ ERP (Plate 16Cb; King et al., 2013a).

14.3 | Arrhythmic Substrate Arising from Spatial Heterogeneities in Action Potential Recovery at the Tissue Level

In the thin-walled atria, excitation and recovery likely uniformly involve the entire myocardial thickness. However, pro-arrhythmic disruptions in the normal temporal sequence of action potential repolarisation and refractoriness can potentially involve the thickness of individual ventricular chambers. The sequence of normal ventricular repolarisation is paradoxically opposite to that of depolarisation, running from epi- to endocardium and apex to base. This reflects longer endo- than epicardial and basal than apical APDs (Yan and Antzelevitch, 1998; Wang et al., 2006; London et al., 2007). These differences result from differing local K^+ channel densities, and the consequent repolarising currents, and APDs and ERPs. This gradient of transmural repolarisation, quantifiable from the difference (ΔAPD) between endocardial and epicardial APDs, and therefore their respective ERPs, normally minimises the likelihood of re-entrant retrograde excitation from epicardium to endocardium (Figure 14.3A).

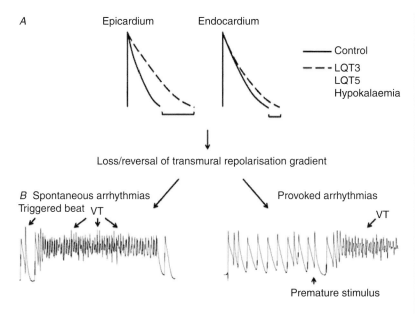

Figure 14.3 Reductions and reversals of transmural repolarisation gradients (ΔAPDs) in murine heart and consequent arrhythmogenesis. (A) Under normal conditions (solid lines), epicardial action potential durations (APDs) are significantly shorter than endocardial APDs. Genetically modified mice with mutations in sarcolemmal ionic currents corresponding to long QT syndrome type 3 and long QT syndrome type 5, and murine hearts under hypokalaemic conditions show preferential increases in epicardial compared to endocardial APDs (dashed lines). These can significantly reduce or even reverse baseline transmural repolarisation heterogeneities. (B) Such effects increase incidences of both spontaneous arrhythmias and arrhythmias provoked by premature extra-systolic stimuli. (From Killeen *et al.*, 2008a.)

Disruptions in ΔAPD potentially generate arrhythmogenic re-entrant circuits where they result in the presence of time windows permitting re-excitation. The latter arise from positive time differences between measures of action potential repolarisation, such as APD_{90}, and recovery from refractoriness exemplified by ventricular ERP (VERP). These critical intervals reflect the extent to which full repolarisation ending the action potential precedes or follows Na^+ channel recovery from refractoriness. They could either occur within a given cardiac region or involve adjoining, electrotonically coupled regions of myocardium (Killeen *et al.*, 2008b; Sabir *et al.*, 2008a). Risks of local re-excitation in either an epi- or endocardial region might be assessed from their respective (APD_{90}–VERP) differences. Risks of transmural re-entry resulting respectively in pro-arrhythmic epicardial or endocardial excitation (Figure 14.3B) would involve critical intervals involving both epicardial and endocardial APD_{90} and VERP terms, allowing for any delays between endocardial and epicardial excitation (Sabir *et al.*, 2007).

Finally, *Phase 2 re-entry* earlier in the action potential recovery phase can also result from retrograde propagation of an action potential wave into regions that have recovered from refractoriness. These have been implicated in Brugada syndrome (BrS) (Section 14.7; Gussak and Antzelevich, 2003).

14.4	Arrhythmic Substrate Arising from Temporal Electrophysiological Heterogeneities at the Tissue Level

Temporal electrophysiological heterogeneities manifest as unstable, often alternating, action potential amplitudes or durations, often

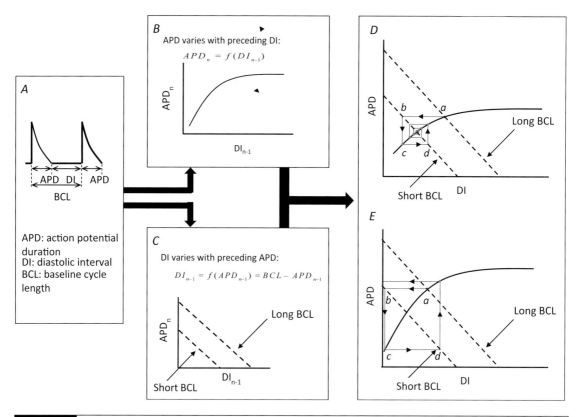

Figure 14.4 Arrhythmic substrate arising from temporal electrophysiological heterogeneities at the tissue level. (A) Action potential showing a basic cycle length (BCL) made up of the action potential duration (APD) and succeeding diastolic interval (DI). (B) A-curve showing normal adjustment of APD_n of an nth action potential in response to its preceding DI_{n-1}, and (C) D-line showing consequent linear effect of variations of APD_n on DI_{n-1} at different BCLs: DI = BCL − APD. (D, E) Intersection between the A-curve and D-line gives steady-state APD and DI at any BCL (point a). Increasing heart rate, thereby decreasing BCL alters the D-line (point b), in turn adjusting the APD (point c), thence affecting the succeeding DI (point d). The resulting cycle converges where the dependence of APD_n on DI_{n-1} has less than unity slope (D), remains constant with unity slope and diverges with greater than unity slope (E). These respectively result in declining, stable or unstable waxing oscillations in APD.

presenting as alternans in ECG signals. T-wave alternans classically precedes breakdown of regular electrophysiological activity leading to ventricular arrhythmias. APD alternans that is discordant between adjacent tissue areas produces spatial APD gradients between them. This potentially induces re-excitation in re-entrant circuits, leading to ventricular arrhythmia.

APD alternans has been attributed to interactions between action potential recovery, often termed restitution, properties, and variations in the basic cycle length BCL, reflecting the heart rate. Altered BCL alters the diastolic interval, DI, during which the membrane potential has regained the resting level following repolarisation. This in turn determines the period left for ion channel recovery prior to the subsequent action potential (Figure 14.4A). Alterations in this recovery period, DI, alter the APD of the subsequent action potential along an A-curve (Figure 14.4B). However, since BCL = APD + DI, so DI = BCL − APD.

Consequently, altered APD at a given BCL linearly alters the DI preceding the subsequent beat along a D-line (Figure 14.4C).

Superimposing the A-curve and D-line gives an intersection defining the baseline steady-state APD and DI at any given heart rate or BCL (point a in Figure 14.4D, E). Alterations in heart rate or BCL, exemplified for the situation of an increased heart rate (point b in Figure 14.4D, E) would result in a transition to an altered D-line representing the new BCL. This predicts an altered subsequent DI. A vertical line drawn from the intersection point b to the corresponding A-curve gives the corresponding subsequent APD (point c in Figure 14.4D, E). This in turn alters the succeeding DI (point d in Figure 14.4D, E), generating a cycle of oscillating APD in successive heartbeats.

The slope of the A-curves (Figure 14.4D, E) in the region where they intersect the D-line determines the persistence and nature of such APD oscillations. An intersection at an A-curve region where its slope is zero results in immediate attainment of the final steady-state APD without oscillations. A slope between zero and unity (Figure 14.4D) gives successive projection lines converging back to the set point, giving a transient alternans. An intersection at a critical DI, DI_{crit}, where the A-curve shows unity slope, predicts projection lines that neither converge nor diverge, reflecting sustained alternans. Finally, an intersection where the A-curve slope exceeds unity produces diverging projection lines, reflecting progressively increasing instability. These veer away from the left-hand limit of the A-curve, producing conduction block and potential re-entry (Figure 14.4E; Matthews et al., 2010; 2012).

14.5 | Arrhythmic Triggers at the Cellular Level

Sustained cardiac arrhythmias can follow *triggered* events arising from repeated abnormally initiated action potentials or extra-systoles. Alternatively, failure of excitability to recover to resting levels during or following an action potential wave can result in re-entrant excitation and an arrhythmogenic substrate sustaining an arrhythmic process following triggering.

Of triggering events, enhanced or compromised *automaticity* in pacemaker cells can arise from accelerated or retarded depolarisation in their pacemaker potentials. Abnormal atrial triggering precipitating AF can also arise from the pulmonary or the superior caval veins. In atrial or ventricular cardiomyocytes, triggering events in the form of extra-systolic membrane depolarisations, if of sufficient magnitude, can generate premature action potentials that follow the normally paced action potentials (Killeen et al., 2008a; Figure 14.5).

First, interruptions in the normally smooth action potential repolarisation timecourse following the Phase 2 plateau can produce *early*

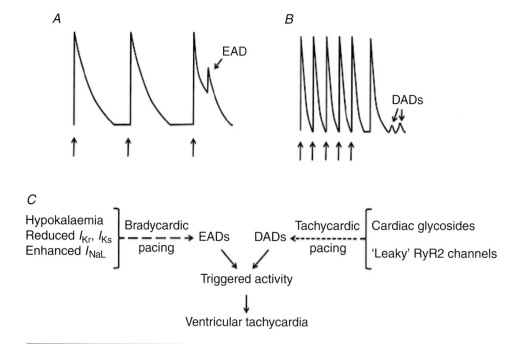

Figure 14.5 Arrhythmogenic triggers and conditions favouring their induction in the murine heart. (A) Early after-depolarisations (EADs) interrupting the repolarisation phase in murine ventricles. (B) Delayed after-depolarisations (DADs), small, transient membrane potential oscillations, following full repolarisation, and (C) conditions promoting their respective occurrence. (From Killeen et al., 2008a.)

after-depolarisations (EADs) (Figure 14.5A). These result from a combination of plateau currents associated with prolonged action potentials, as in long QT syndromes (Section 14.8), and L-type Ca^{2+} channel recovery from inactivation within a critical voltage window (January and Riddle, 1989). The re-activated inward depolarising L-type Ca^{2+} currents, I_{CaL}, possibly accompanied by increased Na^+–Ca^{2+} exchange current, I_{NCX}, (Section 13.16) produces further regenerative depolarisation, potentially triggering a pro-arrhythmic extra-systolic action potential (Killeen *et al.*, 2008a). EADs are frequently induced under conditions of action potential prolongation, through compromised repolarisation in long QT syndrome and bradycardia (Figure 14.5C).

Secondly, *delayed after-depolarisations* (DADs) represent small, transient oscillations in resting membrane potential following full repolarisation (Figure 14.5B). They are associated with conditions of rapid pacing and situations favouring diastolic Ca^{2+} leak through RyR2 channels, typified in catecholaminergic polymorphic VT, (CPVT; Section 14.14), cardiac failure or cardiac glycoside toxicity. The resulting disrupted Ca^{2+} homeostasis causes transient inward currents (I_{ti}) produced by increased NCX activity following excessive SR Ca^{2+} release.

Both EADs and DADs can cause triggered beats initiating arrhythmia, where their associated membrane potential displacements reach the re-excitation threshold (January and Riddle, 1989). In the presence of arrhythmic substrate they can thereby trigger polymorphic VT.

14.6 | Action Potential Activation and Conduction Abnormalities: Impaired Na⁺ Channel Function and Atrial Arrhythmia

An important group of studies on genetically modified murine systems explored associations between genetic abnormalities affecting Nav1.5 function and a range of clinical arrhythmic disorders. These illustrated the pro-arrhythmic consequences of many of the fundamental conduction, depolarisation and repolarisation properties described above (Sections 14.2–14.4). Compromised Nav1.5 function is associated with clinical sinus node dysfunction, AF, Brugada syndrome (BrS) (Sarquella-Brugada *et al.*, 2016) and progressive fibrotic cardiac conduction disease (Lev–Lenegre syndrome; Martin *et al.*, 2010). These provide paradigms for the pro-arrhythmic consequences of slowed action potential conduction. Intact isolated Langendorff-perfused Nav1.5-haploinsufficient murine hearts (Section 14.1) with heterozygotic *Scn5a+/–* or *Scn5a*-1798insD genotypes recapitulated these respective sinus bradycardic, atrial and ventricular arrhythmic, and age-dependent fibrotic phenotypes as well as progressive conduction disorders.

These experimental studies also implicated Nav1.5 in the normally ordered action potential conduction through the SAN and from there to the surrounding atrial myocytes that thereby initiates periodic atrial excitation. Patch clamped murine *Scn5a+/–* SAN cardiomyocytes correspondingly showed reduced I_{Na} relative to wild type (WT) myocytes (Plate 17*Aa,b*). Otherwise, activating and inactivating kinetics, and the ~–70 mV, activation thresholds and ~+20 mV voltages at which currents were maximal, were normal.

This reduced Nav1.5-mediated I_{Na} reduced heart rate by compromising impulse propagation within the SAN and from SAN to atrium (Lei *et al.*, 2005). Action potentials recorded from sites near the centre of the SAN showed longer cycle lengths in *Scn5a+/–* reflecting lower pacemaker rates (Plate 17*Ba,b*). Multiple extracellular recordings mapped the activation sequence in intact atrial preparations that included the SAN. These demonstrated that both WT and *Scn5a+/–* atria showed spontaneous pacemaker activity beginning in the SAN, propagating to the crista terminalis (CT) and interatrial septum (SEP) (Plate 17*Ca*) with similar overall activation patterns. However, *Scn5a+/–* hearts showed significantly slower conduction in regions towards the SAN periphery and its junction with the atria suggesting a peripheral exit block (Plate 17*Cb,c*).

Finally, both multi-electrode array mapping and electrocardiographic PR intervals demonstrated slowed conduction and increased arrhythmic tendency in *Scn5a+/–* relative to WT atria (Guzadhur *et al.*, 2012).

14.7 | Action Potential Activation and Conduction Abnormalities: Impaired Na$^+$ Channel Function and Ventricular Arrhythmia

The autosomal dominant BrS is also associated with increased risks of arrhythmogenic sudden cardiac death, particularly in middle-aged (~40–45 years) males. BrS accounts for 4–12% of unexpected sudden deaths. Although causative mutations are demonstrable only in ~30% of cases, the commonest associated mutation concerns the *SCN5A*-encoded Nav1.5, which accounts for 15–30% of BrS cases. BrS is linked to ~300 different *SCN5A* mutations. Mutations in other genes associated with BrS include other Na$^+$ channel α- (e.g. *SCN10A*) and β-subunits (Section 5.2; *SCN1B*, *SCN2B* and *SCN3B*; Hakim *et al.*, 2008; O'Malley and Isom, 2015), as well as other proteins likely associated with the Na$^+$ channel.

Langendorff-perfused murine *Scn5a+/−* ventricles demonstrated pro-arrhythmic tendencies with extra-systolic stimulation. They recapitulated the clinically observed respective pro- and anti-arrhythmic effects of the Na$^+$ and K$^+$ channel blockers, flecainide and quinidine, reported in the corresponding human condition. Their phenotypes were attributed to reduced I_{Na} and fibrotic change, particularly involving the right ventricular outflow tract (RVOT) especially in male *Scn5a+/−* mice. Such alterations would compromise action potential conduction velocity. *Scn5a+/−* ventricular myocytes showed reduced Nav1.5 mRNA and protein expression. Patch clamped *Scn5a+/−* ventricular myocytes demonstrated correspondingly reduced I_{Na} and decreased action potential $(dV/dt)_{max}$ particularly in the right (RV) compared to the left ventricle (LV) (Plate 18*A–C*). This matched the higher RV than LV Kv4.2, Kv4.3 and KChIP2 expression expected to result in greater RV than LV magnitudes of I_{to} in both *Scn5a+/−* and WT (Martin *et al.*, 2012).

Scn5a+/− hearts similarly recapitulated the clinical BrS phenotypes in showing age- and sex-dependent progressive myocardial fatty replacement and fibrosis, and compromised connexin expression, particularly in male animals. This was reflected in multi-electrode array activation maps of conduction latency (Plate 18*Da,b*) and their frequency distributions of activation times (Plate 18*Ea,b*). Ventricles from aged male *Scn5a+/−* mice (Plate 18*Eb*) showed reduced representations of rapidly conducting components compared to those from young male WT mice (Plate 18*Ea*). Such findings paralleled known age and sex-dependent risks of arrhythmias in human BrS. The latter correlated with histological demonstrations of fibrotic changes. These could also compromise action potential conduction (Sections 6.2–6.4) through altered gap junction expression or function, increased cardiomyocyte capacitance following fibroblast–cardiomyocyte fusion and/or altered tissue geometry dispersing action potential propagation pathlengths. This would add compromised local circuit currents conducting the action potential to the deficiencies in Na^+ current required for its initiation to the resulting abnormal cardiac excitation process (Sections 6.1–6.3; Jeevaratnam *et al.*, 2016).

Further observations in *Scn5a+/–* hearts paralleled clinical findings implicating the RV, in particular, selective action potential conduction delays in the right ventricular outflow tract (RVOT), as the arrhythmia initiating site. RVOT stimulation elicited increased arrhythmic incidences in *Scn5a+/–* but not WT hearts. Multi-array analyses demonstrated lower overall action potential conduction velocities in *Scn5a+/–* than WT, and in RVOT than RV. The magnitudes of such conduction velocities were lower in *Scn5a+/–* than WT but similar between RV and RVOT. However, dispersions in conduction direction were greater in RVOT than RV, but similar between *Scn5a+/–* and WT. These electrophysiological results closely paralleled the reduced Nav1.5 protein expression in the RVs and RVOTs of *Scn5a+/–* hearts compared to WT, in combination with increased fibrotic levels in the RVOT relative to the RV in both WT and *Scn5a+/–* hearts (Martin *et al.*, 2011*a*; Jeevaratnam *et al.*, 2012; Zhang *et al.*, 2014).

Relationships between conduction velocity, and the initiation and maintenance of re-entrant arrhythmia could be dynamically visualised in intact beating hearts. Contact multi-electrode (0.5 mm) array mapping techniques compared recordings from LV and RV epicardial surfaces of spontaneously beating murine *Scn5a+/–* and WT hearts before and following flecainide or quinidine challenge (Plate 19, inset; Martin *et al.*, 2011*b*). Isochronal maps representing timings of action-potential activation and recovery, and the intervening activation–recovery intervals between them implicated slowed action potential conduction, particularly in the RV with arrhythmic abnormalities. *Scn5a+/–* ventricles showed slowed conduction and increased dispersions of activation relative to WT. The greater dispersions in the *Scn5a+/–* were particularly noticeable in the RV, and these were further accentuated by flecainide challenge. *Scn5a+/–* ventricles also showed shorter recoveries, further shortened by flecainide, but these were more dispersed, particularly with flecainide challenge. Both the latter effects were again particularly noticeable in the RV.

These features predisposed to an initiation of VT from the RV through formation of lines of conduction block inducing re-entrant circuits. Plate 19 illustrates the generation of a premature ventricular beat following arrival of activation at the RV epicardium leading to initiation of polymorphic VT in a flecainide-treated, spontaneously beating, *Scn5a+/–* heart. The mapped events (Plate 19*A–F*) are correlated with the recorded ECG (Plate 19*G*). Plate 19*A* shows the close isochronal contours reflecting delayed arrival of epicardial activation in the last normal beat. However, these are accompanied by marked repolarisation heterogeneities in the corresponding repolarisation map (Plate 19*A''*). A superimposed, premature ventricular activation produces a line of block, with impulse propagation flowing around it (Plate 19*B*) and a ventricular ectopic event initiates an anticlockwise running circuit (Plate 19*C*) that persists into the following beat (Plate 19*D*). This initiates VT, particularly when the premature ventricular beats coincide with the T-wave in the ECG (Plate 19*J*). There is now a continually changing line of block that produces a non-stationary vortex that results in the polymorphic character of the VT (Plate 19*E*, *F*, *G*). The VT propagates as a wavefront across the LV from its initiation site in the RV (Plate 19*I*; Martin *et al.*, 2011*b*). In contrast to these findings in spontaneously beating *Scn5a+/−*, WT hearts did not show ventricular ectopic or tachycardic phenomena before, and rarely did so following flecainide challenge.

In contrast to these conduction changes, *Scn5a+/−* ventricular action potential recovery timecourses were actually shortened, with longer endo- than epicardial APDs, giving normally directed ventricular ΔAPD, particularly in the RV. These tendencies were accentuated by flecainide but reduced by quinidine. ERPs were increased rather than decreased, effects also accentuated by flecainide and quinidine challenge. Hence the ΔAPDs and consequent ERP, and changes in these with flecainide challenge would minimise re-entrant, retrograde excitation from epicardium to endocardium. These differences also applied in comparisons between the RVOT and remaining RV. *Scn5a+/−* hearts continued to show greater ERPs in both the RVOT and the remaining RV than WT, but similar ERPs in the remaining RV as WT. This made re-entrant events, including those occurring in Phase 2 (Section 14.3), during action potential recovery less likely, particularly in the RVOTs of murine *Scn5a+/−* hearts (Lukas and Antzelevitch, 1996; Veeraraghavan *et al.*, 2013).

Scn5a+/− and WT ventricles nevertheless showed differing instability properties with progressively incremented steady-state pacing rates (Section 14.4). *Scn5a+/−* ventricles showed APD alternans at longer BCLs and DIs than WT. These effects were most noticeable following flecainide challenge, particularly in the RV epicardium. *Scn5a+/−* ventricles accordingly had steeper restitution functions, particularly in the RV epicardium, than the corresponding WT. Flecainide particularly destabilised the RV of *Scn5a+/−* hearts (Matthews *et al.*, 2010; 2012).

14.8 | Action Potential Repolarisation Abnormalities: Long QT Syndromes and Ventricular Arrhythmia

Arrhythmic substrate also arises from action potential repolarisation and refractoriness abnormalities (Section 14.3). Alterations in Na^+ channel recovery can result from changes in fast or slow inactivation or development of late Na^+ currents, I_{NaL}. Persistent or increased inward depolarising, I_{Na} or plateau I_{CaL}, or compromised outward repolarising I_K all prolong the action potential, resulting in clinical long QT syndrome (LQTS). This predisposes to potentially fatal ventricular, TdP arrhythmias (London, 2001; Killeen *et al.*, 2008*b*).

Table 14.1	*Variant long QT syndromes and their underlying ion channels*		
Variant	**Ion channel/molecule type**	**Encoding gene**	**Comments**
LQTS1	I_{Ks}: Kv7.1	KCNQ1	
LQTS2	I_{Kr}: Kv11.1	KCNH2	
LQTS3	I_{Na}: Nav1.5	SCN5A	
LQTS4	Ankyrin B	ANK2	Anchoring protein
LQTS5	MinK	KCNE1	K$^+$ channel β-subunits
LQTS6	MiRP	KCNE2	K$^+$ channel β-subunits
LQTS7	I_{K1}: Kir2.1	KCNJ2	Anderson-Tawil syndrome
LQTS8	I_{Ca}: Cav1.2	CACNA1C	Timothy Syndrome
LQTS9	caveolin-3	CAV3	cytoskeletal protein
LQTS10	Navβ4	SCN4B	β-subunit
LQTS11	A-kinase anchor protein 9	AKAP9	cytoskeletal protein
LQTS12	α-1-syntrophin	STNA1	cytoskeletal protein
LQTS13	I_{KAch}: Kir3.4	KCNJ5	

Acquired LQTS follows a number of clinical conditions, as well as a wide range of medications (Fermini and Fossa, 2003; Frommeyer and Eckardt, 2016). The latter often produce LQTS through reduced I_{Kr} (Roden, 2004). *Hereditary* LQTS was first described as the autosomal dominant Romano–Ward (RW) and autosomal recessive Jervell and Lange-Nielsen (JLN) syndrome; the latter is accompanied by congenital sensorineural deafness (Keating and Sanguinetti, 2001). Inherited LQTS affects ~0.01% of the population, with ~50% of LQTS patients having known mutations. LQTS is associated with a wide range of genetic modifications (Table 14.1). The commonest, LQTS1, LQTS2 and LQTS3, account for 45%, 45% and 5%, respectively, of inherited genotype-confirmed LQTS cases. All variants can present as Romano–Ward syndrome, with some LQTS1 and LQTS5 also presenting as Jervell and Lange-Nielsen syndrome.

14.9 | Gain of Na$^+$ Channel Function: LQTS3

LQTS3 is associated with increased as opposed to loss of Na$^+$ channel function, in contrast to the Na$^+$ channel deficiency associated with BrS. It shows correspondingly differing arrhythmic and electrophysiological, including ECG, features. The class IA cardiotropic drug, quinidine is pro-arrhythmic and the class IC drug flecainide is anti-arrhythmic in LQTS3. This directly contrasts with their respective anti- and pro-arrhythmic effects in BrS (Brugada, 2000). LQTS3 also differs from other LQTS variants in that 39% of fatal arrhythmias occur during sleep or rest rather than stress or arousal as in LQTS1 and LQTS2. LQTS3 patients also show paradoxical anti-arrhythmic effects with β-adrenoceptor agonists (Section 13.18), and benefit less from β-adrenoceptor antagonist therapy than do LQTS1 and LQTS2 patients.

LQTS3 was modelled in murine hearts with targeted ΔKPQ1505–1507 deletions involving the Nav1.5 inactivation domain (Sections 4.6 and 5.2; Nuyens *et al.*, 2001; Head *et al.*, 2005). Their ventricular myocytes recapitulated the increased I_{NaL} associated with clinical LQTS3 implicated in many of the electrophysiological features of LQTS3 (Belardinelli *et al.*, 2015; Chadda *et al.*, 2017b). Isolated, Langendorff-perfused, murine *Scn5a+/ΔKPQ* hearts similarly showed increased ventricular arrhythmogenecity provoked by extrasystolic stimuli, exacerbated by quinidine and relieved by flecainide. Arrhythmogenesis persisted with challenge by the adrenergic β-receptor antagonist propranolol, paralleling clinical reports (Chadda *et al.*, 2017a).

Murine *Scn5a+/ΔKPQ* hearts proved useful in exploring arrhythmic mechanisms underlying the consequences of increased Nav1.5 function in LQTS3. Epicardial and endocardial MAP recordings assessed action potential recovery timecourses from in situ Langendorff-perfused cardiac preparations (Figure 14.6*A*). MAP recordings measure the potential difference between a positive electrode placed firmly on the myocardial surface and a lightly touching negative electrode in contact with the extracellular fluid. The pressure exerted by the positive electrode opens stretch-sensitive ion channels. This causes a local membrane depolarisation to ~–20 to –30 mV and inactivation of voltage-gated Na⁺ channels. It also creates an electrical pathway exposing the electrode tip to the intracellular potentials of the cardiomyocytes with which it makes contact. It thus senses membrane potential changes resulting from electrical activity, providing valuable in situ measurements of the timecourse of their action potentials.

Ventricular MAP recordings demonstrated increased epicardial APD_{90} (~60 ms vs ~47 ms in WT), recapitulating clinically observed increased QT intervals, and correlating with regional I_{NaL} differences. This could provide the necessary critical voltage conditions permitting I_{NaL} reactivation, leading to EAD-mediated triggering (Section 14.5). *Scn5a+/ΔKPQ* ventricles indeed showed triggering EAD events, often followed by VT. Furthermore, the I_{CaL} antagonist nifedipine at progressively increasing concentrations (10 nM–1 µM) decreased the incidences of both EADs and arrhythmias evoked by programmed electrical stimulation in *Scn5a+/ΔKPQ* ventricles. Yet it did not affect the remaining epicardial or endocardial APD_{90}, ΔAPD_{90} or VERP in either *Scn5a+/ΔKPQ* or WT considered below (Thomas *et al.*, 2007a).

Secondly, in contrast to the increased epicardial APD_{90}, endocardial APD_{90} remained close to values shown by WT (~52–54 ms). *Scn5a+/ΔKPQ* ventricles accordingly showed altered ΔAPD_{90} patterns. Epicardial APD now exceeded endocardial APD, with endocardial repolarisation preceding rather than following epicardial repolarisation (Section 14.3; Thomas *et al.*, 2007a). This caused arrhythmic substrate through mechanisms distinct from those underlying the EADs. Low (100 nM) concentrations of the β-adrenoceptor antagonist, propranolol, suppressed *both* spontaneous and provoked arrhythmias. It also both suppressed EADs and reduced epicardial APD, thereby correcting the repolarisation

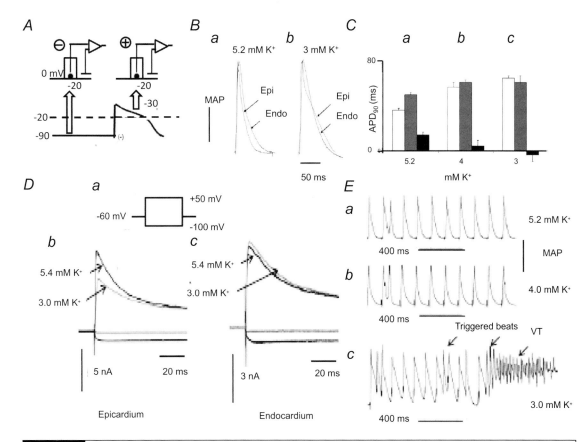

Figure 14.6 Ventricular action potential recovery characteristics in Langendorff-perfused hypokalaemic murine hearts. (A) Monophasic action potential (MAP) recording technique. (B) LV endocardial ('endo') and epicardial ('epi') action potential recordings under (a) control (5.2 mM) $[K^+]_o$, and (b) hypokalaemic conditions (3 mM $[K^+]_o$); hearts were paced at 125 ms intervals. (C) Steady-state epicardial (white columns) and endocardial APD_{90} values (grey columns) and the resulting ΔAPD_{90} (black columns) at $[K^+]_o =$ (a) 5.2 mM, (b) 4 mM and (c) 3 mM, respectively. (D) Epicardial (b) and endocardial (c) outward and inward K^+ currents in response to depolarising and hyperpolarising voltage steps (a) in whole cell patch clamped epicardial (b) and endocardial (c) myocytes under normokalaemic (dark lines) and hypokalaemic (3 mM) conditions (pale lines). (E) Programmed electrical stimulation of isolated, WT Langendorff-perfused mouse hearts under (a) normokalaemic (5.2 mM) and hypokalaemic, (b) 4 mM and (c) 3 mM $[K^+]_o$, conditions. No observed VT in 5.2 mM $[K^+]_o$ (a), VT in two of seven hearts in 4 mM $[K^+]_o$ (b) and 9 of 11 hearts in 3 mM $[K^+]_o$ (c). (From Figures 2, 3, 5, 6 of (Killeen et al., 2007b.)

gradient abnormalities. High (1 mM) propranolol concentrations eliminated both EADs and spontaneous arrhythmias, but increased susceptibility to provoked arrhythmogenesis. This was accompanied by further prolonged epicardial yet reduced endocardial APDs, exacerbating the abnormalities in repolarisation gradient shown by Scn5a+/ΔKPQ hearts (Thomas et al., 2008).

Thirdly, Scn5a+/ΔKPQ ventricular action potential durations showed greater temporal heterogeneities involving their repolarisation as opposed to conduction or activation phases (Section 14.4). They showed increased epicardial and endocardial APD_{90} alternans, further accentuated at increased pacing rates. The resulting epicardial and endocardial restitution curves gave increased values of DI_{crit}. These were increased and decreased respectively by quinidine and flecainide in parallel with their observed pro-arrhythmic and anti-arrhythmic effects (Sabir *et al.*, 2008b).

14.10 | Gain of Ca^{2+} Channel Function: LQTS8 (Timothy Syndrome)

LQTS8 (Timothy syndrome) is associated with autosomal dominant gain-of-function missense *CACNA1C*-G406R mutations involving the junction between DI/S6 and the I-II loop of Cav1.2, with or without an accompanying *CACNA1C*-G402S mutation. Patch clamp recordings reported increased Cav1.2 opening probabilities, open time durations and coupled gating frequencies. Hearts from a cardio-specific LQTS8 model were hypertrophied relative to WT. A slowed inactivation in their ventricular myocyte I_{CaL} likely accounted for their prolonged AP waveforms. Intact hearts showed exercise-induced premature ventricular depolarisations and TdP (Cheng *et al.*, 2011).

14.11 | Loss of K^+ Channel Function: Hypokalaemic Murine Models for Acquired LQTS

Most LQTS-mediated arrhythmias are associated with acquired rather than congenital risk factors. These include metabolic and electrolyte, particularly hypokalaemic (reduced $[K^+]_o$), conditions, bradycardic situations and drug therapies, particularly those affecting I_{Kr} (Vandenberg *et al.*, 2001). Murine hearts replicated the $[K^+]_o$-sensitivity shown by the major voltage-dependent repolarising I_K, in particular I_{K1}, (Section 13.9) in larger mammals.

In common with the LQTS exemplars outlined above, hypokalaemia modified ventricular action potential recovery timecourses in intact, Langendorff-perfused WT murine hearts. It increased endocardial and epicardial MAP durations to differing extents (Figure 14.6*Ba*, *b*) thereby reducing ΔAPD (Figure 14.6*C*). This arose from differential I_{to} and I_{K1} alterations. Under normokalaemic (5.4 mM $[K^+]_o$) conditions, I_{to} was greater in patch clamped epi- than endocardial myocytes subject to depolarising test steps in keeping with their higher epi- than endocardial protein expression levels (Figure 14.6*Da,b*). This explains the normally observed ΔAPDs. However, hypokalaemia decreased epicardial but not endocardial I_{to}, explaining the correspondingly diminished repolarisation gradients. In contrast, epicardial and endocardial myocytes under normokalaemic conditions showed similar I_{K1} obtained in response to hyperpolarising steps

(Figure 14.6Da). Hypokalaemia reduced these to similar extents (Figure 14.6Db,c). This was consistent with increased APD but altered ΔAPDs (Section 14.3; Killeen et al., 2007b).

These reductions in $[K^+]_o$ additionally gave rise to an occurrence of triggered beats ((Figure 14.6E; Section 14.5), that were often followed by episodes of non-sustained VT.

In common with LQTS3-related arrhythmic mechanisms (Section 14.9), hypokalaemic hearts showed pharmacologically separable pro-arrhythmic contributions, arising from triggering events and ΔAPDs, to the generation of triggering and persistent arrhythmia. Both the I_{CaL}-blocker nifedipine (100 nM) and the CaMKII inhibitor KN-93 (Killeen et al., 2007a) reduced EADs and spontaneous arrhythmias. These implicated Ca^{2+} channel reactivation processes in the premature action potentials and triggered beats (January and Riddle, 1989; Fabritz et al., 2003). However, neither produced any further alterations in either epicardial or endocardial APD. Nor did they prevent persistent arrhythmias provoked by programmed electrical stimulation. However, higher (1 μM) nifedipine concentrations abolished both the spontaneous and provoked arrhythmic phenomena. They shortened epicardial APDs, restoring the ΔAPDs to control values.

Finally, hypokalaemia contributed arrhythmic substrate resulting from temporal heterogeneities in action potential recovery. These were particularly noticeable with delivery of pacing stimuli at progressively shortened cycle lengths (Section 14.4). Hearts under hypokalaemic (3 mM) but not normokalaemic (5.2 mM $[K^+]_o$) conditions showed increased incidences of APD alternans ultimately leading to arrhythmias. Their epicardial and endocardial restitution curves both showed increased maximum gradients and DI_{crit} values corresponding to unity gradients, features identified with the onset of arrhythmia (Sabir et al., 2008b).

14.12 | Congenital LQTS Related to Loss of K$^+$ Channel Function

Ventricular action potential recovery in murine hearts contrasts with that in hearts of larger mammals in their increased transient outward I_{to}, and diminished I_{Kr} and I_{Ks}. The resulting more rapid, triangulated decay timecourse often lacks a Phase 3 plateau. Nevertheless, transgenic ($Kv4.2$-DN) mice lacking $I_{to,f}$ or ($Kv4.2$-DN × $Kv1.4$–/–) mice lacking both $I_{to,f}$ and $I_{to,s}$ showed APD and QT prolongation, ventricular tachyarrhythmias with abnormal apex–base, endocardium–epicardium and septum-free wall ventricular repolarisation patterns. However, knockout of murine $Erg1B$ preventing I_{Kr} expression resulted in a predisposition to episodic sinus bradycardia, but normal QT intervals.

Normal adult mouse myocytes similarly lack Kv7.1-mediated I_{Ks}. Langendorff-perfused murine $Kcnq1$–/– or $Kcne1$–/– hearts lacking

Kv7.1 or its β-subunit *Kcne1* showed normal QT intervals and APDs, although these could be prolonged by nicotinic or sympathomimetic stimulation or increased heart rates. The mice continued to recapitulate morphological inner-ear abnormalities, auditory defects and the shaker/waltzer and cardiac phenotypes of Jervell and Lange-Nielsen syndrome. Spontaneously beating isolated *Kcne1–/–* hearts showed frequent EADs, coupled triggered beats, spontaneous VT and VT triggered by premature stimuli (Balasubramaniam *et al.*, 2003; Hothi *et al.*, 2009). They showed prolonged epicardial and endocardial APDs, increased APD_{90}-VERP differences and reduced ΔAPDs consistent with spatial re-entrant substrate, increased APD_{90} alternans and steeper epicardial and endocardial APD_{90} restitution curves during dynamic pacing. Nifedipine suppressed the EADs, triggered beats and repolarisation alternans (Thomas *et al.*, 2007b).

14.13 | Pro-Arrhythmic Perturbations in Intracellular Ca^{2+} Homeostasis

The above exemplars implicate surface ion channel and transporter activity abnormalities as immediate causes of most arrhythmias, through abnormal SA, AVN or Purkinje fibre automaticity (Section 13.2), or pathological activity involving atrial and ventricular cardiomyocytes (Sections 14.2–14.5). However, these electrophysiological processes feed forward into excitation–contraction coupling events, triggering RyR2-mediated SR Ca^{2+} release (Section 13.14). The resulting $[Ca^{2+}]_i$ elevations and alterations in cellular energetic status can exert reciprocal, potentially pro-arrhythmic, feedback effects upon these surface membrane processes. In particular, increased electrogenic I_{NCX} activity (Section 13.16) drives DADs, potentially causing triggered activity (Section 13.10), and a downregulated I_{Na} (Sections 13.19 and 13.20) produces arrhythmic substrate (Section 14.7; Huang, 2017).

Isolated perfused murine hearts proved useful in identifying and characterising this further group of pro-arrhythmic mechanisms. They recapitulated the opposing effects of β-adrenergic stimulation and Epac activation (Section 13.18), and nifedipine-mediated I_{CaL} blockade on ventricular arrhythmogenesis. These were related to altered Ca^{2+} homeostatic and triggering events (Section 14.5; Balasubramaniam *et al.*, 2003; 2004; Hothi *et al.*, 2008). In ventricular cardiomyocytes with normal, or increased, SR Ca^{2+} loading, both Epac activation and challenge by the RyR2 agonists caffeine or FK506 increased the frequencies of propagating Ca^{2+} waves arising from abnormal SR Ca^{2+} release (Section 11.8; Balasubramaniam *et al.*, 2005). Murine atrial preparations similarly demonstrated caffeine-induced atrial arrhythmias related to Ca^{2+} waves resulting from diastolic Ca^{2+}-induced Ca^{2+} release, even from a more depleted atrial SR Ca^{2+} store (Zhang *et al.*, 2009). These changes occurred in the absence of altered spatial or temporal electrophysiological recovery heterogeneities, in contrast to the alterations shown by the congenital or acquired ventricular LQTS models described above (Sections 14.3, 14.9–14.12).

14.14 | Altered Intracellular Ca^{2+} Homeostasis: Catecholaminergic Polymorphic Ventricular Tachycardia

Murine hearts replicating genetic disorders involving Ca^{2+}-mediated arrhythmia could be used to clarify their underlying mechanisms. Of these, catecholaminergic polymorphic ventricular tachycardia (CPVT) is associated with gain of RyR2 or loss of cardiac calsequestrin, CASQ2, function. It presents with a predisposition to acute VT on adrenergic stimulation that may degenerate into VF, leading to ~30–33% mortality by age 35 years. Heterozygotic *RyR2*-P2328S (*RyR2$^{S/+}$*) murine hearts recapitulated the clinically reported ECG episodes of bigeminy and bidirectional ventricular tachycardia showing alternating 180° rotations of the electrocardiographic QRS axis (Sections 13.4 and 14.1) following isoproterenol challenge. Homozygotic, *RyR2$^{S/S}$*, animals showed this both before and following isoproterenol challenge, and additionally recapitulated atrial arrhythmic phenotypes also associated with CPVT (Zhang *et al.*, 2011).

Isolated murine *RyR2$^{S/S}$* channels in lipid bilayer membranes (Section 3.1) studied at +40 or −40 mV transmembrane voltages showed normal opening activity at 1 mM (Figure 14.7*Aa*) but greater opening activity at the ~1 μM $[Ca^{2+}]_i$ than might exist in end-diastolic myocytes, compared to WT channels (Figure 14.7*Ab*). *RyR2$^{S/S}$* open probabilities, P_o, plotted against $[Ca^{2+}]_i$ showed marked, >10-fold, negative shifts in half-maximal activating $[Ca^{2+}]_i$ from ~3.5 μM to ~320 nM, relative to corresponding features in WT channels (Figure 14.7*Ba,b*). *RyR2$^{S/S}$* channels consequently showed increased sensitivity to Ca^{2+}-induced Ca^{2+} release (Figure 14.7*Ca,b*; Salvage *et al.*, 2019a). In contrast, gene and protein expression levels of remaining Ca^{2+} homeostasis proteins (CASQ2, FKBP12, SERCA2A, NCX1, and Cav1.2) remained unchanged (Saadeh *et al.*, 2020).

At the cellular level, regularly stimulated *RyR2$^{S/+}$* and *RyR2$^{S/S}$* ventricular myocytes both showed altered Ca^{2+} release patterns modifiable by 100 nM isoproterenol (Goddard *et al.*, 2008; Zhang *et al.*, 2013b). Isoproterenol-stimulated *RyR2$^{S/S}$* ventricular myocytes additionally showed spontaneous Ca^{2+} propagating waves (Plate 20*A*). In the absence of isoproterenol challenge, spontaneously beating Langendorff-perfused *RyR2$^{S/S}$* though not *RyR2$^{S/+}$* hearts showed spontaneous extra-systolic triggering events often followed by VT (Plate 20*B*, *C*). *RyR2$^{S/+}$* hearts nevertheless showed non-sustained VT with extrasystolic stimulation. Following isoproterenol challenge, *RyR2$^{S/+}$*

Figure 14.7 Properties of mutant (*RyR2-P2328S*) compared to wild-type (WT) ryanodine receptors (RyR2) studied in lipid bilayer membrane. (*A*) Channel conductances under conditions of 1 mM extracellular and (*a*) 1 mM and (*b*) 1 μM cytoplasmic [Ca^{2+}]$_i$. Representative 25 s recordings from one WT RyR2 channel (upper panels) and one RyR2-P2328S channel (lower panels) with (*a*) 1 mM or (*b*) 1μM cytoplasmic Ca^{2+} at –40 mV (upper record) or +40 mV (lower record). Broken lines (labelled 'o') indicate maximum open currents, and solid lines (labeled 'c') indicate closed levels. The *RyR2-P2328S* mutation enhances RyR2 channel activity compared to WT when cytoplasmic [Ca^{2+}] is 1 μM, without altering maximum conductance. Note similar maximum conductances of WT and RyR2-P2328S channels at any given bilayer voltage and cytoplasmic [Ca^{2+}] in (*a*) but greater maximum currents in both channel types with 1 μM cytoplasmic Ca^{2+}, likely reflecting removal of partial pore block by Ca^{2+} binding in (*b*). (*B*) Comparisons of cytoplasmic Ca^{2+} sensitivity of WT (*Ba,b*) and RyR2-P2328S channels (*Ca,b*) at (*a*) –40 and (*b*) +40 mV. (From Figure 1 and 4 of Salvage et al., 2019a.)

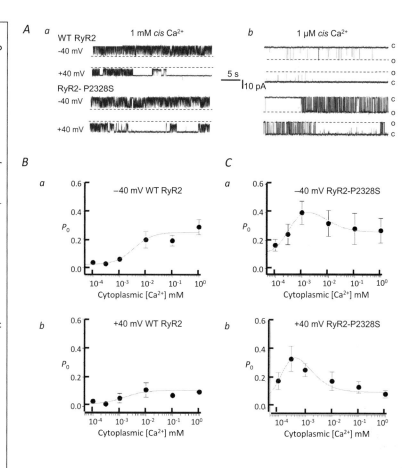

and *RyR2*$^{S/S}$ hearts both showed VT during both regular pacing and following extra-systolic stimulation (Plate 20D, E). Murine *RyR2*$^{S/S}$ hearts similarly recapitulated reported clinical atrial arrhythmic phenotypes, in the form of frequent, sustained tachyarrhythmias, delayed after-depolarisations and ectopic action potentials.

Both *RyR2*$^{S/S}$ atria and ventricles additionally demonstrated evidence for arrhythmic substrate in the form of slowed action potential conduction (Section 14.2). *RyR2*$^{S/S}$ atrial action potential epicardial conduction velocities and (dV/dt)$_{max}$ were reduced to extents comparable to those shown by *Scn5a+/–* hearts (Plate 20F,G; King et al., 2013c). Loose patch clamp measurements (Section 10.2) demonstrated reduced peak I_{Na} in *RyR2*$^{S/S}$ (Plate 20I) relative to WT atria (Plate 20H; King et al., 2013b). These findings could be attributed both to reduced Nav1.5 expression in *RyR2*$^{S/S}$ (King et al., 2013b; Ning et al., 2016) and Na$^+$ channel inhibition by altered [Ca^{2+}]$_i$ (Section 13.19). Thus flecainide (1 μM)-mediated RyR2 block rescued the arrhythmogenic phenotype and restored the compromised I_{Na} in *RyR2*$^{S/S}$ atria, yet was pro-arrhythmic and reduced I_{Na} in WT (Salvage et al., 2015). Untreated WT and *RyR2*$^{S/S}$ ventricles showed similar action potential conduction velocities and (dV/dt)$_{max}$, but these were selectively reduced by isoproterenol and caffeine in *RyR2*$^{S/S}$ (Zhang et al., 2013a). Finally, WT atria replicated these I_{Na} reductions with acute $[Ca^{2+}]_i$ elevations produced by increased $[Ca^{2+}]_o$, or RyR2 activation by caffeine, or SERCA antagonism by cyclopiazonic acid (King *et al.*, 2013b).

Finally, action potential recovery characteristics in both *RyR2*^S/+^ and *RyR2*^S/S^ atria and ventricles, reflected in action potential and ERP durations (Section 14.3; Goddard *et al.*, 2008; King *et al.*, 2013c) and restitution properties, were unchanged relative to WT, whether before or following catecholaminergic challenge (Section 14.4; Sabir *et al.*, 2010).

14.15 | Mechanistic Schemes for Ca²⁺-Mediated Arrhythmias

Similar arrhythmic consequences followed mutations in other biomolecules also involved in intracellular Ca^{2+} homeostasis. Mice with genetic *Casq2* modifications, altering Casq2-dependent SR Ca^{2+} storage, similarly showed CPVT and bidirectional VT phenotypes, and increased tendencies to AF. Similarly, the cardiac serine/threonine protein kinase p2-activated kinase-1 (Pak1) is associated with and likely acts as an upstream cardiac PP2A activator in SAN and ventricular cells. PP2A removes phosphate groups at PKA phosphorylation sites, including those in L-type Ca^{2+} channels, RyR2s and the SERCA regulator phospholamban. *Pak1*-cko conditional knockout and *Pak1*–/– hearts showed increased incidences of polymorphic VT, despite normal ERPs, following either acute or chronic β-adrenergic isoproterenol challenge. *Pak1*-cko myocytes also showed higher incidences of irregular, alternating Ca^{2+} transients with repetitive stimulation, and of spontaneous Ca^{2+} waves, both before and particularly following imposition of chronic β-adrenergic stress. *Pak1*-cko and control, *Pak1*-f/f, ventricular myocytes showed comparable baseline I_{CaL} and NCX activity, similarly reduced by chronic β-adrenergic stress. However, *Pak1*-cko myocytes showed reduced SERCA2 activity, as assessed by the recovery timecourse of stimulated Ca^{2+} transients following initial SR Ca^{2+} depletion produced by caffeine challenge, accentuated by chronic β-adrenergic stress (Wang *et al.*, 2014).

These observations together suggest a hypothetical scheme relating perturbed Ca^{2+} homeostasis, triggering events, arrhythmic substrate and the resulting generation of persistent arrhythmia (Huang, 2017). Both acquired and genetic perturbations could either increase RyR2-mediated SR Ca^{2+} or decrease SERCA-mediated Ca^{2+} reuptake from cytosol to SR store, thereby perturbing $[Ca^{2+}]_i$ (Plate 21A,B). This in turn increases electrogenic NCX activity and I_{ti}, leading to diastolic triggering (Plate 21C). It also either reduces Nav1.5 membrane expression or directly alters Nav1.5 function (Plate 21D). The consequent reduced I_{Na} slows action potential conduction, causing arrhythmic substrate. In contrast, action potential recovery characteristics remain normal. The combination of triggering and re-entrant substrate potentially leads to sustained arrhythmia (Plate 21E).

14.16 | Pro-Arrhythmic Consequences of Compromised Cellular Energetics

Genetic modifications directed at specific molecules often associated with rare pro-arrhythmic conditions thus proved useful for physiological analysis. Murine models also proved useful in electrophysiological analyses of commoner clinical situations. Cardiac failure, ischaemia-reperfusion injury, ageing and metabolic conditions such as obesity, insulin resistance and type 2 diabetes, are associated with energetic, particularly mitochondrial, dysfunction, in turn implicated in both atrial and ventricular arrhythmia. Their accompanying increased reactive oxygen species (ROS) and reduced NAD$^+$/NADH levels decrease I_{Na} and activate RyR2-Ca^{2+} channel activity, increasing cytosolic [Ca^{2+}]$_o$. Both ROS and an accompanying reduced ATP/ADP also activate sarcolemmal ATP-sensitive K$^+$ (K$_{ATP}$) channels (Section 13.20).

Gene expression related to mitochondrial oxidative phosphorylation is modified by peroxisome-proliferator-activated receptor-γ coactivator-1 (PGC-1α and PGC-1β) transcriptional coactivators, abundant in brown adipose tissue and cardiac and skeletal muscle. Such coactivators are involved in energy regulation in both health and disease (Lin *et al.*, 2005; Finck and Kelly, 2006). Compromised cardiomyocyte metabolic energetics and mitochondrial function, modelled in *Pgc-1β$^{-/-}$* mice, resulted in irregular heartbeats and polymorphic VT following isoproterenol challenge. In common with *Scn5a+/−*, Langendorff-perfused *Pgc-1β$^{-/-}$* hearts showed pro-arrhythmic compromised atrial and ventricular action potential (dV/dt)$_{max}$ and conduction velocities, and age-dependent fibrotic phenotypes (Ahmad *et al.*, 2017; Valli *et al.*, 2017). Their isolated ventricular myocytes showed EADs and DADs, and diastolic Ca^{2+} transients exacerbated by isoproterenol challenge (Gurung *et al.*, 2011). Loose patch clamp studies implicated compromised I_{Na} in these pro-arrhythmic phenotypes, with reductions in ventricular I_{Na} with otherwise unchanged voltages at half-maximal current, V^*, and steepness factors, k, in an absence of altered Nav1.5 gene or protein expression (Valli *et al.*, 2018a; Ahmad *et al.*, 2019; Edling *et al.*, 2019).

These observations can be simplified into a scheme relating energetic dysfunction and its arrhythmogenic effects (Huang, 2017). The relatively common clinical conditions of cardiac ischaemia and/or failure, ageing and diabetes (Plate 22A) lead to mitochondrial dysfunction increasing ROS production, compromising NAD$^+$/NADH and ATP/ADP ratios (Plate 22B). These in turn accentuate RyR2-mediated SR Ca^{2+} release, NCX and consequent DAD-mediated triggering activity (Plate 22C), and reduce I_{Na} and increase I_K activity, affecting AP excitation, propagation and recovery (Plate 22D), changes potentially leading to arrhythmic substrate (Plate 22E).

14.17 | Pro-Arrhythmic Structural Remodelling Resulting from Upstream Pathophysiological Changes

Tissue structural, fibrotic and/or hypertrophic, change reduce connexin-mediated cardiomyocyte–cardiomyocyte coupling and increase cardiomyocyte capacitance through their fusion with fibroblast cells. Both potentially compromise conduction of excitation (Sections 6.2 and 6.3). This has been clinically implicated in age-related chronic AF. Both experimental and clinical Nav1.5 haploinsufficiency are also associated with pro-arrhythmic atrial and ventricular fibrotic change (Section 14.7). Murine cardiac-specific TGF-β_1 overexpression results in atrial fibrosis, atrial conduction disturbances and AF (Verheule *et al.*, 2004).

A direct modelling of pro-arrhythmic consequences of such age-dependent changes compared mitogen-activated protein kinase kinase 4, conditional knockout *Mkk4*-acko, mice with *Mkk4*-f/f controls. MKK4 is a kinase for c-Jun N-terminal kinase (JNK) that itself downregulates transforming growth factor-β1 (TGF-β1) gene expression. Knockout, *Mkk4*-acko, hearts showed normal mRNA or protein expression levels of Nav1.5, RyR2, NCX, inositol 1,4,5-*tris*phosphate receptor (IP$_3$R), the transient receptor potential channel proteins, TRPC1, TRPC3 and TRPC6, and the gap junction proteins, Cx40 and Cx43. Yet aging *Mkk4*-acko mice were more susceptible to both atrial and ventricular arrhythmias. They showed slowed and dispersed atrial conduction. Mapping studies demonstrated ventricular activation delays, and mathematical two-dimensional modelling implicated formation of fibroblast–cardiomyocyte couplings in a slowed, fragmented and potentially pro-arrhythmic ventricular conduction. Nevertheless, Langendorff-perfused *Mkk4*-acko and *Mkk4*-f/f hearts showed similar APD$_{90}$ and ERP, and Cx levels and distribution, through all age groups. These findings accompanied widened electrocardiographic QRS durations (~16 vs 10 ms), but normal QT intervals (Davies *et al.*, 2014).

14.18 | Translation of Mechanistic Insights into Therapeutic Strategy

The analysis of selected experimental pro-arrhythmic genetic exemplars can be integrated with different levels of normal and abnormal cardiac function and their underlying surface membrane (Plate 23A, B) and cytosolic excitation processes (Plate 23C–D) using the functional scheme introduced in Figure 13.11. These relationships are translatable to their corresponding clinical arrhythmic conditions (Plate 23E–F).

Figure 14.8 (A) Mapping of physiological processes underlying arrhythmic events (B) at the level of (a) cardiomyocyte membrane, (b) autonomic signalling, (c) excitation–contraction coupling and (d) upstream energetic or structural remodelling targets, onto classes of anti-arrhythmic drugs targeted at their underlying surface and intracellular membrane ion channels, ion exchangers, transporters, autonomic receptors, ionic pumps, and energetic and structural remodelling processes (From Figure 1 of Huang et al., 2020).

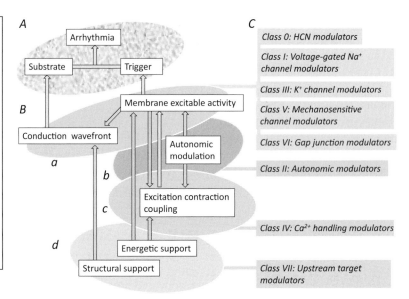

Plate 23 summarises the (A) surface membrane action potential excitation, conduction (Section 14.2) and recovery (Sections 14.3 and 14.4), and ectopic triggering (Section 14.5), and (B) their underlying ion channels (Sections 13.6–13.12). Particular abnormalities in one or more of these constitute the fundamental causes of atrial or ventricular arrhythmic activity. The surface events feed forward successively to intracellular processes of (C) excitation–contraction coupling, resulting in SR Ca^{2+} release and reuptake (Section 13.14) and (D) further upstream effects reflecting cellular energetics (Section 13.20) and pathological remodelling. Outcomes of both (C) and (D), exemplified by increased $[Ca^{2+}]_i$ and other signalling molecules, additionally exert up- (+) or downregulatory (–) feedback effects upon the excitation process (dotted lines), thereby influencing the surface electrophysiological events (Sections 13.18–13.20). (E) The selection of their underlying molecular and potential therapeutic targets, studied through either gain- or loss-of-function mutations, can be mapped onto the mechanistic scheme (A)–(D) as well as to their corresponding (F) human arrhythmic syndromes, showing (G) one or more forms of arrhythmic substrate or triggering.

Pharmacological intervention remains a mainstay in the clinical management of arrhythmic conditions. Much of this depends on systematic drug development and classification, their correlation with their modes of action, clinical indications and therapeutic actions. A classical attempt at such classification of available anti-arrhythmic medications long remained a mainstay of anti-arrhythmic management (Vaughan Williams, 1975). Of subsequent updating attempts incorporating the subsequent developments in cardiac electrophysiology and pharmacology (Task Force of the Working Group on

Arrhythmias of the European Society of Cardiology, 1991), the most recent modernised classification scheme (Lei *et al.*, 2018) retained and extended its original classes of investigational and therapeutic drugs. It mapped their ultimate anti-arrhythmic actions (Figure 14.8*A*) actions on membrane ((Figure 14.8*Ba*), excitation-contraction coupling (Figure 14.8*Bc*) and further upstream levels of cardiomyocyte function (Figure 14.8*Bd*) and their autonomic modulation (Figure 14.8*Bb*):

Class I, comprising drugs acting not only on early I_{Na} but also late I_{NaL} (Sections 14.8 and 14.9).

Class II, comprising not only β-adrenergic inhibitors, but also other modifiers of G-protein-mediated signalling (Section 13.18).

Class III, directed at the large number of K^+ channel species (Sections 14.11 and 14.12).

Class IV, targeting the wide range of physiological mechanisms related to Ca^{2+} homeostasis (Section 14.13).

It introduced new classes of target, in the light of the insights and findings subsequent to introduction of the original Vaughan Williams classification:

Class 0, bearing on cardiac automaticity (Section 13.8);

Class V, acting on mechanically sensitive channels;

Class VI, acting on cell–cell electrotonic coupling (Section 13.8);

Class VII, acting on upstream signalling processes (Sections 13.20, 14.16 and 14.17).

These have identified and classified currently established and investigated pharmacological targets strategic to cardiac electrophysiological activity, relating them to their effects whether at single cell, tissue or organ levels (Huang *et al.*, 2020). The effect of individual agents can be related to their likely therapeutic actions and consequent clinical indications. This will facilitate both the use of existing clinically used drugs and the future development of investigational agents for clinical use.

15

Smooth Muscle

Smooth, unstriated muscle forms the motile component in walls of hollow organs such as the gastrointestinal tract, the trachea, bronchi and bronchioles of the respiratory system, blood vessels in the cardiovascular system, and the ureters and bladder in the urogenital system. Its consequently diverse functions are reflected in its wide variations in structure and detailed physiological properties. Smooth muscle contracts and relaxes more slowly than skeletal muscle, but is better adapted to achieve sustained contractions. The load against which smooth muscle works is typically the pressure within the tubular structures it surrounds. In blood vessels, its tonic contraction produces their steady intraluminal pressure. In the gastrointestinal tract, its phasic contractions propel its luminal contents onward. Smooth muscle also occurs in the iris, ciliary body and nictitating membrane in the eye, and are the erector pili muscles which erect the hairs.

15.1 | Structure of Smooth Muscle Cells

Smooth muscle cells are uninucleate, elongated, often spindle-shaped and much smaller than the multi-nucleate skeletal muscle fibres (Figure 15.1). They are typically 3–5 μm in diameter and up to 400 μm long. Adjacent cells are connected at regions where their membranes are apposed to form gap junctions. These allow propagation of waves of electrical excitation or diffusion of intracellular messengers through the tissue. There is a vesicular sarcoplasmic reticulum (SR) close to the membrane, but no transverse (T-) tubular system. Their thick myosin and thin actin filaments are arranged longitudinally in the cytoplasm, but are not transversely aligned. The cells consequently show no visible striations or sarcomeres. The actin filaments are attached in bundles at dense bodies in the cytoplasm, and to attachment plaques at the membrane. The attachment plaques are analogous to the Z-disks in skeletal muscle and similarly contain α-actinin. These act as anchors for filaments to permit effective cell shortening.

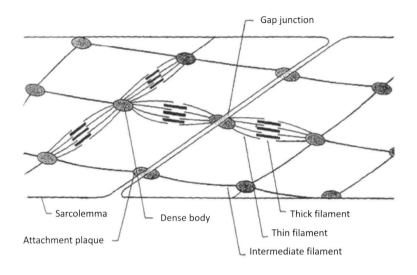

Gap junction

Sarcolemma

Attachment plaque

Dense body

Thick filament

Thin filament

Intermediate filament

Figure 15.1 Basic features of smooth muscle cells showing layout of gap junctions, thick and thin filaments, and dense bodies. (From Koeppen and Stanton, 2009.)

In contrast to skeletal muscle, smooth muscle has a diffuse innervation, with presynaptic terminal varicosities containing vesicles with transmitters and relatively unspecialised postsynaptic membranes. Different smooth muscle types vary in their detailed innervation pattern as well as the extent of electrical coupling between myocytes. Multi-unit smooth muscle cells in the iris and the ciliary body of the eye show little coupling. They are each independently innervated by synaptic endings. In contrast, the more extensively occurring unitary smooth muscle cells found in gastrointestinal tract, peripheral vasculature, bladder and uterus show varying extents of intercellular connection. Smooth muscle cells consequently show varying degrees of independent or synchronised activation.

15.2 | Functional Features of Smooth Muscle

Smooth muscle shows diverse excitable phenomena varying with both species and smooth muscle type. It further possesses a wide range of both hormonal and autonomic, involuntary, direct regulatory neural inputs, and degrees of spontaneous activity. Smooth muscles of the iris, nictitating membrane and vas deferens are not spontaneously active. However, other smooth muscle types show significant spontaneous activity, generating active tension, even in the absence of nerve activity. Intestinal muscles show spontaneous contractions that mix and propel the gut contents. Their electrical activity includes both slow waves of variable amplitude and all-or-nothing action potentials (Figure 15.2). The neuronal input modulates rather than initiates tension generation. Its spontaneous activity is modified by activity in its extrinsic nerves, and by circulating adrenaline (epinephrine). There

Figure 15.2 Simultaneous records of tension (upper trace) and electrical activity (lower trace) in guinea pig taenia coli. (From Bülbring, 1979.)

is often a reciprocal innervation by both sympathetic and parasympathetic fibres that form loosely associated terminals on the membrane surface.

Smooth muscle often shows graded, either depolarising or hyperpolarising, variations in membrane potential in response to circulating or local signal molecules or mechanical stimuli. Uterine muscle activity is modified by the action of hormones of the reproductive cycle. Electrical activity in smooth muscle membranes can show both temporal and spatial summation. If occurring with sufficient magnitude these can nevertheless elicit all-or-none action potentials particularly in some unitary and some multi-unit smooth muscle cells.

15.3 | Interstitial Cell of Cajal Networks

Smooth muscle properties reflect the particular features of their activation mechanisms. Many smooth muscle tissues, here exemplified by gastrointestinal smooth muscle, contain mesenchymal interstitial cells of Cajal (ICC) (Figure 15.3A). These cells have fusiform cell bodies and multiple thin processes contacting with and electrically coupling to other ICCs through connexin 40-, 43- and 45-containing gap junctions to form intricate functional networks. ICCs are also anatomically closely related to both enteric motor neurons and smooth muscle cells. This reflects their pacing role in smooth muscle activity, and their function as initial targets for neurotransmitter released from enteric motor neurons (Figure 15.3B). They express receptors, transduction mechanisms and ionic conductances mediating postjunctional responses to enteric motor neurotransmission. Among others, small intestinal ICCs express vasoactive intestinal peptide type 1 (VIP_1), muscarinic (M_2 and M_3) and neurokinin (NK_1 and NK_3) receptors. Finally, ICCs are electrically coupled to, thereby regulating the excitability of, smooth muscle cells.

ICCs occur throughout the gastrointestinal tract with distribution patterns and detailed morphological features varying with anatomical position (Figure 15.3C). ICC-SMPs and ICC-SMs occur between the submucosal and circular muscle layers in colon and gastric pylorus. ICC-DMPs occur in the deep muscular plexus between small intestinal

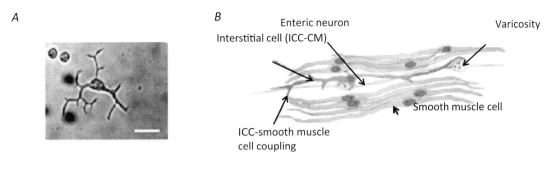

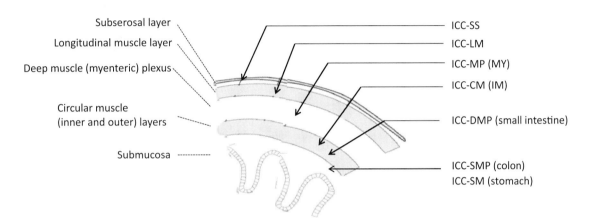

inner and outer circular muscle sublayers. ICC-CMs and ICC-LMs are embedded within gastrointestinal circular and longitudinal muscle. In the myenteric plexus, ICC-MPs (also termed ICC-MYs) occur between the circular and longitudinal muscle layers. Finally, ICC-SSs occur in the subserosal connective tissue space.

Figure 15.3 (A) Phase contrast image of cultured interstitial cell of Cajal (ICC) from tunica muscularis of the proximal colon. Scale bar 25 µm. (B) Anatomical relationship between enteric motor neurons, interstitial cells of Cajal and smooth muscle cells. (C) ICC types located in different tissue layers of the gastrointestinal tract ((A) from Figure 1A of Kim et al., 2002, (B) from Figure 2 of Sanders et al., 2010.)

15.4 | Pacing Properties of Interstitial Cells of Cajal

ICCs provide a diffusely distributed smooth muscle pacemaker system. This contrasts with the highly localised cardiac SAN pacemaker cells that mediate cardiac pacing (Section 13.7). ICCs show intrinsic pacemaker activity mediated by unique ionic currents specifically generating spontaneous electrical slow waves. They also show stimulus-induced pacemaker activity reflecting their autonomic inputs. ICC-IMs transduce inputs from enteric motor neurons that can alter slow wave frequencies, resulting in their becoming the dominant intestinal pacemaker under some circumstances.

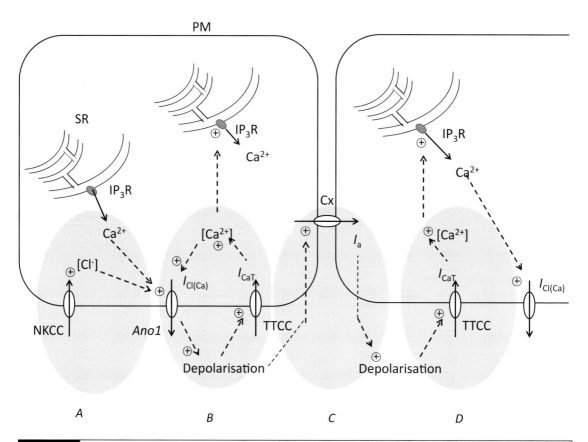

Figure 15.4 Pacing function of interstitial cells of Cajal (ICCs). (A) Primary pacemaker mechanisms: inositol *tris*phosphate receptor (IP$_3$R) Ca^{2+} channel mediated sarcoplasmic reticular (SR) Ca^{2+} release, increasing local cytosolic [Ca^{2+}]$_i$ together with outward Cl$^-$ electrochemical gradient generated by background Na$^+$-K$^+$-2Cl$^-$ (NKCC). These lead to (B) Ca^{2+}-activated Cl$^-$ current ($I_{Cl(Ca)}$) mediated depolarisation. This triggers Ca^{2+} current I_{CaT} through T-type Ca^{2+} channels, resulting in a regenerative increase in [Ca^{2+}]$_i$, in turn further increasing $I_{Cl(Ca)}$, leading to slow potentials. (C) Connexin/gap junction-mediated electrotonic spread of depolarisation to neighbouring ICCs, in turn causing (D) activation of I_{CaT}-mediated Ca^{2+} entry and further IP$_3$R-Ca^{2+} channel mediated SR Ca^{2+} release, continuing and propagating the cycle. TTCC, T-type Ca^{2+} channels; Ano1, Ca^{2+}-activated Cl$^-$ channels; IP$_3$R, inositol *tris*phosphate receptor; SR, endoplasmic reticulum; PM, plasma membrane.

Intestinal ICCs contain extensive intracellular Ca^{2+} stores, from which Ca^{2+} release through inositol *tris*phosphate receptor (IP$_3$R)-Ca^{2+} channels locally elevates [Ca^{2+}]$_i$, thereby providing a primary pacemaker (Figure 15.4A). In urethral smooth muscle, the RyR rather than the IP$_3$R may form the primary oscillator. The RyR inhibitor tetracaine here abolishes [Ca^{2+}]$_i$ oscillatory activity, whereas the IP$_3$R blocker 2-aminoethoxydiphenylborate (2-APB) diminishes its amplitude (Johnston *et al.*, 2005; McHale *et al.*, 2006). The elevated [Ca^{2+}]$_i$ increases opening frequencies in local *Ano1*-encoded Ca^{2+}-activated Cl$^-$ channels (Figure 15.4B; Zhu *et al.*, 2009). The latter opening events produce stochastically occurring spontaneous transient inward currents (STICs), leading to spontaneous transient depolarization events (STD) or unitary potentials (Figure 15.5A, B; Hirst and Edwards, 2001).

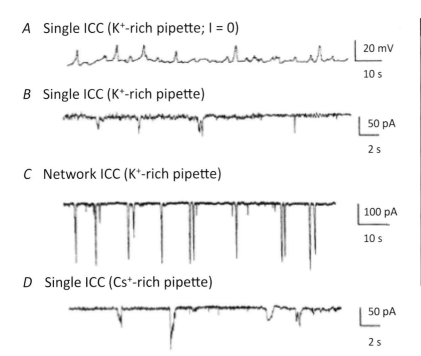

A Single ICC (K⁺-rich pipette; I = 0)

20 mV

10 s

B Single ICC (K⁺-rich pipette)

50 pA

2 s

C Network ICC (K⁺-rich pipette)

100 pA

10 s

D Single ICC (Cs⁺-rich pipette)

50 pA

2 s

Figure 15.5 Spontaneous inward currents from cultured single and networked interstitial cells of Cajal (ICCs) in murine colon. (A, B) Single ICC studies in the presence of nicardipine. Intracellular (pipette) [K⁺]: 140 mM. (A) Spontaneous membrane potential oscillations under current clamp in the absence of injected current. (B) Spontaneous inward currents observed under voltage clamp at a –80 mV holding potential. (C, D) Spontaneous inward currents from ICCs in networks in the presence of 140 mM pipette [K⁺] (C) or [Cs⁺] (D). (From Figure 2 of Kim et al., 2002.)

A summation of STICs results in large (~80 pA/pF) Ca^{2+}-activated Cl^- currents ($I_{Cl(Ca)}$) that produce slow waves depolarising the cell towards ~–10 mV (Zheng et al., 2014). This reflects the large driving force on Cl^- efflux arising from a high $[Cl^-]_i$. The latter results from extensive *Slc12a2*-encoded Na^+-K^+-$2Cl^-$ cotransporter type 1, NKCC1, expression and activity (Figure 15.4A). Gramicidin-permeabilised patch clamped ICCs demonstrated NKCC activity that resulted in reversal potentials of the STICs (~–10.5 mV) close to the peaks of slow waves recorded directly from in situ ICCs. These reversal potentials were shifted to more positive potentials by reducing extracellular $[Cl^-]$ ($[Cl^-]_o$) as expected from the Cl^- Nernst equation, and to more negative potentials by challenge by the NKCC inhibitor bumetanide (Zhu et al., 2016). The latter agent similarly modified E_{Cl} to a lesser degree in skeletal muscle (Ferenczi et al., 2004). In addition, the Ca^{2+}-activated Cl^- channels may be localised within high-$[Cl^-]_i$ microdomains intervening between closely apposed plasma membrane and endoplasmic reticulum (ER).

Summation of STDs drives the membrane voltage to a threshold that results in regenerative activation of *Cacna1h-* and *Cacna1g*-encoded T-type Cav3.2 channels, and Ca^{2+} entry (Figure 15.4B). Patch clamp recordings from freshly isolated murine small intestinal ICCs exhibited both low-voltage-activated T-type and high-voltage-activated L-type Ca^{2+} currents (I_{CaT} and I_{CaL}, respectively). However, slow-wave generation persisted with challenge by dihydropyridine I_{CaL} inhibitors, but was reduced in *Cacna1h–/–* mice, and by the I_{CaT} inhibitors nicardipine, Ni^{2+}, and mibefradil.

The Ca^{2+} entry reinforces the *Ano1* activation and its induced cell depolarisation. This voltage-dependent interaction thereby synchronises and entrains STICs (Figure 15.5C,D) into whole-cell *slow-wave* currents which can similarly be activated or paced by imposed step depolarisations. The resulting slow waves are abolished by genetic deactivation of *Ano1*.

15.5 | Propagation of Interstitial Cell of Cajal Pacing Events

ICCs form electrically coupled networks mediating active regenerative slow wave propagation. These involve voltage-dependent mechanisms in common with those mediating pacing (Figure 15.4C). They involve a gap junction-mediated spread of voltage changes to hitherto quiescent neighbouring cells. These trigger Ca^{2+} entry through voltage-activated I_{CaT} (Figure 15.4D). The connexin blocker β-glycyrrhetinic acid blocks coherent cell-to-cell conduction whilst sparing pacemaker activity within individual ICCs, whereas 2-APB blocks both slow waves and their propagation (Park *et al.*, 2005). The resulting increase in $[Ca^{2+}]_i$ activates IP_3R-Ca^{2+} channel mediated Ca^{2+} release. The further $[Ca^{2+}]_i$ elevation activates $I_{Cl(Ca)}$, continuing the propagation process (Zhu *et al.*, 2009). Alternatively, the depolarisation could itself activate phospholipase C, generating IP_3, causing release of intracellularly stored Ca^{2+} in turn activating $I_{Cl(Ca)}$ (Figure 15.4D).

The spontaneously active inward pacemaker currents generate rhythmically occurring and propagating slow potentials, often with plateau components that can persist, even in the absence of neural inputs. Both murine ICCs in short-term cultures and submucosal ICCs isolated from canine colon showed slow potentials closely resembling the slow waves recorded from intact muscle strips. The slow potentials are propagated through the network of coupled ICC-MY and ICC-DMP cells.

15.6 | Electrical Coupling Between Interstitial Cells of Cajal and Smooth Muscle Cells

The slow potentials are conducted to smooth muscle cells electrically coupled with the ICC-MY network. Slow waves are not actively propagated amongst smooth muscle cells themselves. Limited structural evidence exists for gap junction coupling between pacemaker ICCs and smooth muscle cells. Nevertheless, dual current injection and recording microelectrode experiments in both gastric antral and small intestinal sections containing ICC-MY and circular and longitudinal muscle demonstrated electrical coupling between all three cell types (Figure 15.6). Coupling between the ICC-MY and smooth muscle cells (Figure 15.6B) was weaker than between the smooth muscle cells

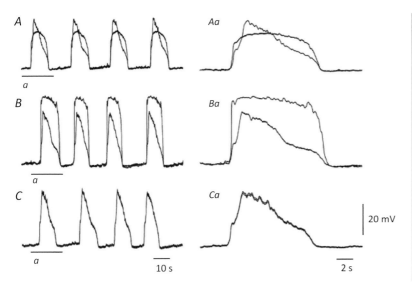

Figure 15.6 Paired intracellular recordings from different guinea pig gastric antral cells. Superimposed records obtained simultaneously from (A) circular and longitudinal muscle cell (of smaller amplitude). (Aa) portion of the same record labelled a on a magnified expanded time base. (B) Circular muscle cell and ICC-NY cell (of larger amplitude). (Ba) portion of the same record labelled a shown on expanded time base. (C) Two independently impaled circular muscle cells. (Ca) portion of the same record labelled a shown on expanded time base. (From Figure 2 of (Cousins et al., 2003.)

(Figure 15.6A, C). Cable analysis nevertheless demonstrated that this coupling would permit passage of sufficient pacemaker current to conduct slow waves into the smooth muscle syncytium. Such an arrangement may reduce the tendency of the smooth muscle syncytium to act as a low-impedance sink draining off pacemaker current from the relatively small and sparse ICCs, and thereby ensure that their threshold for entrainment is attained. It is therefore ICCs that drive active propagation of slow waves; smooth muscle cells lack the ionic mechanisms to do so (Sanders et al., 2006).

15.7 | Membrane Excitation in Smooth Muscle Cells

Smooth muscle cell depolarisation caused by slow waves sums with electrophysiological changes from other stimuli such as distention or neurotransmitters interacting with the numerous smooth muscle membrane receptors. Sufficient depolarisation takes the membrane potentials of the coupled smooth muscle cells to a range in which there is appreciable and steeply voltage-dependent L-type Ca^{2+} channel open probability. This results in a correspondingly periodic I_{CaL}-mediated phasic pattern of depolarisation. Its temporal patterns can reflect the rhythmic depolarisation deriving from a single ICC network, or more complex patterns of electrical activity resulting from interactions between the pacemaker activity in different ICC networks. They vary between gastrointestinal tract regions and between species. The resulting Ca^{2+} entry in turn phasically triggers excitation–contraction coupling. The resulting contraction amplitude is determined by the extent to which the absolute membrane potential exceeds the I_{CaL} threshold.

In some smooth muscle cells, slow-wave depolarisation also elicits regenerative Ca^{2+} action potentials. These cause more intense episodes of Ca^{2+} entry driving strong contractile responses, whose thresholds

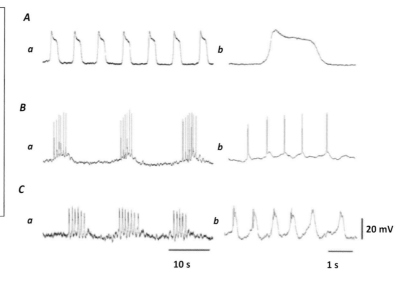

Figure 15.7 Comparison of electrical activity recorded at (*a*) slow- and (*b*) fast-sweep speed from (A) submucosal interstitial cells of Cajal recorded with mucosal layer uppermost and intact submucosal layer attached, (B) circular muscle; tissue with circular smooth muscle layer uppermost and no submucosal layer, (C) longitudinal smooth muscle cells; tissue with longitudinal muscle uppermost, without the submucosal layer. (From Figure 1 of Yoneda *et al.*, 2002.)

vary between smooth muscle types (Ozaki *et al.*, 1991b). These action potentials have slower upstrokes and longer duration than nerve, skeletal or cardiac muscle Na^+ action potentials. Their upstrokes likely reflect regenerative I_{CaL} rather than I_{Na} activation. They are inhibited by L-type Ca^{2+} channel blocking agents, such as nifedipine, but are insensitive to the Na^+ channel blocker, tetrodotoxin. Their detailed waveforms vary between muscle types. They can take the form of simple spikes, or spike trains superimposed upon a prolonged depolarisation. Genitourinary, including ureteric, bladder or uterine smooth muscle additionally show prolonged plateau phases. These events are similarly followed by a relatively gradual recovery, which is suggested to reflect slow Ca^{2+} channel inactivation or delayed activation of Ca^{2+}-activated or voltage-dependent K^+ channels.

The latter activity is related to the slow potentials generated by ICCs (Figure 15.7A*a,b*). In circular (Figure 15.7B*a,b*) and longitudinal intestinal smooth muscle cells (Figure 15.7C*a,b*) this respectively leads to periodic bursts of spike potentials superimposed upon plateau voltage changes (~4–5/min) and bursts of oscillatory potentials. Both were sensitive to the L-type Ca^{2+} channel blocker, nifedipine. The plateau potentials were additionally sensitive to the Ca^{2+}-ATPase inhibitor, cyclopiazonic acid, the IP_3R inhibitor, 2-APB, and the mitochondrial inhibitor, carbonyl cyanide *m*-chlorophenylhydrazone.

15.8 | Excitation–Contraction Coupling in Smooth Muscle Cells

As in other muscle types, increased $[Ca^{2+}]_i$ triggers smooth muscle excitation–contraction coupling (Figure 15.8). However, in smooth muscle this involves multiple overlapping, voltage-dependent and

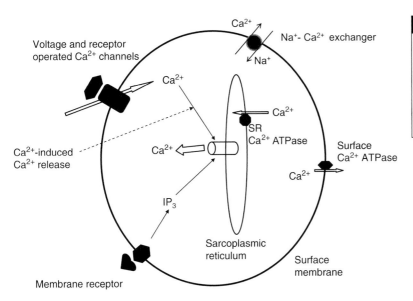

Voltage and receptor operated Ca^{2+} channels

Ca^{2+}

Na$^+$- Ca^{2+} exchanger

Na$^+$

Ca^{2+}

Ca^{2+}

SR Ca^{2+} ATPase

Surface Ca^{2+} ATPase

Ca^{2+}-induced Ca^{2+} release

Ca^{2+}

Ca^{2+}

IP$_3$

Membrane receptor

Sarcoplasmic reticulum

Surface membrane

Figure 15.8 Mechanisms for Ca^{2+} release through Ca^{2+}-induced Ca^{2+} release and inositol 1,3,5-*tris*phosphate pathways, and its subsequent recovery by sarcoplasmic reticular and surface membrane Ca^{2+} transport and Na$^+$-Ca^{2+} exchange.

voltage-independent coupling mechanisms. Of these, electromechanical coupling resembles the corresponding skeletal and cardiac muscle processes in which increased [Ca^{2+}]$_i$ results from activation of voltage-dependent Ca^{2+} channels by membrane depolarisation. Phasic smooth muscle shows a pattern of oscillatory membrane potential changes upon which are superimposed calcium 'spikes' driven by Ca^{2+} channel opening in turn leading to extracellular Ca^{2+} entry. The larger surface–volume ratio of smooth compared to either skeletal or cardiac muscle cells make such Ca^{2+} entry through voltage-gated Ca^{2+} channels a significant means for altering cytosolic [Ca^{2+}].

Nevertheless, SR Ca^{2+} release remains a major source of [Ca^{2+}]$_i$ elevation, as in skeletal and cardiac muscle. Smooth muscle similarly shows electron-dense couplings bridging sarcolemmal and SR membranes, consistent with a Ca^{2+}-induced SR Ca^{2+} release mechanism initiated by Ca^{2+} influx (Section 11.5). However, this mechanism appears effective only at relatively high [Ca^{2+}]$_i$. An alternative route triggering SR Ca^{2+} release involves IP$_3$ production initiated by membrane-associated phospholipase C (PLC) action on membrane phosphoinositide lipid (PIP$_2$) (Section 8.7). This follows activation of a G-protein-coupled seven-transmembrane domain receptor. IP$_3$ binding triggers IP$_3$R-mediated SR Ca^{2+} release into the cytoplasm (cf. Section 13.18). The resulting SR Ca^{2+} store depletion can further lead to activation of store-operated surface membrane Ca^{2+} channels that mediate trans-sarcolemmal influx of extracellular Ca^{2+}, potentially maintaining the [Ca^{2+}]$_i$ elevations (Section 15.10). Following activity, intracellular [Ca^{2+}] is reduced to its low resting level by ATP-utilising Ca^{2+} pumps in the plasma membrane and in the SR, the latter replenishing the SR Ca^{2+} store (cf. Section13.16).

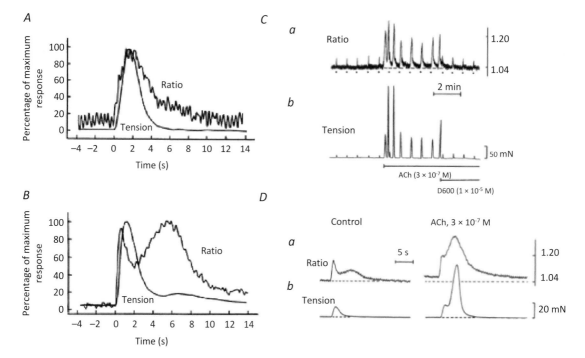

Figure 15.9 Ca^{2+} transients and contractile activity in canine antral circular muscle. (A, B) patterns shown by spontaneous Ca^{2+} transients, measured by fura-2 ratiometric analysis, and contractile force, both normalised to their respective maxima. (C) Effect of 3×10^{-7} M acetylcholine (ACh) followed by 10^{-5} M D600 on (a) $[Ca^{2+}]_i$ and (b) muscle tension during regular stimulation. (D) Spontaneous (a) Ca^{2+} transients and (b) contractions (left) before and (right) following ACh challenge. (From Figures 2 and 3 of Ozaki et al., 1991a.)

Records of spontaneous $[Ca^{2+}]_i$ transients and contractile activity in antral circular muscle demonstrated that the increases in $[Ca^{2+}]_i$ expectedly preceded force generation. However, the resulting tension traces fell more rapidly than did the $[Ca^{2+}]_i$ transient (cf. Section 13.17). Single-phase $[Ca^{2+}]_i$ transients elicited correspondingly single-phase contractions (Figure 15.9A). Where the Ca^{2+} transient showed a second later phase of increased amplitude, this was frequently similar to that of the first. However, the amplitude of the corresponding second tension transient was far less than the first, a phenomenon discussed in Section 15.12 (Figure 15.9B).

15.9 | Pharmacomechanical Coupling in Smooth Muscle Cells

Further, pharmacomechanical coupling activation mechanisms, independent of membrane potential may nevertheless influence membrane potential once activated (Figure 15.10; Guibert et al., 2008). These also elicit both Ca^{2+} entry and release of intracellularly stored Ca^{2+}. One group of such processes involve a group of receptor-operated mechanisms (Figure 15.10A) acting on particular G-protein-coupled receptors, frequently seven transmembrane domain receptors (Section 8.7; Figure 15.10B) whose activation mediates strategic cellular responses (Figure 15.10C–G). Their agonists include the classical

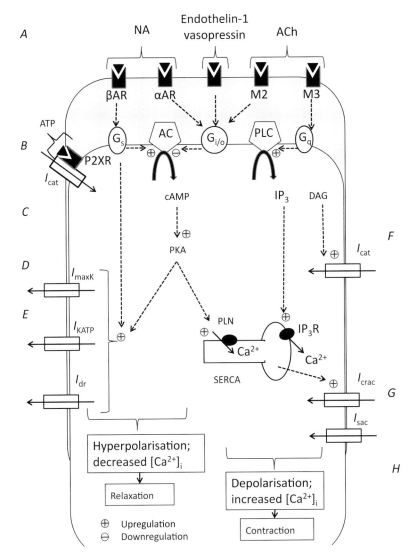

Figure 15.10 Voltage-independent mechanisms influencing contractile activity in smooth muscle. (*A*) Transmitters and receptors: autonomic transmitters: NA: noradrenaline; ACh: acetylcholine. Local transmitters: exemplified by endothelin I and vasopressin. Receptors: αAR: α-adrenergic, βAR: β-adrenergic receptors; M2 and M3: muscarinic receptors; P2XR: P2X purinergic receptors. (*B*) Receptor mechanisms: G_s, $G_{i/o}$ and G_q: G-proteins; AC: adenylyl cyclase; PLC: phospholipase C. (*C*) Messengers: cAMP, cyclic 3′,5′-adenosine monophosphate; PKA: protein kinase A; IP_3: inositol *tris*phosphate; DAG: diacylglycerol. (*D*) Intracellular targets: PLN: phospholamban, SERCA: sarcoplasmic reticular Ca^{2+}-ATPase; IP_3R: IP_3 receptor; (*E–G*) Surface membrane targets: (*E*) K$^+$ channels: I_{maxK}: Ca^{2+}-activated, I_{KATP}: ATP-activated, I_{dr}, delayed rectifying K$^+$ currents; (*F*) Cationic channels: I_{cat}: cation current; (*G*) Store-operated and mechanosensitive ion channels: I_{crac}: capacititative Ca^{2+} current; I_{sac}: stretch-activated current. (*H*) Physiological outcomes: cell hyperpolarisation and relaxation, and cell depolarisation and contraction.

neurotransmitters adrenaline (Adr; epinephrine), noradrenaline (NA; norepinephrine) and acetylcholine (ACh), as well as local messengers and hormones including ATP, histamine, endothelin-1, neurokinins, substance P and vasopressin (Figure 15.10*A*).

Of these receptors, the three known β-adrenoceptor subtypes each have particular relative specificities for isoproterenol, adrenaline and noradrenaline binding. Their occurrences vary with smooth muscle subtypes: β2-adrenoceptors are abundant in trachea, bronchi and microvasculature, β3-adrenoceptors are abundant in the gastrointestinal tract. The β-adrenoceptors are coupled to a choleratoxin-sensitive G_s-receptor whose activation increases adenylyl cyclase (AC)-mediated $[cAMP]_i$ elevation, in turn activating protein kinase A (PKA) (Figure 15.10C). PKA phosphorylates a range of membrane and cytosolic proteins, including phospholamban (PLN) associated with muscle relaxation (Figure 15.10D; Tanaka *et al.*, 2005). G_s-receptor activation also mediates direct, cAMP-independent activation of large-conductance, Ca^{2+}-activated (maxiK, BK), ATP-sensitive (K_{ATP}) or delayed rectifier (K_{dr}) K^+ channels. These would be expected to hyper-polarise the surface membrane and cause smooth muscle relaxation (Figure 15.10E).

Smooth muscle also co-expresses α-adrenoceptor and M_2 acetyl-choline (ACh) receptor (AChR) subtypes that activate $G_{i/o}$, and M_3 AChR subtypes that activate G_q proteins. The pertussis toxin-sensitive G_i/G_o occurs particularly in visceral and G_q in vascular smooth muscle These respectively inhibit AC activity and activate phospholipase C (PLC) (Eglen *et al.*, 1994). Vasopressin receptors and endothelin-1 receptors are also linked to $G_{i/o}$-protein.

PLC activation increases IP_3 and diacylglycerol (DAG) production. IP_3 induces SR Ca^{2+} release. Elevated DAG activates receptor-operated cation channels (ROCCs) resulting in a non-selective cation current (I_{cat}) carrying both mono- and divalent cations, with varying degrees of Ca^{2+} selectivity (Figure 15.10F). ROCCs showed similar unit channel conductances consistent with the different receptor types ultimately acting on a similar or common channel species. Recent evidence attributes I_{cat} to a family of transient receptor potential (TRP) surface channels. These show some structural simi-larities to membrane-spanning domains of voltage-dependent ion channels (Section 5.2). They are homo- or heterotetramers whose subunits each contain contains six transmembrane domains, S1–S6. All show a hydrophobic loop contributing to a channel pore between S5 and S6 and cytoplasmic N- and C-termini, but they lack voltage-sensing S4 regions (Section 5.5). Together these processes have a net effect of causing membrane depolarisa-tion, increasing $[Ca^{2+}]_i$ and enhancing contractile activation (Figure 15.10H).

Noradrenaline exerts its differing actions through distinct α- and β-adrenergic receptor subtypes. These exert excitatory effects at sites such as the vas deferens, and inhibitory effects at others, such as intestinal muscle and the iris of the eye. ACh, acting on muscarinic receptors, is an excitatory transmitter in much intestinal muscle and in the iris.

Figure 15.9C, D illustrates the interactions between activation by voltage-dependent and voltage-independent mechanisms in smooth muscle. In regularly stimulated smooth muscle, ACh challenge increased resting $[Ca^{2+}]_i$ but not resting tension. It also increased the amplitudes of Ca^{2+} transients (Figure 15.9Ca) and contractions, and resulted in the appearance of some spontaneous transients between applied stimuli (Figure 15.9Cb). Spontaneous Ca^{2+} transients and contractions that demonstrated two-phase Ca^{2+} transients and mono-phasic mechanical responses showed increased amplitudes of the second phase of the Ca^{2+} transient (Figure 15.9Da) and the appearance of a second contraction phase with ACh challenge (Figure 15.9Db). The Ca^{2+} channel blocker, D600, decreased the steady-state [Ca^{2+}]$_i$ to its to the original level and decreased the amplitudes of both Ca^{2+} transients and phasic contractions (Figure 15.9Ca,b).

Finally, ATP binding to the P2X1 receptor exceptionally results in a direct, ligand-gated ROCC opening process within the same macro-molecular complex (Figure 15.10B).

15.10 Store-Operated and Mechanosensitive Ion Channels in Smooth Muscle Cells

A second group of voltage-independent processes involve store-operated Ca^{2+} channels (SOCCs), similarly reflecting relatively non-selective cation conductances, activated by depletion of SR Ca^{2+}. Such activation might take place through a chemical signal, reflecting store depletion, fusion between SR and SOCC channels or fusion between vesicle membranes normally containing the SOCC and sur-face membrane. The resulting Ca^{2+} influx carried by the resulting I_{crac} current tends to maintain the raised [Ca^{2+}]$_i$ despite prolonged agonist application and aids in SR store refilling on agonist withdrawal (Figure 15.10G; Putney et $al.$, 2001).

A final group of voltage-independent processes involve mechano-sensitive stretch-activated channels (SACs) sensitive to membrane stretch resulting from the conditions of mechanical stimulation to which smooth muscle cells are exposed. Their open probabilities varied with the pressure applied through a recording pipette. Particu-lar, Ca^{2+} permeable, SAC types might act to increase local [Ca^{2+}]$_i$ that in turn may activate Ca^{2+}-induced SR Ca^{2+} release or exert other effects, such as opening Ca^{2+}-dependent K$^+$ channels (BKCa). Their cation influx may also produce a sustained membrane depolarisation, activating other voltage-dependent processes (Figure 15.10G,H).

15.11 Ca^{2+}-Mediated Contractile Activation

Smooth muscles contain the major contractile proteins actin and myosin, together with tropomyosin. There is a lower relative propor-tion of myosin than in skeletal muscle. However, in contrast to skel-etal muscle, the myosin molecules within the thick filaments are oriented in opposite directions on the two faces of a filament. This permits a thin filament to be pulled over its entire length by a thick filament, so that the muscle can operate at near maximum tension over a wide range of lengths (Figure 15.11).

Figure 15.11 A contractile unit of smooth muscle, showing how the filaments could slide past each other during contraction. (From Squire, 1986.)

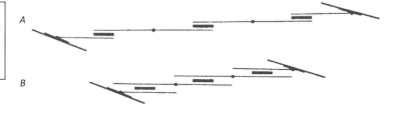

Figure 15.12 Control of contractile activation at the level of actin filaments by caldesmon.

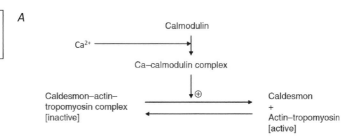

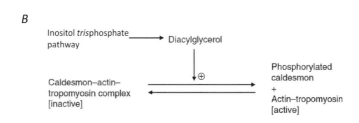

Contractile activation in smooth muscle is more complex and more subject to modulation than in either skeletal or cardiac muscle. Cross-bridge activity is controlled by a range of different $[Ca^{2+}]_i$-dependent mechanisms. These involve an initial Ca^{2+} combination with calmodulin rather than troponin (Section 9.10), which is absent in smooth muscle. There follows a range of biochemical cascades entailing considerably slower kinetics of mechanical activation.

At the level of the actin filaments, the Ca^{2+}–calmodulin complex binds to the protein caldesmon, dissociating it from the actin–tropomyosin thin filaments. This permits their interaction with myosin, leading to cross-bridge cycling (Figure 15.12A). Caldesmon dissociation from actin–tropomyosin can also directly result from its phosphorylation by protein kinase C (PKC). The latter is activated by diacylglycerol (DAG), a product of phospholipase C activation (Figure 15.12B).

At the level of the myosin filaments, the Ca^{2+}–calmodulin complex activates myosin light chain kinase (MLCK). This catalyses phosphorylation of the myosin light chain (MLC). This phosphorylation involves formation of a covalent bond and constitutes a form of

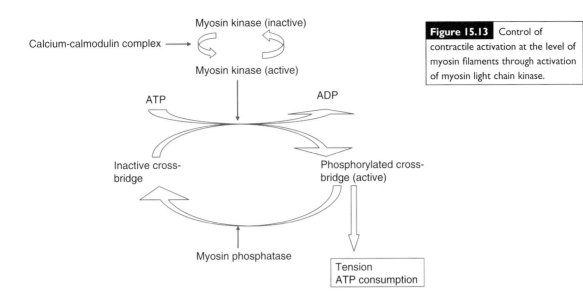

Figure 15.13 Control of contractile activation at the level of myosin filaments through activation of myosin light chain kinase.

covalent regulation. The degree of MLC phosphorylation modifies the relationship between $[Ca^{2+}]_i$ and muscle tension generation. This phosphorylation permits cross-bridge cycling and associated splitting of ATP (Figure 15.13). Conversely, MLC dephosphorylation by myosin phosphatase terminates the cross-bridge cycling. Direct Ca^{2+} binding to MLC also increases cross-bridge cycling. There are also mechanisms that modulate contraction strength by varying the Ca^{2+} sensitivity of the proteins regulating contraction.

15.12 | Effects of Myosin Light Chain Phosphorylation Levels

A second level of control of contractile activity involves modulation of MLC phosphorylation. The balance between MLC phosphorylation and dephosphorylation strongly influences contractile force for any given $[Ca^{2+}]_i$.

Increased MLCK-mediated MLC phosphorylation or diminished phosphatase-mediated MLC dephosphorylation increases force generation for any given $[Ca^{2+}]_i$. This can extend to the point where muscle contraction can actually occur in the absence of altered $[Ca^{2+}]_i$. One means of altering this balance could involve inhibiting or activating MLCK. Conversely, the phosphatase inhibitor, calyculin-A (10^{-6} M) induced tonic contraction, even in the absence of increased $[Ca^{2+}]_i$. Electrical activity produced superimposed $[Ca^{2+}]_i$ and phasic contractions (Ozaki *et al.*, 1991a). This phosphatase action was inhibited by protein kinase C.

In contrast, MLC dephosphorylation reduces force generation at any given $[Ca^{2+}]_i$. This may underly the contractile relaxation

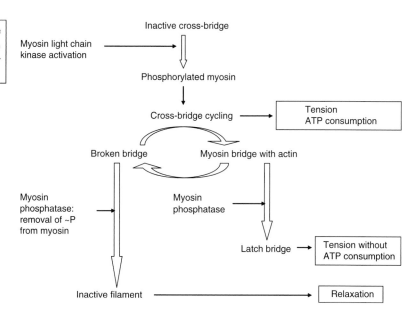

Figure 15.14 Termination of the process of cross-bridge cycling with the formation of either detached or latch bridges.

preceding restoration of resting $[Ca^{2+}]_i$ and the lack of contractile responses to secondary increases in $[Ca^{2+}]_i$ under some conditions (Figure 15.9B). ACh promotes MLC phosphorylation and may thereby increase Ca^{2+} transient amplitudes and contractile force (Figure 15.9C). Some neurotransmitters produce a G-protein linked inhibition of the phosphatase.

Finally, it is possible to modify the Ca^{2+} sensitivity of MLCK itself. MLCK is phosphorylated by several protein kinases. PKA, PKC and Ca^{2+}–CaM-dependent kinases decrease MLCK sensitivity to activation by the Ca^{2+}–CaM complex.

Myosin dephosphorylation at a stage when the myosin and the thin filament are detached permits muscle relaxation and an end of contractile activity (Figure 15.14). If dephosphorylation occurs when the myosin is attached to the thin filament, it remains bound with high affinity. Cross-bridges in this state are termed latch bridges. They allow maintained tension without a requirement for cross-bridge cycling or ATP consumption. This mechanism explains the greater (~300-fold) energy efficiency of smooth compared to skeletal muscle during maintained contractions. The cellular levels of the kinases and phosphatase involved in this regulation can be modulated to make long-term adjustments to the contractile properties of smooth muscle cells.

15.13 | Mechanical Properties of Smooth Muscle

Despite its lack of obvious sarcomere structure, isometric contraction in smooth muscle shows a force–length relationship resembling that of skeletal muscle (Section 12.3), reflecting similar, fundamental, sliding filament mechanisms (Section 9.3). However, maximal isotonic contraction velocity is much lower than in skeletal muscle, whilst retaining its inverse, Hill, relationship with load. In addition, shortening velocity in smooth muscle can be increased by increasing levels of cross-bridge phosphorylation (Section 15.11). However, a near-maximal isometric force can be generated, even at low phosphorylation levels. In a maintained contraction, Ca^{2+} and the rate of cross-bridge phosphorylation first rise to their peak levels, to produce rapid shortening, and then subside to lower levels, while tension is maintained.

Many smooth muscles, such as those in the bladder, uterus and gut, contract phasically in response to stretch. This is the result of mechanically induced depolarisation, brought about by stretch-activated surface membrane ion channels (Section 15.10). This plays a role in peristaltic movements in the intestine. Other smooth muscles show a tonic stretch-induced contraction that allows compensatory adjustment of tension. A constant muscle fibre length is consequently maintained, despite variations in load, as occurs in the response of arteriolar smooth muscle to raised blood pressure.

15.14 | Propagation of Excitation and Contraction in Smooth Muscle

Excitable and contractile activity initiated in a particular region of small intestine, often at its pyloric end, spreads to remaining regions along the length of the organ through slow-wave propagation (Section 15.5) in experimental feline intestinal preparations. Whereas slow waves propagate to the limit of the organ or tissue border studied, spikes resulting from slow-wave initiation of I_{CaL} activity (Section 15.7) propagate only for limited, 5–25 mm, distances before terminating spontaneously. The slow-wave activity propagates rapidly around a localised ring of bowel, and when the resulting membrane depolarisation reaches its threshold potential, the spike potential causes circumferential contraction. Slow-wave propagation drives sequential, coordinated peristaltic contractions along the longitudinal axis, leading to regions of high and low lumina pressures that move and mix the intestinal contents.

Multi-electrode array recordings demonstrated such slow-wave conduction in cut-open isolated small-intestinal segments (Plate 24Aa). 121-electrode arrays were placed on their serosal surfaces (Plate 24Ab), giving electrogram recordings exemplified here for successive lengthwise columns of electrodes (Plate 24Ac). At each recording site, the timing of the downstrokes of the electrical deflections gives the slow-wave latencies. Results from all the recording sites could be reconstituted into isochronal maps representing wave latencies over the recording surface (Plate 24Ad). Electrograms and propagation maps from duodenal, jejunum and ileal segments (Plate 24*B*(i)–(iii), respectively) all demonstrated coherently propagating slow waves. Their propagation velocity was relatively slow, around 1–10 cm/s (cf. canine atria, ~100 cm/s). Together with their long refractory periods, slow waves gave propagation wavelengths of 1–2 cm, falling within distances compatible with typical dimensions of the intestinal wall.

Deficiently expressed, poorly developed or distributed, or abnormally associated ICCs have been demonstrated in mice deficient in neuronal nitric oxide synthase (nNOS). The deficiency compromises nitric oxide (NO) production known to promote ICC formation. Abnormal ICC function leads to disrupted intestinal motility, leading to dysrhythmias, gastroparesis and slow intestinal transit, and even chronic intestinal pseudo-obstruction. Some such abnormalities have been associated with diabetes mellitus, possibly through deficiencies in a stem cell factor (SCF) thought to act as an ICC trophic factor (Ördög, 2008). Local ischaemia also caused areas of acute conduction blocks and the appearance of ectopic pacemakers.

Slow-wave excitation is susceptible to experimental disruption. In common with findings bearing on cardiac arrhythmias (Chapter 14; Plate 16), it also possible to induce episodes of pro-arrhythmic circus movements of excitation following ethanol-induced reductions in slow-wave propagation velocities in rat small intestine (cf. Section 14.2). These preparations can show a variety of propagation disturbances, varying from very small circuits to intertwined re-entrant loops. Plate 24*Cb*(i)–(iii) shows three successive episodes of slow wave propagation within a small tissue area. Note the clockwise circus pattern re-entering the region of their originating excitation sites (Panels (i) and (ii)), which shifted their location (Panel (i)) consistent with their arising from focal excitation. Furthermore, recorded electrograms (Plate 24*Ca*) demonstrate that the second and third impulses were initiated while the previous wave was still propagating, giving rise to two simultaneously propagating wavelets.

Further Reading

I. Historical background studies on cell excitability

Prosser CL, Curtis BA and Esmail M (2010). *A History of Nerve, Muscle and Synapse Physiology*. USA: Stipes Pub Llc. 572 pp.

Rapport R (2005). *Nerve Endings, the Discovery of the Synapse: The Quest to Find How Brain Cells Communicate*. New York: W. W. Norton & Company. 224 pp.

Hodgkin AL (1977). *The Pursuit of Nature: Informal Essays on the History of Physiology*. Cambridge: Cambridge University Press. 190 pp.

Cole KS (1968). *Membranes, Ions and Impulses: A Chapter of Classical Biophysics*. Berkeley, USA: University of California Press. 580 pp.

Noble D (1979). *The Initiation of the Heartbeat 2/e*. Oxford: Oxford University Press. 200 pp.

2. Biophysical and mathematical background

Glaser R (2012). *Biophysics: An Introduction 2/e*. Heidelberg: Springer. 428 pp.

Aitken M, Broadhurst W and Hladky S (2009). *Mathematics for Biological Scientists*. Abingdon, UK: Garland Science. 482 pp.

Keener J and Sneyd J (2008). *Mathematical Physiology 2/e*. Heidelberg and Berlin: Springer. Vol. I: Cellular Physiology. 576. Vol. II. Systems Physiology. 608 pp.

3. Fundamental properties of membranes

Dillon PF (2012). *Biophysics: A Physiological Approach*. Cambridge: Cambridge University Press. 314 pp.

Jackson MB (2006). *Molecular and Cellular Biophysics*. Cambridge: Cambridge University Press. 528 pp.

4. Nerve and ion channel function

Heimburg T (2020). *The Biophysics of Nerve Cells*. London: Wiley VCH. 350 pp.

Hille B (2001). *Ionic Channels of Excitable Membranes 3/e*. Sunderland, USA: Sinauer Associates Inc. 814 pp.

Jack JJB, Noble D and Tsien RW (1983). *Electric Current Flow in Excitable Cells*. Oxford: Oxford University Press. 534 pp.

Scolding N and Wilkins A. (2012). *Multiple Sclerosis (Oxford Neurology Library)*. Oxford: Oxford University Press. 90 pp.

5. | Synaptic and neuromuscular physiology

Pickel VM and Segal M (Eds.). (2014). *The Synapse: Structure and Function*. Oxford: Academic Press/Elsevier. 512 pp.

Hell JW and Ehlers MD (Eds.). (2008). *Structural and Functional Organization of the Synapse*. NY: Springer. 820 pp.

Cowan WM, Südhof TC and Stevens CF (2000). *Synapses*. Baltimore: The Johns Hopkins University Press. 792 pp.

Amato A and Russell JA (2016). *Neuromuscular Disorders 2/e*. NY: McGraw-Hill Education/Medical. 960 pp.

6. | Skeletal muscle

MacIntosh B, Gardiner PF and McComas AJ (2005). *Skeletal Muscle: Form and Function 2/e*. Champaign, Ill, USA: Human Kinetics. 432 pp.

Sugi H (2005). *Sliding Filament Mechanism in Muscle Contraction: Fifty Years of Research (Advances in Experimental Medicine and Biology)*. Heidelberg and Berlin: Springer. 448 pp.

Huang CL-H (1993). *Intramembrane Charge Movements in Striated Muscle (Monographs of the Physiological Society)*. Oxford: Oxford University Press. 302 pp. [Available at: https://www.oxfordscholarship.com/view/10.1093/acprof:oso/9780198577492.001.0001/acprof-9780198577492].

Ruegg JC (1992). *Calcium in Muscle Contraction: Cellular and Molecular Physiology 2/e*. Berlin and Heidelberg: Springer. 376 pp.

Campbell AK (2018). *Fundamentals of Intracellular Calcium*. London: Wiley-Blackwell. 464 pp.

Kaminski HJ and Kusner LL (Eds.). (2018). *Myasthenia Gravis and Related Disorders 3/e*. Cham, Switzerland: Humana/Springer. 353 pp.

Takahashi MP and Matsumura T (Eds.). (2018). *Myotonic Dystrophy: Disease Mechanism, Current Management and Therapeutic Development*. Singapore: Springer-Nature. 214 pp.

Ellis FR and Hopkins PM (Eds.). (1996). *Hyperthermic and Hypermetabolic Disorders: Exertional Heat-Stroke, Malignant Hyperthermia and Related Syndromes*. Cambridge: Cambridge University Press. 302 pp.

7. | Cardiac muscle

Macleod KT (2013). *An Essential Introduction to Cardiac Electrophysiology*. London: Imperial College Press. 286 pp.

Katz AM (2010). *Physiology of the Heart 5/e*. Philadelphia: Wolters Kluwer/Lippincott Williams and Wilkins. 576 pp.

Bers DM (2001). *Excitation-Contraction Coupling and Cardiac Contractile Force 2/e*. Heidelberg: Springer. 456 pp.

Zipes DP, Jalife J and Stephenson WG (2017). *Cardiac Electrophysiology: From Cell to Bedside 7/e*. Philadelphia: Elsevier. 1424 pp.

Callans D (2020). *Josephson's Clinical Cardiac Electrophysiology: Techniques and Interpretations 6/e*. Philadelphia: Wolters Kluwer. 850 pp.

8. | Smooth muscle

Trebak M and Earley S (Eds.). (2018). *Signal Transduction and Smooth Muscle*. Boca Raton, USA: CRC Press/Taylor & Francis. 432 pp.

Hai C–M (Ed.). (2016). *Vascular Smooth Muscle: Structure and Function in Health and Disease*. Singapore: World Scientific Publishing Company. 308 pp.

Barany M (1996). *Biochemistry of Smooth Muscle Contraction*. San Diego, USA: Academic Press Inc/Elsevier Science Publishing Co Inc. 418 pp.

Kao CY and Carsten ME (Eds.). (1997). *Cellular Aspects of Smooth Muscle Function*. Cambridge: Cambridge University Press. 312 pp.

9. | Historic Nobel Prize Lectures

The chapters referring to material related to each lecture are indicated with each award.

Physiology or Medicine; https://www.nobelprize.org/prizes/medicine:

1906: C Golgi and S. Ramon y Cajal: "work on the structure of the nervous system" (Chapter 1).

1922: AV Hill and OF Meyerhof; "the production of heat in the muscle; the fixed relationship between the consumption of oxygen and the metabolism of lactic acid in the muscle" (Chapter 12).

1924: W. Einthoven; "the mechanism of the electrocardiogram" (Chapter 13).

1932: C Sherrington and ED Adrian; "the functions of neurons" (Chapter 2).

1936: HH Dale and O Loewi; "chemical transmission of nerve impulses" (Chapter 7, 8).

1944: J Erlanger and HS Gasser; "the highly differentiated functions of single nerve fibres" (Chapter 6).

1963: JC Eccles, AL Hodgkin and AF Huxley; "the ionic mechanisms involved in excitation and inhibition in the peripheral and central portions of the nerve cell membrane" (Chapters 4, 8).

1970: B Katz, U von Euler and J. Axelrod; "the humoral transmitters in the nerve terminals and the mechanism for their storage, release and inactivation" (Chapter 7, 8).

1971: EW Sutherland, Jr; " the mechanisms of the action of hormones" (Chapter 8).

1988: JW Black, GB Elion and GH Hitchings "discoveries of important principles for drug treatment" (Chapter 13, 14)

1991: E Neher and B Sakmann; "discoveries concerning the function of single ion channels in cells" (Chapter 7).

1994: AF Gilman and M Rodbell; "G-proteins and the role of these proteins in signal transduction in cells" (Chapter 8).

Chemistry; https://www.nobelprize.org/prizes/chemistry/:

1997: JC Skou; "first discovery of an ion-transporting enzyme, Na^+, K^+-ATPase" (Chapter 3).

2008: O Shimomura, M Chalfie and RY Tsien; "discovery and development of the green fluorescent protein, GFP" (Chapter 11).

2017: J Dubochet, J Frank and R Henderson; "cryo-electron microscopy for the high-resolution structure determination of biomolecules in solution" (Chapter 5).

References

Adams BA and Beam KG (1990). Muscular dysgenesis in mice: a model system for studying excitation-contraction coupling. *FASEB J* 4, 2809–2816.

Adams DJ and Gage PW (1979). Sodium and calcium gating currents in an *Aplysia* neurone. *J Physiol* 291, 467–481.

Adrian RH (1956). The effect of internal and external potassium concentration on the membrane potential of frog muscle. *J Physiol* 133, 631–658.

Adrian RH (1964). The rubidium and potassium permeability of frog muscle membrane. *J Physiol* 175, 134–159.

Adrian ED and Lucas K (1912). On the summation of propagated disturbances in nerve and muscle. *J Physiol* 44, 68–124.

Adrian RH and Almers W (1976a). Charge movement in the membrane of striated muscle. *J Physiol* 254, 339–360.

Adrian RH and Almers W (1976b). The voltage dependence of membrane capacity. *J Physiol* 254, 317–338.

Adrian RH and Bryant SH (1974). On the repetitive discharge in myotonic muscle fibres. *J Physiol* 240, 505–515.

Adrian RH and Peachey LD (1973). Reconstruction of the action potential of frog sartorius muscle. *J Physiol* 235, 103–131.

Adrian RH, Chandler WK and Hodgkin AL (1970). Voltage clamp experiments in striated muscle fibres. *J Physiol* 208, 607–644.

Adrian RH, Costantin LL and Peachey LD (1969). Radial spread of contraction in frog muscle fibres. *J Physiol* 204, 231–257.

Ahmad S, Valli H, Edling CE, et al. (2017). Effects of ageing on pro-arrhythmic ventricular phenotypes in incrementally paced murine Pgc-1β–/– hearts. *Pflügers Arch* 469, 1579–1590.

Ahmad S, Valli H, Smyth R, et al. (2019). Reduced cardiomyocyte Na$^+$ current in the age-dependent murine Pgc-1β–/– model of ventricular arrhythmia. *J Cell Physiol* 234, 3921–3932.

Aidley D (1998). *The Physiology of Excitable Cells 4/e*. Cambridge: Cambridge University Press.

Allen DG, Lamb GD and Westerblad H (2008). Skeletal muscle fatigue: cellular mechanisms. *Physiol Rev* 88, 287–332.

Almers W, Stanfield PR and Stühmer W (1983). Lateral distribution of sodium and potassium channels in frog skeletal muscle: measurements with a patch-clamp technique. *J Physiol* 336, 261–284.

Anderson BR and Granzier HL (2012). Titin-based tension in the cardiac sarcomere: molecular origin and physiological adaptations. *Prog Biophys Mol Biol* 110, 204–217.

Armstrong CM and Bezanilla F (1973). Currents related to movement of the gating particles of the sodium channels. *Nature* 242, 459–461.

Armstrong CM and Bezanilla F (1974). Charge movement associated with the opening and closing of the activation gates of the Na channels. *J Gen Physiol* 63, 533–552.

Armstrong CM and Bezanilla F (1977). Inactivation of the sodium channel. II. Gating current experiments. *J Gen Physiol* 70, 567–590.

Ashley CC and Ridgway EB (1968). Simultaneous recording of membrane potential, calcium transient and tension in single muscle fibres. *Nature* 219, 1168–1169.

Ashley CC and Ridgway EB (1970). On the relationships between membrane potential, calcium transient and tension in single barnacle muscle fibres. *J Physiol* 209, 105–130.

Bagshawe C (1993). *Muscle Contraction 2/e*. London: Chapman & Hall.

Baker PF, Hodgkin AL and Shaw TI (1962). The effects of changes in internal ionic concentrations on the electrical properties of perfused giant axons. *J Physiol* 164, 355–374.

Balasubramaniam R, Chawla S, Grace AA and Huang CL-H (2005). Caffeine-induced arrhythmias in murine hearts parallel changes in cellular Ca^{2+} homeostasis. *Am J Physiol Heart Circ Physiol* 289, H1584–H1593.

Balasubramaniam R, Chawla S, Mackenzie L, et al. (2004). Nifedipine and diltiazem suppress ventricular arrhythmogenesis and calcium release in mouse hearts. *Pflügers Arch* 449, 150–158.

Balasubramaniam R, Grace A, Saumarez R, Vandenberg J and Huang CL-H (2003). Electrogram prolongation and nifedipine-suppressible ventricular arrhythmias in mice following targeted disruption of KCNE1. *J Physiol* 552, 535–546.

Barnard EA, Miledi R and Sumikawa K (1982). Translation of exogenous messenger RNA coding for nicotinic acetylcholine receptors produces functional receptors in Xenopus oocytes. *Proc R Soc London - Biol Sci* 215, 241–246.

Baruscotti M, Bottelli G, Milanesi R, DiFrancesco JC and DiFrancesco D (2010). HCN-related channelopathies. *Pflügers Arch* 460, 405–415.

Baylor SM, Chandler WK and Marshall MW (1983). Sarcoplasmic reticulum calcium release in frog skeletal muscle fibres estimated from arsenazo III calcium transients. *J Physiol* 344, 625–666.

Belardinelli L, Giles W, Rajamani S, Karagueuzian H and Shryock J (2015). Cardiac late Na^+ current: Proarrhythmic effects, roles in long QT syndromes, and pathological relationship to CaMKII and oxidative stress. *Heart Rhythm* 12, 440–448.

Ben-Johny M, Yang PS, Niu J, et al. (2014). Conservation of Ca^{2+}/Calmodulin Regulation across Na and Ca^{2+} channels. *Cell* 157, 1657–1670.

Bers DM (2001). *Excitation-Contraction Coupling and Cardiac Contractile Force 2/e*. Dordrecht, The Netherlands: Kluwer Academic Publishers.

Bezanilla F (2018). Gating currents. *J Gen Physiol* 150, 911–932.

Blaauw B, Schiaffino S and Reggiani C (2013). Mechanisms modulating skeletal muscle phenotype. *Compr Physiol* 3, 1645–1687.

Bliss TVP and Cooke SF (2011). Long-term potentiation and long-term depression: a clinical perspective. *Clinics* 66, 3–17.

Bliss TVP and Lomo T (1973). Long-lasting potentiation of synpatic transmission in the dentate area of the anaesthetized rabbit following stimulation of the perforant path. *J Physiol* 232, 331–356.

Block BA, Imagawa T, Campbell KP and Franzini-Armstrong C (1988). Structural evidence for direct interaction between the molecular components of the transverse tubule/sarcoplasmic reticulum junction in skeletal muscle. *J Cell Biol* 107, 2587–2600.

Boyle PJ and Conway EJ (1941). Potassium accumulation in muscle and associated changes. *J Physiol* 100, 1–63.

Brock LG, Coombs JS and Eccles JC (1952). The recording of potentials from motoneurones with an intracellular electrode. *J Physiol* 117, 431–460.

Brodal P (2016). *The Central Nervous System 5/e*. New York: Oxford University Press.

Brugada R (2000). Use of intravenous antiarrhythmics to identify concealed Brugada syndrome. *Curr Control Trials Cardiovasc Med* 1, 45–47.

Bruns D and Jahn R (1995). Real-time measurement of transmitter release from single synaptic vesicles. *Nature* 377, 62–65.

Bülbring E (1979). Post junctional adrenergic mechanisms. *Brit Med Bull* 35, 285–294.

Buller A (1975). *The Contractile Behaviour of Mammalian Skeletal Muscle (Oxford Biology Reader No. 36)*. London: Oxford University Press.

Cain DF, Infante AA and Davies RE (1962). Chemistry of muscle contraction: adenosine triphosphate and phosphorylcreatine as energy supplies for single contractions of working muscle. *Nature* 196, 214–217.

Caldwell PC and Keynes R (1957). The utilization of phosphate bond energy for sodium extrusion from giant axons. *J Physiol* 137, 12P–13P.

Caldwell PC, Hodgkin AL, Keynes RD and Shaw TI (1960). The effects of injecting "energy-rich" phosphate compounds on the active transport of ions in the giant axons of *Loligo*. *J Physiol* 152, 561–590.

Cannon SC (2018). Sodium channelopathies of skeletal muscle In: *Handbook of Experimental Pharmacology*. Berlin: Springer, 309–330.

Catterall WA (1992). Cellular and molecular biology of voltage-gated sodium channels. *Physiol Rev* 72, S15–S48.

Chadda KR, Ahmad S, Valli H et al. (2017a). The effects of ageing and adrenergic challenge on electrocardiographic phenotypes in a murine model of long QT syndrome type 3. *Sci Rep* 7, 11070.

Chadda KR, Jeevaratnam K, Lei M, Huang CL-H (2017b). Sodium channel biophysics, late sodium current and genetic arrhythmic syndromes. *Pflugers Arch* 469, 629–641.

Chandler WK, Rakowski RF and Schneider MF (1976). A non-linear voltage dependent charge movement in frog skeletal muscle. *J Physiol* 254, 245–283.

Chawla S, Skepper JN, Hockaday AR and Huang CL-H (2001). Calcium waves induced by hypertonic solutions in intact frog skeletal muscle fibres. *J Physiol* 536, 351–359.

Chawla S, Skepper JN and Huang CL-H (2002). Differential effects of sarcoplasmic reticular Ca^{2+}-ATPase inhibition on charge movements and calcium transients in intact amphibian skeletal muscle fibres. *J Physiol* 539, 869–882.

Chawla S, Vanhoutte P, Arnold FJL, Huang CL-H and Bading H (2003). Neuronal activity-dependent nucleocytoplasmic shuttling of HDAC4 and HDAC5. *J Neurochem* 85, 151–159.

Cheng EP, Yuan C, Navedo MF, et al. (2011). Restoration of normal L-type Ca^{2+} channel function during Timothy syndrome by ablation of an anchoring protein. *Circ Res* 109, 255–261.

Cheng H, Lederer WJ and Cannell MB (1993). Calcium sparks: elementary events underlying excitation-contraction coupling in heart muscle. *Science* 262, 740–744.

Chiamvimonvat N, Kargacin ME, Clark RB and Duff HJ (1995). Effects of intracellular calcium on sodium current density in cultured neonatal rat cardiac myocytes. *J Physiol* 483, 307–318.

Clausen T (2003). Na^+-K^+ pump regulation and skeletal muscle contractility. *Physiol Rev* 83, 1269–1324.

Cole KS (1941). Rectification and inductance in the squid giant axon. *J Gen Physiol* 25, 29–51.

Cole KS and Curtis HJ (1939). Electric impedance of the squid giant axon during activity. *J Gen Physiol* 22, 649–670.

Colquhoun D and Sakmann B (1985). Fast events in single-channel currents activated by acetylcholine and its analogues at the frog muscle end-plate. *J Physiol* 369, 501–557.

Conway EJ (1957). Nature and significance of concentration relations of potassium and sodium ions in skeletal muscle. *Physiol Rev* 37, 84–132.

Coombs JS, Eccles JC and Fatt P (1955a). Excitatory synaptic action in motoneurones. *J Physiol* 130, 374–395.

Coombs JS, Eccles JC and Fatt P (1955b). The specific ionic conductances and the ionic movements across the motoneuronal membrane that produce the inhibitory post-synaptic potential. *J Physiol* 130, 326–373.

Corrias A, Giles W and Rodriguez B (2011). Ionic mechanisms of electrophysiological properties and repolarization abnormalities in rabbit Purkinje fibers. *Am J Physiol - Heart Circ Physiol* 300, H1806–H1813.

Costantin LL (2011). Activation in striated muscle. *Compr Physiol* 215–259.

Cousins HM, Edwards FR, Hickey H, Hill CE and Hirst GDS (2003). Electrical coupling between the myenteric interstitial cells of Cajal and adjacent muscle layers in the guinea-pig gastric antrum. *J Physiol* 550, 829–844.

Dale HH, Feldberg W and Vogt M (1936). Release of acetylcholine at voluntary motor nerve endings. *J Physiol* 86, 353–380.

Davies L, Jin J, Shen W, et al. (2014). Mkk4 is a negative regulator of the transforming growth factor beta 1 signaling associated with atrial remodeling and arrhythmogenesis with age. *J Am Heart Assoc* 3, 1–19.

Del Castillo J and Katz B (1954). Quantal components of the end-plate potential. *J Physiol* 124, 560–573.

Del Castillo J and Katz B (1955). On the localization of acetylcholine receptors. *J Physiol* 128, 157–181.

Del Castillo J and Moore JW (1959). On increasing the velocity of a nerve impulse. *J Physiol* 148, 665–670.

DiFrancesco D (1993). Pacemaker mechanisms in cardiac tissue. *Annu Rev Physiol* 55, 455–471.

DiFrancesco D and Noble D (1985). A model of cardiac electrical activity incorporating ionic pumps and concentration changes. *Philos Trans R Soc Lond B Biol Sci* 307, 353–398.

Doyle DA, Cabral JM, Pfuetzner RA, et al. (1998). The structure of the potassium channel: Molecular basis of K^+ conduction and selectivity. *Science* 280, 69–77.

Dutka TL and Lamb GD (2000). Effect of lactate on depolarization-induced Ca^{2+} release in mechanically skinned skeletal muscle fibers. *Am J Physiol Cell Physiol* 278, C517–C525.

Ebashi S and Endo M (2003). Calcium and muscle contraction. *Prog Biophys Mol Biol* 18, 123–183.

Eccles JC (1964). *The Physiology of Synapses*. Berlin: Springer-Verlag.

Edling CE, Fazmin IT, Saadeh K, et al. (2019). Molecular basis of arrhythmic substrate in ageing murine peroxisome proliferator-activated receptor γ co-activator deficient hearts modelling mitochondrial dysfunction. *Biosci Rep.* 39, BSR20190403.

Eglen RM, Reddy H, Watson N and Challiss RAJ (1994). Muscarinic acetylcholine receptor subtypes in smooth muscle. *Trends Pharmacol Sci* 15, 114–119.

Einthoven W (1924). The string galvanometer and the measurement of the action currents of the heart, Nobel lecture 1925 In: *Nobel Lectures, Physiology or Medicine 1921–41. Republished 1965*. Amsterdam: Elsevier 287–306.

Emslie-Smith D, Paterson C, Scratcherd T and Read NW (1988). *Textbook of Physiology 11/e*. Edinburgh: Churchill-Livingstone.

Erlanger J and Gasser HS (1937). *Electrical Signs of Nervous Activity*. Philadelphia: University of Pennsylvania Press.

Fabritz L, Kirchhof P, Franz MR, et al. (2003). Prolonged action potential durations, increased dispersion of repolarization, and polymorphic ventricular tachycardia in a mouse model of proarrhythmia. *Basic Res Cardiol* 98, 25–32.

Fatt P and Katz B (1951). An analysis of the end-plate potential recorded with an intra-cellular electrode. *J Physiol* 115, 320–370.

Fatt P and Katz B (1952). Spontaneous subthreshold activity at motor nerve endings. *J Physiol* 117, 109–128.

Fawcett DW and McNutt NS (1969). The ultrastructure of the cat myocardium. I. Ventricular papillary muscle. *J Cell Biol* 42, 1–45.

Ferenczi EA, Fraser JA, Chawla S, et al. (2004). Membrane potential stabilization in amphibian skeletal muscle fibres in hypertonic solutions. *J Physiol* 555, 423–438.

Ferenczi EA, Tan X and Huang CL-H (2019). Principles of optogenetic methods for application to cardiac experimental systems. *Front Physiol* 10, 1096.

Fermini B and Fossa AA (2003). The impact of drug-induced QT interval prolongation on drug discovery and development. *Nat Rev Drug Discov* 2, 439–447.

Field AC, Hill C and Lamb GD (1988). Asymmetric charge movement and calcium currents in ventricular myocytes of neonatal rat. *J Physiol* 406, 277–297.

Filatov GN, Pinter MJ and Rich MM (2009). Role of Ca^{2+} in injury-induced changes in sodium current in rat skeletal muscle. *Am J Physiol Cell Physiol* 297, C352–C359.

Finck BN and Kelly DP (2006). PGC-1 coactivators: inducible regulators of energy metabolism in health and disease. *J Clin Invest* 116, 615–622.

Frankenhaeuser B and Hodgkin AL (1957). The action of calcium on the electrical properties of squid axons. *J Physiol* 137, 218–244.

Fraser JA and Huang CL-H (2004). A quantitative analysis of cell volume and resting potential determination and regulation in excitable cells. *J Physiol* 559, 459–478.

Fraser JA and Huang CL-H (2007). Quantitative techniques for steady-state calculation and dynamic integrated modelling of membrane potential and intracellular ion concentrations. *Prog Biophys Mol Biol* 94, 336–372.

Fraser JA, Huang CL-H and Pedersen TH (2011). Relationships between resting conductances, excitability, and t-system ionic homeostasis in skeletal muscle. *J Gen Physiol* 138, 95–116.

Fraser JA, Middlebrook CE, Usher-Smith JA, Schwiening CJ and Huang CL-H (2005). The effect of intracellular acidification on the relationship between cell volume and membrane potential in amphibian skeletal muscle. *J Physiol* 563, 745–764.

Fraser JA, Skepper JN, Hockaday AR and Huang CL-H (1998). The tubular vacuolation process in amphibian skeletal muscle. *J Muscle Res Cell Motil* 19, 613–629.

Frommeyer G and Eckardt L (2016). Drug-induced proarrhythmia: risk factors and electrophysiological mechanisms. *Nat Rev Cardiol* 13, 36–47.

Gattuso JM, Davies AH, Glasby MA, Gschmeissner SE and Huang CL-H (1988). Peripheral nerve repair using muscle autografts: recovery of transmission in primates. *J Bone Joint Surg Br* 70, 524–529.

Glasby MA, Gschmeissner S, Hitchcock R and Huang CL-H (1986a). Regeneration of the sciatic nerve in rats: the effect of muscle basement membrane. *J Bone Joint Surg Br* 68, 829–833.

Glasby MA, Gschmeissner SG, Hitchcock RJI and Huang CL-H (1986b). The dependence of nerve regeneration through muscle grafts in the rat on the availability and orientation of basement membrane. *J Neurocytol* 15, 479–510.

Goddard CA, Ghais NS, Zhang Y, et al. (2008). Physiological consequences of the P2328S mutation in the ryanodine receptor (RyR2) gene in genetically modified murine hearts. *Acta Physiol* 194, 123–140.

Goldman DE (1943). Potential, impedance and rectification in membranes. *J Gen Physiol* 27, 37–60.

Gordon AM, Huxley AF and Julian FJ (1966). The variation in isometric tension with sarcomere length in vertebrate muscle fibres. *J Physiol* 184, 170–192.

Grabner M, Dirksen RT, Suda N and Beam KG (1999). The II-III loop of the skeletal muscle dihydropyridine receptor is responsible for the bidirectional coupling with the ryanodine receptor. *J Biol Chem* 274, 21913–21919.

Granzier H and Labeit S (2002). Cardiac titin: an adjustable multi-functional spring. *J Physiol* 541, 335–342.

Granzier H and Labeit S (2007). Structure-function relations of the giant elastic protein titin in striated and smooth muscle cells. *Muscle and Nerve* 36, 740–755.

Greeff NG, Keynes RD and Van Helden DF (1982). Fractionation of the asymmetry current in the squid giant axon into inactivating and non-inactivating components. *Proc R Soc London - Biol Sci* 215, 375–389.

Guibert C, Ducret T and Savineau JP (2008). Voltage-independent calcium influx in smooth muscle. *Prog Biophys Mol Biol* 98, 10–23.

Gulbis JM and Doyle DA (2004). Potassium channel structures: do they conform? *Curr Opin Struct Biol* 14, 440–446.

Gundersen K (2011). Excitation-transcription coupling in skeletal muscle: the molecular pathways of exercise. *Biol Rev* 86, 564–600.

Gurung I, Medina-Gomez G, Kis A, et al. (2011). Deletion of the metabolic transcriptional coactivator PGC1-β induces cardiac arrhythmia. *Cardiovasc Res* 92, 29–38.

Gussak I and Antzelevich C (2003). *Cardiac Repolarization: Bridging Basic and Clinical Science*. Totowa, NJ: Humana Press, Inc.

Guzadhur L, Jiang W, Pearcey SM, et al. (2012). The age-dependence of atrial arrhythmogenicity in Scn5a+/- murine hearts reflects alterations in action potential propagation and recovery. *Clin Exp Pharmacol Physiol* 39, 518–527.

Hakim P, Gurung IS, Pedersen TH, et al. (2008). Scn3b knockout mice exhibit abnormal ventricular electrophysiological properties. *Prog Biophys Mol Biol* 98, 251–266.

Hamill OP, Marty A, Neher E, Sakmann B and Sigworth FJ (1981). Improved patch-clamp techniques for high-resolution current recording from cells and cell-free membrane patches. *Pflügers Arch* 391, 85–100.

Head CE, Balasubramaniam R, Thomas G, et al. (2005). Paced electrogram fractionation analysis of arrhythmogenic tendency in DeltaKPQ Scn5a mice. *J Cardiovasc Electrophysiol* 16, 1329–1340.

Heatwole CR and Moxley RT 3rd (2007). The nondystrophic myotonias. *Neurotherapeutics* 4, 238–251.

Hernandez-Ochoa EO, Pratt SJP, Lovering RM and Schneider MF (2016). Critical role of intracellular RyR1 calcium release channels in skeletal muscle function and disease. *Front Physiol* 6, 1–11.

Hibino H, Inanobe A, Furutani K, et al. (2010). Inwardly rectifying potassium channels: their structure, function, and physiological roles. *Physiol Rev* 90, 291–366.

Hill AV (1938). The heat of shortening and the dynamic constants of muscle. *Proc R Soc London - Biol Sci* 126, 136–195.

Hill AV (1950a). The dimensions of animals and their muscular dynamics. *Sci Prog Lond* 38, 209–230.

Hill AV (1950b). A challenge to biochemists. *Biochim Biophys Acta* 4, 4–11.

Hill AV and Hartree W (1920). The four phases of heat production of muscle. *J Physiol* 54, 84–128.

Hille B (1971). The hydration of sodium ions crossing the nerve membrane. *Proc Natl Acad Sci* 68, 280–282.

Hirst GDS and Edwards FR (2001). Generation of slow waves in the antral region of guinea-pig stomach - a stochastic process. *J Physiol* 535, 165–180.

Hodgkin AL (1939). The relation between conduction velocity and the electrical resistance outside a nerve fibre. *J Physiol* 94, 560–570.

Hodgkin AL (1951). The ionic basis of electrical activity in nerve and muscle. *Biol Rev* 26, 339–409.

Hodgkin AL (1958). The Croonian Lecture: ionic movements and electrical activity in giant nerve fibres. *Proc R Soc London - Biol Sci* 148, 1–37.

Hodgkin AL and Horowicz P (1957). The differential action of hypertonic solutions on the twitch and action potential of a muscle fibre. *J Physiol* 137, 17P–18P.

Hodgkin AL and Horowicz P (1959). The influence of potassium and chloride ions on the membrane potential of single muscle fibres. *J Physiol* 148, 127–160.

Hodgkin AL and Huxley AF (1952). A quantitative description of membrane current and its application to conduction and excitation in nerves. *J Physiol* 117, 500–544.

Hodgkin AL and Katz B (1949). The effect of sodium ions on the electrical activity of the giant axon of the squid. *J Physiol* 108, 37–77.

Hodgkin AL and Keynes RD (1955a). The potassium permeability of a giant nerve fibre. *J Physiol* 128, 61–88.

Hodgkin AL and Keynes RD (1955b). Active transport of cations in giant axons from *Sepia* and *Loligo*. *J Physiol* 128, 28–60.

Hodgkin AL, Huxley AF and Katz B (1949). The effect of sodium ions on the electrical activity of the giant axon of the squid. *J Physiol* 108, 37–77.

Hodgkin AL, Huxley AF and Katz B (1952). Measurement of current-voltage relations in the membrane of the giant axon of Loligo . *J Physiol* 116, 424–448.

Hoffman B and Cranefield P (1960). *Electrophysiology of the Heart*. New York: McGrawHill.

Hollingworth S and Marshall MW (1981). A comparative study of charge movement in rat and frog skeletal muscle fibres. *J Physiol* 321, 583–602.

Hollingworth S, Peet J, Chandler WK and Baylor SM (2002). Calcium sparks in intact skeletal muscle fibers of the frog. *J Gen Physiol* 118, 653–678.

Homsher E (1987). Muscle enthalpy production and its relationship to actomyosin ATPase. *Annu Rev Physiol* 49, 673–690.

Hothi SS, Gurung IS, Heathcote JC, et al. (2008). Epac activation, altered calcium homeostasis and ventricular arrhythmogenesis in the murine heart. *Pflügers Arch* 457, 253–270.

Hothi SS, Thomas G, Killeen MJ, Grace AA and Huang CL-H (2009). Empirical correlation of triggered activity and spatial and temporal re-entrant substrates with arrhythmogenicity in a murine model for Jervell and Lange-Nielsen syndrome. *Pflügers Arch* 458, 819–835.

Huang CL-H (1982). Pharmacological separation of charge movement components in frog skeletal muscle. *J Physiol* 324, 375–387.

Huang CL-H (1984). Analysis of "off" tails of intramembrane charge movements in skeletal muscle of Rana temporaria. *J Physiol* 356, 375–390.

Huang CL-H (1988). Intramembrane charge movements in skeletal muscle. *Physiol Rev* 68, 1197–1247.

Huang CL-H (1990). Voltage-dependent block of charge movement components by nifedipine in frog skeletal muscle. *J Gen Physiol* 96, 535–557.

Huang CL-H (1993). Intramembrane Charge Movements in Striated Muscle. *Monographs of the Physiological Society, No. 44*. Oxford: Clarendon Press.

Huang CL-H (1994). Charge conservation in intact frog skeletal muscle fibres in gluconate-containing solutions. *J Physiol* 474, 161–171.

Huang CL-H (1996). Kinetic isoforms of intramembrane charge in intact amphibian striated muscle. *J Gen Physiol* 107, 515–534.

Huang CL-H (1998a). The influence of caffeine on intramembrane charge movements in intact frog striated muscle. *J Physiol* 512, 707–721.

Huang CL-H (1998b). The influence of perchlorate ions on complex charging transients in amphibian striated muscle. *J Physiol* 506, 699–714.

Huang CL-H (2017). Murine electrophysiological models of cardiac arrhythmogenesis. *Physiol Rev* 97, 283–409.

Huang CL-H (2018). Roger Yonchien Tsien. 1 February 1952 – 24 August 2016. *Biogr Mem Fellows R Soc* 65, 405–428.

Huang CL-H and Hockaday AR (1988). Development of myotomal cells in Xenopus laevis larvae. *J Anat* 159, 129–136.

Huang CL-H and Peachey LD (1989). Anatomical distribution of voltage-dependent membrane capacitance in frog skeletal muscle fibers. *J Gen Physiol* 93, 565–584.

Huang CL-H, Pedersen TH and Fraser JA (2011). Reciprocal dihydropyridine and ryanodine receptor interactions in skeletal muscle activation. *J Muscle Res Cell Motil* 32, 171–202.

Huang CL-H, Wu L, Jeevaratnam K and Lei M (2020). Update on antiarrhythmic drug pharmacology. *J Cardiovasc Electrophysiol* 31, 579–592.

Huxley AF and Niedergerke R (1954). Structural changes in muscle during contraction: interference microscopy of living muscle fibres. *Nature* 173, 971–973.

Huxley AF and Stämpfli R (1949). Evidence for saltatory conduction in peripheral myelinated nerve fibres. *J Physiol* 108, 315–339.

Huxley AF and Taylor RE (1958). Local activation of striated muscle fibres. *J Physiol* 144, 426–441.

Huxley HE (1963). Electron microscope studies on the structure of natural and synthetic protein filaments from striated muscle. *J Mol Biol* 7, 281–308.

Huxley HE (1976). The structural basis of contraction and regulation in skeletal muscle In: *Molecular Basis of Motility*. (Eds.). Heilmeyer Jr L, Ruegg J and Wieland T, Berlin: Springer Verlag. 9–25.

Huxley HE (1990). Sliding filaments and molecular motile systems. *J Biol Chem* 265, 8347–8350.

Huxley HE and Hanson J (1954). Changes in the cross-striations of muscle during contraction and stretch and their structural interpretation. *Nature* 173, 973–976.

Jack J, Noble D and Tsien R (1983). *Electric Current Flow in Excitable Cells*. Oxford: Oxford University Press.

James MF, Smith MI, Bockhorst KHJ, et al. (1999). Cortical spreading depression in the gyrencephalic feline brain studied by magnetic resonance imaging. *J Physiol* 519, 415–425.

January CT and Riddle JM (1989). Early afterdepolarizations: mechanism of induction and block. A role for L-type Ca^{2+} current. *Circ Res* 64, 977–990.

Jeevaratnam K, Guzadhur L, Goh YM, Grace AA and Huang CL-H (2016). Sodium channel haploinsufficiency and structural change in ventricular arrhythmogenesis. *Acta Physiol* 216, 186–202.

Jeevaratnam K, Rewbury R, Zhang Y, et al. (2012). Frequency distribution analysis of activation times and regional fibrosis in murine Scn5a+/- hearts: the effects of ageing and sex. *Mech Ageing Dev* 133, 591–599.

Johnston L, Sergeant GP, Hollywood MA, Thornbury KD and McHale NG (2005). Calcium oscillations in interstitial cells of the rabbit urethra. *J Physiol* 565, 449–461.

Jurkat-Rott K, Holzherr B, Fauler M and Lehmann-Horn F (2010). Sodium channelopathies of skeletal muscle result from gain or loss of function. *Pflügers Arch* 460, 239–248.

Karbat I and Reuveny E (2019). Voltage sensing comes to rest. *Cell* 178, 776–778.

Kato G (1936). On the excitation, conduction, and narcotization of single nerve fibers. *Cold Spr Harb Symp Quant Biol* 4, 202–213.

Katz B (1962). The Croonian Lecture - the transmission of impulses from nerve to muscle, and the subcellular unit of synaptic action. *Proc R Soc London - Biol Sci* 155, 455–477.

Katz B and Miledi R (1969). Spontaneous and evoked activity of motor nerve endings in calcium Ringer. *J Physiol* 203, 689–706.

Keating MT and Sanguinetti MC (2001). Molecular and cellular mechanisms of cardiac arrhythmias. *Cell* 104, 569–580.

Keynes RD (1951). The ionic movements during nervous activity. *J Physiol* 114, 119–150.

Keynes RD (1963). Chloride in the squid giant axon. *J Physiol* 169, 690–705.

Keynes RD and Elinder F (1998). On the slowly rising phase of the sodium gating current in the squid giant axon. *Proc R Soc London - Biol Sci* 265, 255–262.

Keynes RD and Elinder F (1999). The screw-helical voltage gating of ion channels. *Proc R Soc London - Biol Sci* 266, 843–852.

Keynes RD and Kimura JE (1983). Kinetics of activation of the sodium conductance in the squid giant axon. *J Physiol* 336, 621–634.

Keynes RD and Lewis PR (1951). The sodium and potassium content of cephalopod nerve fibers. *J Physiol* 114, 151–182.

Keynes RD and Martins-Ferreira H (1953). Membrane potentials in the electroplates of the electric eel. *J Physiol* 119, 315–351.

Keynes RD and Ritchie JM (1984). On the binding of labelled saxitoxin to the squid giant axon. *Proc R Soc London - Biol Sci* 222, 147–153.

Keynes RD and Rojas E (1973). Characteristics of the sodium gating current in the squid giant axon. *J Physiol* 233, 28P–30P.

Keynes RD and Rojas E (1974). Kinetics and steady-state properties of the charged system controlling sodium conductance in the squid giant axon. *J Physiol* 239, 393–434.

Keynes RJ, Hopkins WG and Huang CL-H (1984). Regeneration of mouse peripheral nerves in degenerating skeletal muscle: guidance by residual muscle fibre basement membrane. *Brain Res* 295, 275–282.

Killeen MJ, Gurung IS, Thomas G, et al. (2007a). Separation of early afterdepolarizations from arrhythmogenic substrate in the isolated perfused hypokalaemic murine heart through modifiers of calcium homeostasis. *Acta Physiol* 191, 43–58.

Killeen MJ, Sabir IN, Grace AA and Huang CL-H (2008a). Dispersions of repolarization and ventricular arrhythmogenesis: lessons from animal models. *Prog Biophys Mol Biol* 98, 219–229.

Killeen MJ, Thomas G, Gurung IS, et al. (2007b). Arrhythmogenic mechanisms in the isolated perfused hypokalaemic murine heart. *Acta Physiol* 189, 33–46.

Killeen MJ, Thomas G, Sabir IN, Grace AA and Huang CL-H (2008b). Mouse models of human arrhythmia syndromes. *Acta Physiol* 192, 455–469.

Kim YC, Koh SD and Sanders KM (2002). Voltage-dependent inward currents of interstitial cells of Cajal from murine colon and small intestine. *J Physiol* 541, 797–810.

King J, Huang CL-H and Fraser JA (2013a). Determinants of myocardial conduction velocity: implications for arrhythmogenesis. *Front Physiol* 4, 154.

King J, Wickramarachchi C, Kua K, et al. (2013b). Loss of Nav1.5 expression and function in murine atria containing the RyR2-P2328S gain-of-function mutation. *Cardiovasc Res* 99, 751–759.

King J, Zhang Y, Lei M, et al. (2013c). Atrial arrhythmia, triggering events and conduction abnormalities in isolated murine RyR2-P2328S hearts. *Acta Physiol* 207, 308–323.

Koeppen B and Stanton B (2009). *Berne and Levy: Principles of Physiology 6/e*. New York: Mosby.

Kovács L, Ríos E and Schneider MF (1979). Calcium transients and intramembrane charge movement in skeletal muscle fibres. *Nature* 279, 391–396.

Kovacs L, Rios E and Schneider MF (1983). Measurement and modification of free calcium transients in frog skeletal muscle fibres by a metallochromic indicator dye. *J Physiol* 343, 161–196.

Kuffler S (1980). Slow synaptic responses in autonomic ganglia and the pursuit of a peptidergic transmitter. *J Exp Biol* 89, 257–286.

Kyte J and Doolittle RF (1982). A simple method for displaying the hydropathic character of a protein. *J Mol Biol* 157, 105–132.

Lammers WJEP (2015). Normal and abnormal electrical propagation in the small intestine. *Acta Physiol* 213, 349–359.

Lammers WJEP, Al-Bloushi HM, Al-Eisaei SA, et al. (2011). Slow wave propagation and plasticity of interstitial cells of Cajal in the small intestine of diabetic rats. *Exp Physiol* 96, 1039–1048.

Laver DR (2018). Regulation of the RyR channel gating by Ca^{2+} and Mg^{2+}. *Biophys Rev* 10, 1087–1095.

Lei M, Goddard C, Liu J, et al. (2005). Sinus node dysfunction following targeted disruption of the murine cardiac sodium channel gene Scn5a. *J Physiol* 567, 387–400.

Lei M, Wu L, Terrar DA and Huang CL-H (2018). Modernized classification of cardiac antiarrhythmic drugs. *Circulation* 138, 1879–1896.

Lei M, Zhang H, Grace AA and Huang CL-H (2007). SCN5A and sinoatrial node pacemaker function. *Cardiovasc Res* 74, 356–365.

Lichtman JW, Magrassi L and Purves D (1987). Visualization of neuromuscular junctions over periods of several months in living mice. *J Neurosci* 7, 1215–1222.

Lillie RS (1925). Factors affecting transmission and recovery in the passive iron nerve model. *J Gen Physiol* 7, 473–507.

Lin J, Handschin C and Spiegelman BM (2005). Metabolic control through the PGC-1 family of transcription coactivators. *Cell Metab* 1, 361–370.

Liu M, Sanyal S, Gao G, et al. (2009). Cardiac Na$^+$ current regulation by pyridine nucleotides. *Circ Res* 105, 737–745.

Loewi O (1921). Über humorale Übertragbarkeit der Herzner-venwirkung. *Pflügers Arch* 189, 239–242.

London B (2001). Cardiac arrhythmias: from (transgenic) mice to men. *J Cardiovasc Electrophysiol* 12, 1089–1091.

London B, Baker LC, Petkova-Kirova P, et al. (2007). Dispersion of repolarization and refractoriness are determinants of arrhythmia phenotype in transgenic mice with long QT. *J Physiol* 578, 115–129.

Lu Y, Mahaut-Smith MP, Huang CL-H and Vandenberg JI (2003). Mutant MiRP1 subunits modulate *HERG* K$^+$ channel gating: a mechanism for pro-arrhythmia in long QT syndrome type 6. *J Physiol* 551, 253–262.

Lu Y, Mahaut-Smith MP, Varghese A, et al. (2001). Effects of premature stimulation on *HERG* K$^+$ channels. *J Physiol* 537, 843–851.

Lukas A and Antzelevitch C (1996). Phase 2 reentry as a mechanism of initiation of circus movement reentry in canine epicardium exposed to simulated ischemia. *Cardiovasc Res* 32, 593–603.

Lüttgau HC and Spiecker W (1979). The effects of calcium deprivation upon mechanical and electrophysiological parameters in skeletal muscle fibres of the frog. *J Physiol* 296, 411–429.

MacLennan DH (2000). Ca^{2+} signalling and muscle disease. *Eur J Biochem* 267, 5291–5297.

Magleby KL and Stevens CF (1972). A quantitative description of end-plate currents. *J Physiol* 223, 173–197.

Makowski L, Caspar DL, Phillips WC, Baker TS and Goodenough DA (1984). Gap junction structures. VI. Variation and conservation in connexon conformation and packing. *Biophys J* 45, 208–218.

Mangoni ME and Nargeot J (2008). Genesis and regulation of the heart automaticity. *Physiol Rev* 88, 919–982.

Marmont G (1949). Studies on the axon membrane. I. A new method. *J Cell Comp Physiol* 34, 351–382.

Martin CA, Grace AA and Huang CL-H (2011a). Spatial and temporal heterogeneities are localized to the right ventricular outflow tract in a heterozygotic Scn5a mouse model. *Am J Physiol Heart Circ Physiol* 300, H605–H616.

Martin CA, Guzadhur L, Grace AA, Lei M and Huang CL-H (2011b). Mapping of reentrant spontaneous polymorphic ventricular tachycardia in a Scn5a+/- mouse model. *Am J Physiol Heart Circ Physiol* 300, H1853–H1862.

Martin CA, Huang CL-H and Grace AA (2010). Progressive conduction diseases. *Card Electrophysiol Clin North Am* 2, 509–519.

Martin CA, Siedlecka U, Kemmerich K, et al. (2012). Reduced Na$^+$ and higher K$^+$ channel expression and function contribute to right ventricular origin of arrhythmias in Scn5a+/- mice. *Open Biol* 2, 120072.

Martins-Ferreira H, Nedergaard M and Nicholson C (2000). Perspectives on spreading depression. *Brain Res Rev* 32, 215–234.

Matthews GDK, Guzadhur L, Grace AA and Huang CL-H (2012). Nonlinearity between action potential alternans and restitution, which both predict ventricular arrhythmic properties in Scn5a+/- and wild-type murine hearts. *J Appl Physiol* 112, 1847–1863.

Matthews GDK, Guzadhur L, Sabir IN, Grace AA and Huang CL-H (2013). Action potential wavelength restitution predicts alternans and arrhythmia in murine Scn5a+/- hearts. *J Physiol* 591, 4167–4188.

Matthews GDK, Huang CL-H, Sun L and Zaidi M (2011). Translational musculoskeletal science: is sarcopenia the next clinical target after osteoporosis? *Ann N Y Acad Sci* 1237, 95–105.

Matthews GDK, Martin CA, Grace AA, Zhang Y and Huang CL-H (2010). Regional variations in action potential alternans in isolated murine Scn5a+/- hearts during dynamic pacing. *Acta Physiol* 200, 129–146.

Matthews HR, Tan SRX, Shoesmith JA, et al. (2019). Sodium current inhibition following stimulation of exchange protein directly activated by cyclic-3′,5′-adenosine monophosphate (Epac) in murine skeletal muscle. *Sci Rep* 9, 1927.

Maylie J, Irving M, Sizto N and Chandler WK (1987). Calcium signals recorded from cut frog twitch fibers containing antipyrylazo III. *J Gen Physiol* 89, 83–143.

McHale NG, Hollywood M, Sergeant G and Thornbury K (2006). Origin of spontaneous rhythmicity in smooth muscle. *J Physiol* 570, 23–28.

Mickelson JR and Louis CF (1996). Malignant hyperthermia: excitation-contraction coupling, Ca^{2+} release channel, and cell Ca^{2+} regulation defects. *Physiol Rev* 76, 537–592.

Miller DJ (2004). Sydney Ringer; physiological saline, calcium and the contraction of the heart. *J Physiol* 555, 585–587.

Miyawaki A, Llopis J, Heim R, et al. (1997). Fluorescent indicators for Ca^{2+} based on green fluorescent proteins and calmodulin. *Nature* 388, 882–887.

Morad M and Orkand RK (1971). Excitation-concentration coupling in frog ventricle: evidence from voltage clamp studies. *J Physiol* 219, 167–89.

Mulroy AD, Glasby MA and Huang CL-H (1990). Regeneration of unmyelinated nerve fibers through skeletal muscle autografts in rat sciatic nerve. *Neuro-Orthopedics* 10, 1–14.

Namadurai S, Yereddi NR, Cusdin FS, et al.(2015). A new look at sodium channel β subunits. *Open Biol* 5, 140192.

Neher E and Sakmann B (1976). Single-channel currents recorded from membrane of denervated frog muscle fibres. *Nature* 260, 799–802.

Nicholls CG, Fuchs PA, Martin AR and Wallace B (2001). *From Neuron to Brain: A Cellular Approach to the Function of the Nervous System 3/e.* Sunderland, USA: Sinauer Associates Inc.

Nielsen OB, De Paoli F and Overgaard K (2001). Protective effects of lactic acid on force production in rat skeletal muscle. *J Physiol* 536, 161–166.

Ning F, Luo L, Ahmad S, et al. (2016). The RyR2-P2328S mutation downregulates Nav1.5 producing arrhythmic substrate in murine ventricles. *Pflügers Arch* 468, 655–665.

Noble D (1979). *The Initiation of the Heartbeat 2/e.* Oxford: Oxford University Press.

Noda M, Shimizu S, Tanabe T, et al. (1984). Primary structure of electrophorus electricus sodium channel deduced from cDNA sequence. *Nature* 312, 121–127.

Noda M, Takahashi H, Tanabe T, et al. (1982). Primary structure of alpha-subunit precursor of *Torpedo californica* acetylcholine receptor deduced from cDNA sequence. *Nature* 299, 793–797.

Nuyens D, Stengl M, Dugarmaa S, et al. (2001). Abrupt rate accelerations or premature beats cause life-threatening arrhythmias in mice with long-QT3 syndrome. *Nat Med* 7, 1021–1027.

O'Malley HA and Isom LL (2015). Sodium channel β subunits: emerging targets in channelopathies. *Annu Rev Physiol* 77, 481–504.

Offer G (1974). The molecular basis of muscular contraction In: *Companion to Biochemistry*. (Eds.). Bull A, Lagnado J, Thomas J and Tipton K, London: Longman. 623–671.

Ördög T (2008). Interstitial cells of Cajal in diabetic gastroenteropathy. *Neurogastroenterol Motil* 20, 8–18.

Ostrowski J, Kjelsberg MA, Caron MG and Lefkowitz RJ (1992). Mutagenesis of the beta 2-adrenergic receptor: how structure elucidates function. *Annu Rev Pharmacol Toxicol* 32, 167–183.

Ozaki H, Gerthoffer WT, Publicover NG, Fusetani N and Sanders KM (1991a). Time-dependent changes in Ca^{2+} sensitivity during phasic contraction of canine antral smooth muscle. *J Physiol* 440, 207–224.

Ozaki H, Stevens RJ, Blondfield DP, Publicover NG and Sanders KM (1991b). Simultaneous measurement of membrane potential, cytosolic Ca^{2+}, and tension in intact smooth muscles. *Am J Physiol Physiol* 260, C917–C925.

Padmanabhan N and Huang CL-H (1990). Separation of tubular electrical activity in amphibian skeletal muscle through temperature change. *Exp Physiol* 75, 721–724.

Pan X, Li Z, Zhou Q, et al. (2018). Structure of the human voltage-gated sodium channel Nav1.4 in complex with β1. *Science* 362, eaau2486.

Park KJ, Hennig GW, Lee H-T, et al. (2005). Spatial and temporal mapping of pacemaker activity in interstitial cells of Cajal in mouse ileum in situ. *Am J Physiol Cell Physiol* 290, C1411–C1427.

Peachey LD (1965). The sarcoplasmic reticulum and transverse tubules of the frog's sartorius. *J Cell Biol* 25, 209–231.

Peachey LD and Eisenberg BR (1978). Helicoids in the T system and striations of frog skeletal muscle fibers seen by high voltage electron microscopy. *BiophysJ* 22, 145–154.

Pedersen TH, Huang CL-H and Fraser JA (2011). An analysis of the relationships between subthreshold electrical properties and excitability in skeletal muscle. *J Gen Physiol* 138, 73–93.

Pedersen TH, Macdonald WA, de Paoli FV, Gurung IS and Nielsen OB (2009). Comparison of regulated passive membrane conductance in action potential-firing fast- and slow-twitch muscle. *J Gen Physiol* 134, 525–525.

Pedersen TH, Nielsen OB, Lamb GD and Stephenson DG (2004). Intracellular acidosis enhances the excitability of working muscle. *Science* 305, 1144–1147.

Porter KR and Palade GE (1957). Studies on the endoplasmic reticulum: III. Its form and distribution in striated muscle cells. *J Cell Biol* 3, 269–300.

Powell T, Terrar DA and Twist VW (1980). Electrical properties of individual cells isolated from adult rat ventricular myocardium. *J Physiol* 302, 131–153.

Putney JW, Broad LM, Braun FJ, Lievremont JP and Bird GS (2001). Mechanisms of capacitative calcium entry. *J Cell Sci* 114, 2223–2229.

Querol L and Illa I (2013). Myasthenia gravis and the neuromuscular junction. *Curr Opin Neurol* 26, 459–465.

Rayment I and Holden HM (1994). The three-dimensional structure of a molecular motor. *Trends Biochem Sci* 19, 129–134.

Rayment I (1996). The active site of myosin. *Annu Rev Physiol* 58, 671–702.

Rios E and Brum G (1987). Involvement of dihydropyridine receptors in excitation-contraction coupling in skeletal muscle. *Nature* 325, 717–720.

Ritchie JM and Rogart RB (1977). The binding of saxitoxin and tetrodotoxin to excitable tissue. *Rev Physiol Biochem Pharmacol* 79, 1–50.

Robertson D (1960). The molecular structure and contact relationships of cell membranes. *Prog Biophys Mol Biol* 10, 343–418.

Roden D (2004). Drug-induced prolongation of the QT interval. *N Engl J Med* 350, 1013–1022.

Rushton WAH (1933). Lapicque's theory of curarization. *J Physiol* 77, 337–364.

Ryall RW (1979). *Mechanisms of Drug Action on the Nervous System*. Cambridge: Cambridge University Press.

Saadeh K, Achercouk Z, Fazmin IT, et al. (2020). Protein expression profiles in murine ventricles modeling catecholaminergic polymorphic ventricular tachycardia: effects of genotype and sex. *Ann NY Acad Sci*. DOI: 10.1111/nyas.14426.

Sabir IN, Fraser JA, Cass TR, Grace AA and Huang CL-H (2007). A quantitative analysis of the effect of cycle length on arrhythmogenicity in hypokalaemic Langendorff-perfused murine hearts. *Pflügers Arch* 454, 925–936.

Sabir IN, Killeen MJ, Grace AA and Huang CL-H (2008a). Ventricular arrhythmogenesis: insights from murine models. *Prog Biophys Mol Biol* 98, 208–218.

Sabir IN, Li LM, Grace AA and Huang CL-H (2008b). Restitution analysis of alternans and its relationship to arrhythmogenicity in hypokalaemic Langendorff-perfused murine hearts. *Pflügers Arch* 455, 653–666.

Sabir IN, Ma N, Jones VJ, et al. (2010). Alternans in genetically modified Langendorff-perfused murine hearts modeling catecholaminergic polymorphic ventricular tachycardia. *Front Physiol* 1, 126.

Salvage SC, Gallant E, Beard N, et al. (2019a). Ion channel gating in cardiac ryanodine receptors from the arrhythmic RyR2-P2328S mouse. *J Cell Sci* 132, jcs229039.

Salvage SC, Huang CL-H, and Jackson AP (2020a). Cell-adhesion properties of auxiliary β-subunits in the regulation of cardiomyocyte sodium channels. Special Issue: Trafficking of Cardiac Ion Channels – Mechanisms and Alterations Leading to Disease. Ed. M. Verges. *Biomolecules* 10, 989.

Salvage SC, King JH, Chandrasekharan KH, et al. (2015). Flecainide exerts paradoxical effects on sodium currents and atrial arrhythmia in murine RyR2-P2328S hearts. *Acta Physiol* 214, 361–375.

Salvage SC, Rees J, McStea A, et al. (2020b). Supramolecular clustering of the cardiac sodium channel Nav1.5 in HEK293F cells, with and without the auxiliary β3-subunit. *FASEB J* 34, 3537–3553.

Salvage SC, Zhu W, Habib Z, et al. (2019b). Gating control of the cardiac sodium channel Nav1.5 by its β3-subunit: distinct roles for a transmembrane glutamic acid and the extracellular domain. *J Biol Chem* 94, 19752 –19763

Samsó M (2017). A guide to the 3D structure of the ryanodine receptor type 1 by cryoEM. *Protein Sci* 26, 52–68.

Sanchez JA and Stefani E (1978). Inward calcium current in twitch muscle fibres of the frog. *J Physiol* 283, 197–209.

Sanders KM, Hwang SJ and Ward SM (2010). Neuroeffector apparatus in gastrointestinal smooth muscle organs. *J Physiol* 588, 4621–4639.

Sanders KM, Koh SD and Ward SM (2006). Interstitial cells of Cajal as pacemakers in the gastrointestinal tract. *Annu Rev Physiol* 68, 307–343.

Sanguinetti MC, Jiang C, Curran ME and Keating MT (1995). A mechanistic link between an inherited and an acquired cardiac arrhythmia: *HERG* encodes the I_{Kr} potassium channel. *Cell* 81, 299–307.

Sarbjit-Singh S, Matthews HR and Huang CL-H (2020). Ryanodine receptor modulation by caffeine challenge modifies Na^+ current properties in intact murine skeletal muscle fibres. *Sci Rep* 10, 2199.

Sarquella-Brugada G, Campuzano O, Arbelo E, Brugada J and Brugada R (2016). Brugada syndrome: clinical and genetic findings. *Genet Med* 18, 3–12.

Scher A (1965). Mechanical events in the cardiac cycle In: *Physiology and Biophysics*. (Eds.). Ruch T and Patton H, Philadelphia, USA: Saunders.

Scheuer T and Gilly WF (1986). Charge movement and depolarization-contraction coupling in arthropod vs. vertebrate skeletal muscle. *Proc Natl Acad Sci U S A* 83, 8799–8803.

Schmidt-Nielsen K (1990). *Animal Physiology 4/e*. Cambridge: Cambridge University Press.

Schneider MF and Chandler WK (1973). Voltage dependent charge movement in skeletal muscle: a possible step in excitation-contraction coupling. *Nature* 242, 244–246.

Schwartz LM, McCleskey EW and Almers W (1985). Dihydropyridine receptors in muscle are voltage-dependent but most are not functional calcium channels. *Nature* 314, 747–751.

Sejersted OM and Sjøgaard G (2000). Dynamics and consequences of potassium shifts in skeletal muscle and heart during exercise. *Physiol Rev* 80, 1411–1481.

Sheikh SM, Skepper JN, Chawla S, et al. (2001). Normal conduction of surface action potentials in detubulated amphibian skeletal muscle fibres. *J Physiol* 535, 579–590.

Shotton DM, Heuser JE, Reese BF and Reese TS (1979). Postsynaptic membrane folds of the frog neuromuscular junction visualized by scanning electron microscopy. *Neuroscience* 4, 427–435.

Singer SJ and Nicolson GL (1972). The fluid mosaic model of the structure of cell membranes. *Science* 175, 720–731.

Skou J (1998). The identification of the sodium pump. *Biosci Rep* 24, 436–451.

Smith JM, Bradley DP, James MF and Huang CL-H (2006). Physiological studies of cortical spreading depression. *Biol Rev Camb Philos Soc* 81, 457–481.

Smith JM, James MF, Bockhorst KHJ, et al. (2001). Investigation of feline brain anatomy for the detection of cortical spreading depression with magnetic resonance imaging. *J Anat* 198, 537–554.

Smith PL, Baukrowitz T and Yellen G (1996). The inward rectification mechanism of the HERG cardiac potassium channel. *Nature* 379, 833–836.

Spudich JA (1994). How molecular motors work. *Nature* 372, 515–518.

Spudich JA, Finer J, Simmons B, et al. (1995). Myosin structure and function. *Cold Spr Harb Symp Quant Biol* 60, 783–791.

Squire J (1986). *Muscle: Design, Diversity and Disease*. California: Benjamin/Cummings, Menlo Park.

Takeshima H, Nishimura S, Matsumoto T, et al. (1989). Primary structure and expression from complementary DNA of skeletal muscle ryanodine receptor. *Nature* 339, 439–445.

Takeuchi A and Takeuchi N (1959). Active phase of frog's end-plate potential. *J Neurophysiol* 22, 395–411.

Takeuchi A and Takeuchi N (1960). On the permeability of end-plate membrane during the action of transmitter. *J Physiol* 154, 52–67.

Takla M, Huang CL-H, Jeevaratnam K (2020). The cardiac CaMKII-Nav1.5 relationship: From physiology to pathology. *J Mol Cell Cardiol* 139, 190–200.

Tanabe T, Beam KG, Adams BA, Niidome T and Numa S (1990). Regions of the skeletal muscle dihydropyridine receptor critical for excitation-contraction coupling. *Nature* 346, 567–569.

Tanaka Y, Horinouchi T and Koike K (2005). New insights into β-adrenoceptors in smooth muscle: distribution of receptor subtypes and molecular mechanisms triggering muscle relaxation. *Clin Exp Pharmacol Physiol* 32, 503–514.

Tasaki I (1953). *Nervous Transmission*. Springfield, Illinois: Charles C. Thomas.

Task Force of the Working Group on Arrhythmias of the European Society of Cardiology (1991). The Sicilian gambit: a new approach to the classification of antiarrhythmic drugs based on their actions on arrhythmogenic mechanisms. *Circulation* 84, 1831–1851.

Thomas G, Gurung IS, Killeen MJ, et al. (2007a). Effects of L-type Ca^{2+} channel antagonism on ventricular arrhythmogenesis in murine hearts containing a modification in the Scn5a gene modelling human long QT syndrome 3. *J Physiol* 578, 85–97.

Thomas G, Killeen MJ, Grace AA and Huang CL-H (2008). Pharmacological separation of early afterdepolarizations from arrhythmogenic substrate in DeltaKPQ Scn5a murine hearts modelling human long QT 3 syndrome. *Acta Physiol* 192, 505–517.

Thomas G, Killeen MJ, Gurung IS, et al. (2007b). Mechanisms of ventricular arrhythmogenesis in mice following targeted disruption of KCNE1 modelling long QT syndrome 5. *J Physiol* 578, 99–114.

Thornton CA (2014). Myotonic dystrophy. *Neurol Clin* 32, 705–719.

Tsien RW and Malinow R (1991). Changes in presynaptic function during long-term potentiation. *Ann NY Acad Sci* 635, 208–220.

Turner C and Hilton-Jones D (2014). Myotonic dystrophy. *Curr Opin Neurol* 27, 599–606.

Unwin N (2003). Structure and action of the nicotinic acetylcholine receptor explored by electron microscopy. *FEBS Lett* 555, 91–95.

Unwin N (2014). Nicotinic acetylcholine receptor and the structural basis of neuromuscular transmission: insights from *Torpedo* postsynaptic membranes. *Q Rev Biophys* 46, 283–322.

Usher-Smith JA, Fraser JA, Bailey PSJ, Griffin JL and Huang CL-H (2006a). The influence of intracellular lactate and H^+ on cell volume in amphibian skeletal muscle. *J Physiol* 573, 799–818.

Usher-Smith JA, Huang CL-H and Fraser JA (2009). Control of cell volume in skeletal muscle. *Biol Rev* 84, 143–159.

Usher-Smith JA, Skepper JN, Fraser JA and Huang CL-H (2006b). Effect of repetitive stimulation on cell volume and its relationship to membrane potential in amphibian skeletal muscle. *Pflügers Arch* 452, 231–239.

Valli H, Ahmad S, Chadda K, et al. (2017). Age-dependent atrial arrhythmic phenotype secondary to mitochondrial dysfunction in Pgc-1β deficient murine hearts. *Mech Ageing Dev* 167, 30–45.

Valli H, Ahmad S, Jiang AY, et al. (2018a). Cardiomyocyte ionic currents in intact young and aged murine Pgc-1β–/–atrial preparations. *Mech Ageing Dev* 169, 1–9.

Valli H, Ahmad S, Sriharan S, et al. (2018b). Epac-induced ryanodine receptor type 2 activation inhibits sodium currents in atrial and ventricular murine cardiomyocytes. *Clin Exp Pharmacol Physiol* 45, 278–292.

Vandenberg JI, Varghese A, Lu Y, et al. (2006). Temperature dependence of human ether-a-go-go-related gene K^+ currents. *Am J Physiol Cell Physiol* 291, C165–C175.

Vandenberg JI, Walker BD and Campbell TJ (2001). HERG K^+ channels: friend and foe. *Trends Pharmacol Sci* 22, 240–246.

Vaughan Williams E (1975). Classification of antidysrhythmic drugs. *Pharmacol Ther B* 1, 115–138.

Veeraraghavan R, Gourdie RG, Poelzing S (2014). Mechanisms of cardiac conduction: a history of revisions. *Am J Physiol Heart Circ Physiol* 306, H619–627.

Veeraraghavan R, Larsen AP, Torres NS, Grunnet M and Poelzing S (2013). Potassium channel activators differentially modulate the effect of sodium channel blockade on cardiac conduction. *Acta Physiol* 207, 280–289.

Verheule S, Sat T, Everett IV T, et al. (2004). Increased vulnerability to atrial fibrillation in transgenic mice with selective atrial fibrosis caused by overexpression of TGF-β1. *Circ Res* 94, 1458–1465.

Verschuuren J, Strijbos E and Vincent A (2016). Neuromuscular junction disorders In: *Handbook of Clinical Neurology. Vol 133*. (Eds.). Pittock SJ and Vincent A, New York: Elsevier. 447–466.

Vinogradova TM, Lyashkov AE, Zhu W, et al. (2006). High basal protein kinase A-dependent phosphorylation drives rhythmic internal Ca^{2+} store oscillations and spontaneous beating of cardiac pacemaker cells. *Circ Res* 98, 505–514.

Wang Y, Cheng J, Joyner RW, Wagner MB and Hill JA (2006). Remodeling of early-phase repolarization: a mechanism of abnormal impulse conduction in heart failure. *Circulation* 113, 1849–1856.

Wang Y, Tsui H, Ke Y, et al. (2014). Pak1 is required to maintain ventricular Ca^{2+} homeostasis and electrophysiological stability through SERCA2a regulation in mice. *Circ Arrhythm Electrophysiol* 7, 938–948.

Waxman S (2017). *Clinical Neuroanatomy 28/e*. New York: McGrawHill.

Weidmann S (1956). *Elektrophysiologie der Herzmuskelfaser*. Berne: Huber.

Weidmann S (1952). The electrical constants of Purkinje fibres. *J Physiol* 118, 348–360.

Wilkie DR (1968). Heat work and phosphorylcreatine break-down in muscle. *J Physiol* 195, 157–183.

Wisedchaisri G, Tonggu L, McCord E, et al. (2019). Resting-state structure and gating mechanism of a voltage-gated sodium channel. *Cell* 178, 993–1003.

Yan GX and Antzelevitch C (1998). Cellular basis for the normal T wave and the electrocardiographic manifestations of the long-QT syndrome. *Circulation* 98, 1928–1936.

Yan Z, Zhou Q, Wang L, et al. (2017). Structure of the Nav1.4-β1 complex from electric eel. *Cell* 170, 470–482.

Yarov-Yarovoy V, DeCaen PG, Westenbroek RE, et al. (2012). Structural basis for gating charge movement in the voltage sensor of a sodium channel. *Proc Natl Acad Sci* 109, E93–E102.

Yoneda S, Takano H, Takaki M and Suzuki H (2002). Properties of spontaneously active cells distributed in the submucosal layer of mouse proximal colon. *J Physiol* 542, 887–897.

Young J (1949). Factors influencing the regeneration of nerves. *Advan Surg* 1, 165–220.

Zaleska M, Salvage S, Thompson A, et al. (2018). The voltage-dependent sodium channel family In: *Oxford Handbook of Neuronal Ion Channels*. (Ed.). Bhattacharjee A, Oxford: Oxford University Press.

Zhang MM, Wilson MJ, Azam L, et al. (2013a). Co-expression of NaVβ subunits alters the kinetics of inhibition of voltage-gated sodium channels by pore-blocking μ-conotoxins. *Br J Pharmacol* 168, 1597–1610.

Zhang Y, Fraser JA, Jeevaratnam K, et al. (2011). Acute atrial arrhythmogenicity and altered Ca^{2+} homeostasis in murine RyR2-P2328S hearts. *Cardiovasc Res* 89, 794–804.

Zhang Y, Guzadhur L, Jeevaratnam K, et al. (2014). Arrhythmic substrate, slowed propagation and increased dispersion in conduction direction in the right ventricular outflow tract of murine Scn5a+/- hearts. *Acta Physiol* 211, 559–573.

Zhang Y, Schwiening C, Killeen MJ, et al. (2009). Pharmacological changes in cellular Ca^{2+} homeostasis parallel initiation of atrial arrhythmogenesis in murine Langendorff-perfused hearts. *Clin Exp Pharmacol Physiol* 36, 969–980.

Zhang Y, Wu J, Jeevaratnam K, et al. (2013b). Conduction slowing contributes to spontaneous ventricular arrhythmias in intrinsically active murine RyR2-P2328S hearts. *J Cardiovasc Electrophysiol* 24, 210–218.

Zheng H, Park KS, Koh SD and Sanders KM (2014). Expression and function of a T-type Ca^{2+} conductance in interstitial cells of Cajal of the murine small intestine. *Am J Physiol Physiol* 306, C705–C713.

Zhou J, Brum G, González A, et al. (2003). Ca^{2+} sparks and embers of mammalian muscle. Properties of the sources. *J Gen Physiol* 122, 95–114.

Zhu MH, Kim TW, Ro S, et al. (2009). A Ca^{2+}-activated Cl^- conductance in interstitial cells of Cajal linked to slow wave currents and pacemaker activity. *J Physiol* 587, 4905–4918.

Zhu MH, Sung TS, Kurahashi M, et al. (2016). Na^+-K^+-Cl^- cotransporter (NKCC) maintains the chloride gradient to sustain pacemaker activity in interstitial cells of Cajal.. *Am J Physiol Gastrointest Liver Physiol* 311, G1037–G1046.

Index

background conductances,
135–136
ionic currents, 137–140
sliding-filament theory, 124
slow synaptic potentials, 111–114
slow-twitch muscles, 121, 181
slow-wave propagation, 258–259, 270
smooth muscle, 121, 252, *See also*
interstitial cells of Cajal
contractile activation, 265–267
electrical coupling, 258–259
excitation–contraction coupling,
260–262, 269–270
functional features, 253
ion channels, 265
mechanical properties, 268
membrane excitation, 259–260
pharmacomechanical coupling,
262–263
smooth muscle cells, 252–253
sodium channels, 32–37, 208
gain of function, 239–242
gating mechanism, 66–68
impaired function, 235–238
ryanodine receptor and,
176–179
voltage-gated, 57, 59–60
sodium hypothesis, 41–44
predictions of, 43–46
sodium pump, 33–37
soleus, 121, 150, 181
soma, 106, 110
space constant, 76
spatial buffering, 119
spatial heterogeneities, 228–231
spatial summation, 108
spike. *See* excitation, *See* action
potential
squid giant axon, 3, 5
ST segment, 205–206
steady-state space constant, 76
stilbene chromophores, 157–158
store-operated channels, 265
strength–duration curve, 20
striated muscle, 121–123
stroke, 226
supernormal period, 21
surface-membrane excitation, 148
surface monophasic action potential,
227
suxamethonium, 174
swinging lever model, 130
sympathetic nervous system, 1, 218
synapses, 2, 88

electrotonic, 116
serial, 111
synaptic cleft, 89, 91, 107
synaptic delay, 95
synaptic excitation, 106
synaptic transmission, 106
synaptic vesicles, 89–90, 96, 107
syncytium, 200–202, 259

T-wave, 205
tachycardias. *See* ventricular
tachycardias
temporal electrophysiological
heterogeneities, 231–233, 241,
243
temporal summation, 108
tension, 127, 182, 185–186
tetanus, 182
tetracaine, 168, 172–173
tetrads, 164, 170
tetrodotoxin (TTX), 51, 56, 94
thin filament model, 131
threonine, 56
threshold stimulus, 18–19
Timothy syndrome, 242
titin, 132–134
tonic muscle fibres, 88
Torpedo (electric ray), 102–103
torsades de pointes, 225–226
transfer impedance, 149
transmural repolarisation, 230
transverse (T-) tubular membrane,
141–143
action potential, 143–145
altered properties of, 153
changes during exercise, 152–153
electrophysiology, 145–149
physiological modulation, 150–151
triad complexes, 160–162
transverse (T-) tubules, 122
transverse current, 82
triads, 160–162
tropomyosin, 124, 130–132, 265
troponin, 124, 130–131
trypsin, 128
tryptophan, 56
tubocurarine, 104
tubular membrane, 141–143
action potential, 143–145
electrophysiology, 145–149
physiological modulation, 150–151
tubular voltage detection, 164–167
twitch muscle fibres, 88
fast and slow, 121, 181

isometric, 181–183
tyrosine, 56
tyrosine kinase, 104

unmyelinated axons, 2–3
unmyelinated nerves
conduction velocity, 84
excitation along, 76–78
voltages changes along, 75

vagal nerve stimulation, 90
valine, 56
velocity
action potential conduction, 78–79
conduction, 84–85
force–velocity curve, 184–186
ventricular arrhythmias, 225–226,
236–239, 250
ventricular myocytes, 214
ventricular tachycardias, 226, 234,
245–247
verapamil, 209
vesicular sarcoplasmic reticulum, 252
voltage changes
cable system, 73–74
calcium ion release, 158–160
tubular voltage detection, 164–167
unmyelinated nerve, 75
voltage-sensing modules (VSMs), 58
sodium channels, 66–67
voltage clamp experiments, 12,
46–47, 86
end-plate potentials, 98
ionic current, 137–138
ionic current components, 49–52
nerve membrane, 47–49
voltage-gated ion channels, 55
cDNA sequencing, 55, 57
ion occupancy, 69–72
ionic selectivity, 68–69
sodium gating mechanism, 66–68
structure of, 55–60

Wallerian degeneration, 8
Wheatstone bridge, 41
white matter, 6
work output, 186

Xenopus, 57, 103
X-ray crystallography, 71
X-ray diffraction, 4, 102, 116, 128,
130, 132

Z-line, 123, 126, 132, 141